T0180436

METHODS IN MOLECULAR BIOLOGY™

Series Editor
John M. Walker
School of Life Sciences
University of Hertfordshire
Hatfield, Hertfordshire, AL10 9AB, UK

For other titles published in this series, go to
www.springer.com/series/7651

Glycomics

Methods and Protocols

Edited by

Nicolle H. Packer* and Niclas G. Karlsson†

**Department of Chemistry and Biomolecular Sciences, Macquarie University, Sydney, NSW, Australia*

† Centre for Bioanalytical Science, Chemistry Department, NUI Galway, Galway, Ireland

Humana Press

Editors
Nicolle H. Packer
Department of Chemistry and
Biomolecular Sciences
Macquarie University
Sydney, NSW
Australia

Niclas G. Karlsson
Centre for Bioanalytical Science
Chemistry Department
NUI Galway
Galway
Ireland

ISSN: 1064-3745 e-ISSN: 1940-6029
ISBN: 978-1-61737-769-3 ISBN: 978-1-59745-022-5 (eBook)
DOI: 10.1007/978-1-59745-022-5

Printed on acid-free paper

springer.com

Preface

As the number of -omics words proliferate in the analytical life sciences, glycomics is a word increasingly being used as biological researchers realize that carbohydrates may significantly contribute to the functional diversity in the cell. So what is in a word? Glycomics uses the chemical prefix for a sugar, "glyco-", and follows the naming convention established by genomics (which deals with genes) and proteomics (which deals with proteins). If we define *glycome* as the glycan complement of the cell or tissue as expressed by a genome in time and space, then glycomics is the analysis of the structure and function of these glycans or oligosaccharides (chains of sugars) attached to biomolecules. By this definition, the field includes the analysis of glycoproteins, proteoglycans, glycolipids, peptidoglycans, and lipopolysaccharides. *Glycoproteomics*, the study of the glycome attached to proteins, is gaining increasing interest on the back of the proteomics revolution as it becomes obvious that many of the changes associated with disease and differentiation are due to the modifications to the proteins rather than only to the regulation of the expression of the gene.

Historically, the main difficulty which has slowed the understanding of the biological role of glycans has been the perception that the analysis of sugars is too hard and is best left to the experts in the field. This book tries to remedy this misconception as there is an increasing availability of sample preparation, chromatographic, electrophoretic, mass spectrometric, and bioinformatic tools specifically designed for the analysis of glycosylation. In addition, approaches to investigate the interaction between these glycans and a variety of carbohydrate-recognizing proteins are presented so that the functional significance of the oligosaccharides can be explored. We have assembled the experts in the field, and this book presents a compendium of detailed protocols that they use routinely in their laboratories. The methods described can all be readily implemented using the current technologies already in use in research laboratories, especially those established for proteomics research.

The protocols in **Sect. I** concentrate on glycoprotein and proteoglycan analysis and include different approaches to determine the structure of both the N-linked and O-linked glycans released from glycoproteins and the glycosaminoglycans (GAGS) released from proteoglycans (**Subsect. A**). With the limiting quantities of biological material, as in proteomics, mass spectrometry has emerged as the detector of choice for glycan analysis and a wide variety of sample preparation, derivatization, chromatography, and mass spectrometric techniques are currently being used. We have included some methods which may be similar in principle but which differ in their implementation in the different author's laboratories. Current glycomic analytical methods have evolved based on available instrumentation, and a selection of protocols is provided here for the suite of instruments already existing in research laboratories that want to go down the glycomics path without investment in expensive new instrumentation. In much the same way as kitchen recipes, the reader can choose which detailed method works best "in their hands"!

Some glycosylation site determination methods are also described (**Subsect. B**) and usually involve enrichment of the glycoprotein or glycopeptide before analysis by mass spectrometry. The "holy grail" of glycoanalysis is to be able to characterize the oligosaccharide heterogeneity at each glycosylation site on a protein or proteoglycan — this has not yet been achieved by a generic approach. A separate section (**Subsect. C**) reviews the current practices used for the analysis of the specific single O-GlcNAc found on nuclear and cytoplasmic proteins.

As in proteomics, the development of bioinformatic tools for the analysis of the mass spectrometric data holds the key for widespread adoption of the challenge of glycan analysis (**Subsect. D**). Many of the software tools available for the analysis of proteomics data have the potential to be adapted for use in glycomics research and the development and adoption of these tools will enable high-throughput glycomics analysis to be carried out. At this point, we would like to acknowledge the significant contribution to the area of glycoinformatics of one of our authors, Dr. Claus-Wilhelm (Willi) von der Lieth, who unexpectedly passed away on November 24, 2007. Dr. von der Lieth, of the German Cancer Research Center, Molecular Structure Analysis Group in Heidelberg, was a global leader in the development of informatics systems for glycobiology and was the inspiration behind the EUROCarbDB project. His vision, coupled with his enthusiasm, professional and personal skills, will be sadly missed.

Of course, the reason for carrying out the analysis of what and where glycans are in the cell is so the researcher can ultimately determine what they do. Methods for measuring their diverse biomolecular interactions are described in **Sect. II** and cover glycan arrays, mass spectrometry, NMR, antibodies, and small molecule inhibitors.

As more experimenters take up the challenge of glycan analysis, and as more technologies are developed and used, we are confident that the results obtained will provide valuable insights into the biology of cell-cell communication and interaction. It seems that whenever we perturb the cell a change in glycosylation occurs — the function of these changes is largely unknown and tantalizes the researcher, but it is our belief that the more is known about the structure and function of these ubiquitous molecules the closer we will get to the understanding of the complexity of the glycome and its part in the complexity of biology. We hope that this book, *Glycomics: Methods and Protocols*, contributes to this goal, and we sincerely thank all the contributors for sharing their knowledge with the wider community in such a detailed and explanatory format.

Nicolle H. Packer
Sydney, NSW

Contents

Contributors

Hyun Joo An • *Department of Chemistry, University of California, Davis, CA, USA*
Abdel Atrih • *School of Biological Sciences, University of Liverpool, Liverpool, UK*
Parastoo Azadi • *Complex Carbohydrate Research Centre, The University of Georgia, Athens, GA, USA*
Malin Bäckström • *Department of Medical Biochemistry and Cell Biology, Göteborg University, Gothenburg, Sweden*
Inka Brockhausen • *Department of Medicine, Department of Biochemistry, Queen's University, Kingston, ON, Canada*
Ishan Capila • *Division of Biological Engineering, Massachusetts Institute of Technology, Cambridge, MA, USA*
Frédéric Chirat • *Unité Mixte de Recherche CNRS/USTL, Glycobiologie Structurale et Fonctionnelle, Bâtiment, Université des Sciences et Technologies de Lille, Villeneuve d'Ascq Cedex, France*
Catherine Cooper-Liddell • *Apollo Cytokine Research, Beaconsfield, Sydney, NSW, Australia*
Michael Cukan • *GlycoFi Inc., Lebanon, NH, USA*
André M. Deelder • *Biomolecular Mass Spectrometry Unit, Department of Parasitology, Leiden University Medical Center, Leiden, The Netherlands*
Ruby P. Estrella • *Graduate School of Biomedical Engineering, The University of New South Wales, Sydney, NSW, Australia*
Valegh Faid • *Unité Mixte de Recherche CNRS/USTL, Glycobiologie Structurale et Fonctionnelle, Bâtiment, Université des Sciences et Technologies de Lille, Villeneuve d' Ascq Cedex, France*
Richard Fisher • *GlycoFi Inc., Lebanon, NH, USA*
Tillman Gerngross • *GlycoFi Inc., Lebanon, NH, USA*
Bing Gong • *GlycoFi Inc., Lebanon, NH, USA*
Franz-Georg Hanisch • *Institute of Biochemistry II, Medical Faculty, University of Cologne, Köln, Germany Central Bioanalytics, Center for Molecular Medicine Cologne, University of Cologne, Köln, Germany*
Gunnar C. Hansson • *Department of Medical Biochemistry and Cell Biology, Göteborg University, Gothenburg, Sweden*
Iris Härd • *Department of Medical Biochemistry and Cell Biology, Göteborg University, Gothenburg, Sweden*
Thomas Haselhorst • *Institute for Glycomics, Griffith University, Gold Coast Campus, Gold Coast, QLD, Australia*
Ko Hayama • *Research Center for Glycoscience, National Institute of Advanced Industrial Science and Technology, Ibaraki, Japan*
Jerry L. Hedrick • *Department of Animal Science, University of California, Davis, CA, USA*

JAMIE HEIMBURG-MOLINARO • *School of Medicine, Department of Biochemistry, Emory University, Atlanta, GA, USA*
CHRISTIAN HEISS • *Complex Carbohydrate Research Centre, The University of Georgia, Athens, GA, USA*
JUN HIRABAYASHI • *Research Center for Glycoscience, National Institute of Advanced Industrial Science and Technology, Ibaraki, Japan*
JESSICA M. HOLMÉN LARSSON • *Department of Medical Biochemistry and Cell Biology, Göteborg University, Gothenburg, Sweden*
HIROMI ITO • *Research Center for Glycoscience, National Institute of Advanced Industrial Science and Technology (AIST), Open Space Laboratory C-2, Ibaraki, Japan*
SHIGEYASU ITO • *Research Center for Glycoscience, National Institute of Advanced Industrial Science and Technology, Ibaraki, Japan*
SATSUKI ITOH • *Division of Biological Chemistry and Biologicals, National Institute of Health Sciences, Tokyo, Japan*
KAZUAKI KAKEHI • *Faculty of Pharmaceutical Sciences, Kinki University, Higashi-Osaka, Japan*
AKIHIKO KAMEYAMA • *Research Center for Glycoscience, National Institute of Advanced Industrial Science and Technology (AIST), Open Space Laboratory C-2, Ibaraki, Japan*
PILSOO KANG • *Department of Chemistry, Indiana University, Bloomington, IN, USA*
HASSE KARLSSON • *Department of Medical Biochemistry and Cell Biology, Göteborg University, Gothenburg, Sweden*
NICLAS G. KARLSSON • *Centre for Bioanalytical Science, Chemistry Department, NUI Galway, Galway, Ireland*
NANA KAWASAKI • *Division of Biological Chemistry and Biologicals, National Institute of Health Sciences, Tokyo, Japan*
MITSUHIRO KINOSHITA • *Faculty of Pharmaceutical Sciences, Kinki University, Higashi-Osaka, Japan*
CAROLIEN A.M. KOELEMAN • *Biomolecular Mass Spectrometry Unit, Department of Parasitology, Leiden University Medical Center, Leiden, The Netherlands*
ANNE-CHRISTIN LAMERZ • *Institute for Glycomics, Griffith University, Gold Coast Campus, Gold Coast, QLD, Australia*
ERIKA LATTOVÁ • *Chemistry Department, University of Manitoba, Winnipeg, MB, Canada The Institute of Chemistry, Slovak Academy of Sciences, Bratislava, Slovakia*
CARLITO B. LEBRILLA • *Department of Chemistry, University of California, Davis, CA, USA*
URS LEWANDROWSKI • *Protein Mass Spectrometry and Functional Proteomics Group, Rudolf-Virchow-Center for Experimental Biomedicine, Wuerzburg, Germany*
BENSHENG LI • *Department of Chemistry, University of California, Davis, CA, USA*
HUIJUAN LI • *GlycoFi Inc., Lebanon, NH, USA*
THOMAS LÜTTEKE • *Bijvoet Center for Biomolecular Research, Utrecht University, Utrecht, The Netherlands*

YEHIA MECHREF • *Department of Chemistry, Indiana University, Bloomington, IN, USA*
JEAN-CLAUDE MICHALSKI • *Unité Mixte de Recherche CNRS/USTL, Glycobiologie Structurale et Fonctionnelle, Bâtiment, Université des Sciences et Technologies de Lille, Villeneuve d'Ascq Cedex, France*
WILLY MORELLE • *Unité Mixte de Recherche CNRS/USTL, Glycobiologie Structurale et Fonctionnelle, Bâtiment, Université des Sciences et Technologies de Lille, Villeneuve d'Ascq Cedex, France*
STEFAN MÜLLER • *Central Bioanalytics, Center for Molecular Medicine Cologne, University of Cologne, Köln, Germany*
HISASHI NARIMATSU • *Research Center for Glycoscience, National Institute of Advanced Industrial Science and Technology (AIST), Open Space Laboratory C-2, Ibaraki, Japan*
MILOS V. NOVOTNY • *Department of Chemistry, Indiana University, Bloomington, IN, USA*
HÉLÈNE PERREAULT • *Chemistry Department, University of Manitoba, Winnipeg, MB, Canada*
ANDREW K. POWELL • *School of Biological Sciences, University of Liverpool, Liverpool, UK*
VIKAS PRABHAKAR • *Division of Biological Engineering, Massachusetts Institute of Technology, Cambridge, MA, USA*
KATE RITTENHOUSE-OLSON • *Departments of Microbiology and Immunology, Biotechnical and Clinical Laboratory Sciences, The University at Buffalo, Buffalo, NY, USA*
REBECCA H. ROUBIN • *Faculty of Pharmacy, The University of Sydney, Sydney, NSW, Australia*
NICOLLE H. PACKER • *Department of Chemistry and Biomolecular Sciences, Macquarie University, Sydney, NSW, Australia*
RAM SASISEKHARAN • *Division of Biological Engineering, Massachusetts Institute of Technology, Cambridge, MA, USA*
TAKASHI SATO • *Research Center for Glycoscience, National Institute of Advanced Industrial Science and Technology (AIST), Open Space Laboratory C-2, Ibaraki, Japan*
JOHN SCHUTZBACH • *Department of Medicine, Department of Biochemistry, Queen's University, Kingston, ON, Canada*
ALBERT SICKMANN • *Protein Mass Spectrometry and Functional Proteomics Group, Rudolf-Virchow-Center for Experimental Biomedicine, Wuerzburg, Germany*
RAINA SIMPSON • *Apollo Cytokine Research, Beaconsfield, Sydney, NSW, Australia*
MARK SKIDMORE • *School of Biological Sciences, University of Liverpool, Liverpool, UK*
TERRANCE A. STADHEIM • *GlycoFi Inc., Lebanon, NH, USA*
KRISTINA A. THOMSSON • *Department of Medical Biochemistry and Cell Biology, Göteborg University, Gothenburg, Sweden*
JEREMY E. TURNBULL • *School of Biological Sciences, University of Liverpool, Liverpool, UK*
CLAUS-W. VON DER LIETH • *German Cancer Research Center (DKFZ), Central Spectroscopic Department (B090), Heidelberg, Germany*

MARK VON IZSTEIN • *Institute for Glycomics, Griffith University, Gold Coast Campus, Gold Coast, QLD, Australia*
JOHN M. WHITELOCK • *Graduate School of Biomedical Engineering, The University of New South Wales, Sydney, NSW, Australia*
NICOLE WILSON • *Apollo Cytokine Research, Beaconsfield, Sydney, NSW, Australia*
MANFRED WUHRER • *Biomolecular Mass Spectrometry Unit, Department of Parasitology, Leiden University Medical Center, Leiden, The Netherlands*
TERUHIDE YAMAGUCHI • *Division of Biological Chemistry and Biologicals, National Institute of Health Sciences, Tokyo, Japan*
ED YATES • *School of Biological Sciences, University of Liverpool, Liverpool, UK*
NATASHA E. ZACHARA • *Department of Biological Chemistry, The Johns Hopkins University School of Medicine, Baltimore, MD, USA*
ZHENG-LIANG ZHI • *School of Biological Sciences, University of Liverpool, Liverpool, UK*

Section I

Glycan Analysis

Subsection A

Released Glycans

Chapter 1

Analysis of *N*- and *O*-Linked Glycans from Glycoproteins Using MALDI-TOF Mass Spectrometry

Willy Morelle, Valegh Faid, Frédéric Chirat, and Jean-Claude Michalski

Summary

Glycosylation represents the most common of all known protein post-translational modifications. Carbohydrates can modulate the biological functions of a glycoprotein, protect a protein against hydrolysis via protease activity, and reduce or prevent aggregation of a protein. The determination of the carbohydrate structure and function in glycoproteins remains one of the most challenging tasks given to biochemists, as these molecules can exhibit complex branched structures that can differ in linkage and in the level of branching. In this review, we will present the approach followed in our laboratory for the elucidation of *N*- and *O*-glycan chains of glycoproteins. First, reduced/carboxamidomethylated glycoproteins are digested with a protease or a chemical reagent. *N*-Glycans are then released from the resulting peptides/glycopeptides via digestion with peptide *N*-glycosidase F (PNGase F). Oligosaccharides released by PNGase F are separated from peptides and glycopeptides using a C18 Sep-Pak, and their methylated derivatives are characterized by matrix-assisted laser desorption/ionization-time of flight mass spectrometry (MALDI-TOF-MS). *O*-Glycans are released by reductive elimination, which are permethylated, purified on a Sep-Pak C18 cartridge, and analyzed with MALDI-TOF-MS. Finally, to confirm the structures *N*-glycans released by PNGase F are characterized using MALDI-TOF-MS following on-plate sequential exoglycosidase digestions. The clean-up procedures of native and permethylated oligosaccharides for an efficient MALDI-TOF-MS analysis will also be described. This strategy was applied to calf fetuin and glycoproteins present in human serum.

Key words: Mass spectrometry, Glycomics, Structure analysis, Glycoproteins, Glycans.

1. Introduction

The glycosylation process is the covalent attachment of oligosaccharide chains on the protein backbone and is considered as the most common post-translational modification of proteins. There are two main types of protein glycosylation. The first

Nicolle H. Packer and Niclas G. Karlsson (eds.), *Methods in Molecular Biology, Glycomics: Methods and Protocols, vol. 534*

DOI: 10.1007/978-1-59745-022-5_1

one, called O-glycosylation, corresponds to the attachment of an oligosaccharide chain to the oxygen of a hydroxylated amino acid, more commonly Ser or Thr *(1)*. The second one, called N-glycosylation, corresponds to the attachment of an oligosaccharide chain via the amide group of an Asn residue present in the tripeptide consensus sequence – Asn-X-Thr/Ser (where X can be any amino acid except Pro) *(2)*. Each glycosylated site may contain many different glycan structures. For example, more than 100 different glycan structures have been identified on a single glycosylation site of the CD59 glycoprotein (microheterogeneity) *(3)*. These glycan chains play key functions in biological processes, and the challenge of glycobiologists is to establish the structure/function relationship for each glycan chain. Protocols for the structural analysis of the glycan chains of glycoproteins usually involve, as a first step, the release of the carbohydrate chains from the protein backbone.

Both chemical and enzymatic methods can be used for the release of *N*-glycans. Anhydrous hydrazine can release unreduced N-linked glycans from glycoproteins *(4)*. This chemical release suffers from several major disadvantages. First, as peptide bonds are destroyed, information relating to the protein is lost. Second, the acyl groups are cleaved from the *N*-acylamino sugars and sialic acids, calling for a reacetylation step, assuming that the sialic acid residues were originally acetylated. Third, the reacetylation step can also add a small amount of *O*-acetyl substitution. For these reasons, it is often desirable to perform glycan release under mild conditions using an enzyme. Several enzymes are commercially available and the two most popular ones are PNGase F and endoglycosidase H (Endo H). The first one cleaves the intact glycan chain as glycosylamine, which is readily converted into reducing glycan. With few exceptions, PNGase F releases practically all protein-bound N-linked carbohydrates except those with fucose attached to the 3 position of the Asn-linked GlcNAc residue. Their relevant glycoproteins are commonly found in plants. Endo H is more specific than PNGase F and cleaves the glycosidic bonds between the two GlcNAc residues of the chitobiose core of oligomannose and hybrid-type glycans. It is inactive on complex-type glycans.

The release of *O*-glycan chains can be achieved using chemical or enzymatic methods. However, enzymatic methods are restricted by the few enzymes available and their limited substrate specificities *(5,6)*. A generic enzyme that cleaves all O-linked sugar structures from any protein is missing. For these reasons, *O*-glycans of glycoproteins are usually released using chemical reagents. Anhydrous hydrazine can release unreduced O-linked glycans. The release of O-linked glycans occurs with lower temperature dependence than the release of N-linked glycans. However, release of *O*-glycans may be incomplete if many glycans are clustered in short regions of the protein. Alkaline β-elimination is the most commonly used chemical cleavage method for the release of *O*-glycans. Among the different protocols described *(7–9)*, the protocol suggested by Carlson is considered as the most reliable method for the purpose *(10)*. Sugars O-glycosidically linked to serine or threonine residues are released from *O*-glycopeptides or proteins in the reduced form containing GalNAc-ol by alkaline β-elimination in the presence of high concentrations of sodium borohydride, which prevents "peeling" of the released oligosaccharides by reducing the terminal GalNAc residues to its alditol.

After being released, glycans are subjected to structural analysis. Mass spectrometry has become one of the most powerful and versatile techniques for the structural analysis of glycans *(11, 12)*. Mass spectrometry provides many advantages over traditional analytical methods, such as low sample consumption and high sensitivity. MALDI-TOF-MS is unique in its capacity to analyze complex mixtures, producing spectra uncomplicated by multiple charging. Besides, MALDI-TOF-MS provides significant tolerance against contaminants such as salts *(13)* or common buffers *(14)*, and allows fast and easy sample preparation. For these reasons, MALDI-TOF-MS is one of the most popular techniques for glycan analysis.

In a MALDI experiment, the sample is cocrystallized with a 1,000-fold excess of a low molecular weight matrix which absorbs at the wavelength of the laser. The energy of the laser is first absorbed by the matrix molecules. Stimulated matrix molecules transfer their excess energy to the sample molecules, which become ionized. For this reason, MALDI is a soft ionization method that produces predominantly molecular ions from the released glycans with little or no fragmentation. Consequently, MALDI-TOF-MS is the pre-eminent technique for screening complex mixtures of glycans from biological extracts, thereby revealing the types of glycans present.

The most common matrix for the analysis of glycans is 2,5-dihydroxybenzoic acid (DHB). Native *N*- and/or *O*-glycans generate $[M + Na]^+$ species as the major ions, often accompanied by minor ions $[M + K]^+$. The different *m/z* peaks observed on the mass spectrum provide information on the glycan composition in terms of Hex, dHex, HexNAc, and NeuAc. In some cases, if we take into account the rules of *N*- and *O*-glycans biosynthesis, it is also possible to deduce the glycan structure. **Table 1** lists the masses of the common monosaccharides found in *N*- and *O*-glycans. Sialylated glycans are more difficult to analyze since they can lose a significant amount of sialic acid either in the source or after ion extraction from the ion source. Moreover, sialylated ions can give a mixture of cation adducts such as $[M + Na]^+$, $[M + K]^+$, $[M - nH + (n + 1)Na]^+$, and $[M - nH + (n + 1)K]^+$, making the interpretation of spectra more difficult. In order to stabilize

Table 1
Residue masses of common monosaccharides

	Native		Permethylated	
Monosaccharides	**Accurate mass**	**Average mass**	**Accurate mass**	**Average mass**
Pentose	132.042	132.116	160.074	160.170
Deoxyhexose	146.078	146.143	174.089	174.197
Hexose	162.053	162.142	204.099	204.223
N-Acetylhexosamine	203.079	203.179	245.126	245.275
Hexuronic acid	176.032	176.126	218.079	218.206
N-Acetyl-neuraminic acid	291.095	291.258	361.173	361.392
N-Glycolyl-neuraminic acid	307.090	307.257	391.184	391.418

sialic acid residues, several chemical treatments such as methyl esterification of the carboxyl group of sialic acid residues *(15)* or amidation *(16)* have been suggested. In our lab, we have chosen to stabilize sialylated oligosaccharides using permethylation. This derivatization, which consists in converting all hydroxyl groups into methoxy groups, and carboxyl groups into methyl ester, allows the simultaneous analysis of neutral and sialylated oligosaccharides in the positive ion mode. In the procedure developed by Ciucanu and Kerek, which we applied, this conversion is made in two steps *(17)*. In the first one, alcohols are transformed into alcoholate ions in the presence of sodium hydroxide in dimethylsulfoxide (DMSO). In the second step, alcoholates are methylated by methyl iodide. This approach offers several advantages. Permethylated oligosaccharides are detected with higher sensitivity, compared with their native forms. Besides, since the full structural analysis of a glycan includes the elucidation of all interglycosidic linkages, it is possible to complete the MALDI-TOF MS analysis by a gas chromatography/mass spectrometry analysis of partially acetylated and methylated monosaccharides *(18)*.

In this review, we will describe the procedure we used in our laboratory to study the *N*- and *O*-glycan chains of a glycoprotein. Briefly, after glycoprotein proteolysis, *N*- and *O*-glycopeptides are first submitted to the PNGase F digestion to release the *N*-glycan chains. *O*-glycopeptides are subsequently treated in an alkali reductive medium to free *O*-glycan chains. Aliquots of *N*- and *O*-glycan chains are permethylated and analyzed using MALDI-TOF MS. Their primary sequence as well as the anomeric configuration is confirmed by on-plate sequential exoglycosidase digestions of the native oligosaccharide chains. As an illustration, this methodology has been applied to the N- and O-glycoprofiling of calf fetuin (200 μg) and to glycoproteins isolated from 20 μL of normal human serum.

2. Materials

2.1. Reduction and Carboxymethylation of Glycoproteins

1. *Denaturing buffer.* 0.6 M Tris–HCl, 6 M guanidinium chloride, pH 8.4. Degas for 30 min prior use.
2. *Reduction buffer (10×).* 500 mM dithiothreitol (DDT) solution in denaturing buffer.
3. *Alkylating buffer (10×).* 3.0 M iodoacetamide solution in denaturing buffer. This solution must be kept in the dark.

2.2. Dialysis

Spectra/Por regenerated cellulose membrane No.1 (cut-off 6,000–8,000 Da). The dialysis is carried out against a 50 mM ammonium hydrogen carbonate solution containing 10 mM ethylenediaminetetraacetic acid (EDTA) (*see* **Note 1**), except for the last bath which is a 50 mM ammonium hydrogen carbonate solution.

2.3. Trypsin Digestion of Glycoproteins

1. Trypsin (Sigma, St. Louis, MO).
2. *Trypsin buffer.* 50 mM ammonium hydrogen carbonate. Adjust to pH 8.4 with 10% (v/v) ammonia solution.
3. *Trypsin solution.* 10 mg/mL trypsin solution in the trypsin buffer just before the experiment.

2.4. PNGase F Digestion

1. Recombinant peptidyl-*N*-glycosidase F (PNGase F; EC 3.5.1.52) from *Escherichia coli* (Roche, Mannheim, Germany) 1 U/μL in water (stability: several months at 4°C) (*see* **Note 2**).
2. *PNGase F buffer.* 50 mM ammonium hydrogen carbonate. Adjust to pH 8.4 with 10% (v/v) ammonia solution.

2.5. Clean-Up Procedures of the Released N-Glycans

1. Methanol, acetic acid, acetonitrile, and trifluoroacetic acid (TFA) of the highest purity available.
2. Sep-Pak classic C18 cartridges (Waters, Milford, MA), and 150 mg of nonporous graphitized carbon columns (Alltech, Deerfield, IL).
3. *Equilibrium solvent.* 0.1% (v/v) TFA in water (stability: 1 week at room temperature).
4. *Sep-Pak C18 elution solvent.* Acetonitrile/water (80:20; v/v), 0.1% (v/v) TFA, for the elution of peptides and glycopeptides.
5. *Graphitized carbon column elution solvent.* Acetonitrile/water (25:75; v/v), 0.1% (v/v) TFA for the elution of *N*-glycans.

The two elution solvents are stable for 1 week at room temperature.

2.6. Release of O-Glycans Using Reductive Elimination

1. *Release solution.* 50 mM NaOH solution with 1 M sodium borohydride ($NaBH_4$) prepared just before the experiment.

2.7. Desalting of the Released O-Glycans

1. Ion exchange chromatography Dowex 50 W-X8 beads (H^+ form; 50–100 mesh, BioRad, Hercules, CA) (*see* **Note 3**).
2. Acetic acid (5% v/v) solution in water.
3. Coevaporation solvent: methanol/acetic acid (95:5; v/v).
4. Pasteur pipettes (230 × 50 mm) (*see* **Note 4**).

2.8. Permethylation of Glycans

1. DMSO and iodomethane from Fluka (St. Gallen, Switzerland) kept anhydrous in an argon atmosphere.
2. NaOH pellets (Carlo Erba, Rodano, Italy) kept anhydrous in an argon atmosphere.
3. Mortar with pestle (*See* **Note 5**).

2.9. Clean-Up Procedure of Permethylated Glycans Using a Sep Pak C18

1. Methanol, water, and acetonitrile of the highest purity available.
2. *Elution solvent 1.* Acetonitrile/water (10:90; v/v).
3. *Elution solvent 2.* Acetonitrile/water (80:20; v/v).

2.10. On-Plate Exoglycosidase Digestions

1. *Digest buffer.* 50 mM ammonium formiate. Adjust to pH 4.6 with a 25% (v/v) formic acid solution.
2. α-Sialidase (E.C. 3.2.1.18), from *Arthrobacter ureafaciens* (Roche).
3. β-Galactosidase (EC 3.2.1.23) from bovine testis and α-fucosidase (EC 3.2.1.51) from bovine kidney (Sigma).
4. β-*N*-Acetylhexosaminidase (EC 3.2.1.30) from Jack bean (Calbiochem, San Diego, CA).

2.11. Mass Spectrometry Analysis

1. The MALDI-TOF-MS instrument:Voyager Elite DE-STR Pro from PersSeptive Biosystem (Framingham, MA, USA) equipped with a pulsed nitrogen laser (337 nm) and a gridless delayed extraction ion source.
2. *Matrix solution.* 10 mg/mL DHB in methanol/water (1:1; v/v).

3. Methods

The protocol suggested routinely generates good quality data from the MALDI-TOF mass spectrometer in our laboratory. These data can easily be obtained from several micrograms of glycans. The glycoprotein quantity must be chosen depending on its glycosylation level, which typically ranges between 2 and 40% (w/w). For a lower carbohydrate content, the glycoprotein quantity can be increased up to 2 mg. Over this limit, the retention capacity of the Dowex 50 × 8 column becomes insufficient (*see* **Subheading 3.5**). Moreover, oligosaccharide profiling can also be obtained from a mixture of glycoproteins.

3.1. Reduction and Alkylation of the Glycoprotein

Even though PNGase F is a very effective enzyme for the release of *N*-glycans chains, its use requires prior denaturation of the glycoprotein to increase its accessibility to the different glycosylation sites. The first step consists in opening all glycoprotein-containing disulfide bridges by reduction with DDT and blocking the free thiol groups by alkylation with iodoacetamide.

1. The glycoprotein is dissolved at a concentration of 5–10 μg/μL in the degassed denaturing buffer and incubated for 1 h at 50°C.
2. The reduction buffer is added to obtain a final concentration of 20 mM DTT. The sample is flushed with argon and incubated at 50°C for 4 h.
3. The alkylating buffer is added to obtain a final concentration of 110 mM IAA. The sample is flushed with argon and incubated at room temperature overnight in the dark.
4. In order to destroy the proteases which could contaminate the cellulose dialysis tube, the tube is cleaned by boiling for 10 min prior to use and extensively rinsing with water.
5. The glycoprotein solution is then transferred into the dialysis tubing, and the tube containing the glycoprotein is rinsed twice with water. The sample is dialyzed against 4 × 2 L of dialysis solution at 4°C, with each change occurring after stirring for 12 h. For the last bath, EDTA is omitted from the dialysis buffer.
6. The dialyzed solution is transferred into a glass tube and freeze-dried.

3.2. Proteolytic Digestion of the Glycoprotein

To facilitate the enzymatic de-N-glycosylation by increasing the accessibility of the PNGase F to its substrate, the reduced and carboxamidomethylated glycoprotein is submitted to a proteolytic digestion to generate small peptides and glycopeptides (*see* **Note 6**).

1. The freeze-dried, reduced, and carboxamidomethylated glycoprotein is dissolved in trypsin solution with an enzyme/substrate ratio of 1:10 (by mass) and incubated for 24 h at 37°C under gently agitation.
2. Trypsin is then destroyed by boiling the solution for 10 min before the lyophlization step.

3.3. PNGase F Digestion

1. The de-N-glycosylation step is carried out by dissolving the dried peptides and glycopeptides in 500 μL of the PNGase F buffer.

2. PNGase F is added at the final concentration of 5 U/mg of glycoprotein and the digestion is carried out at 37 °C for 16 h (*see* **Note 7**). The reaction is terminated by lyophilization.

3.4. Clean-Up Procedure of the Released N-Glycans

1. *N*-Glycan chains are separated from the peptides and *O*-glycopeptides using a Sep-Pak C18 cartridge. This clean-up procedure relies on the adsorption of peptides and *O*-glycopeptides on the hydrophobic C18 phase while *N*-glycans pass through this phase (*see* **Note 8**).
2. Prior to use, the Sep-Pak C18 cartridge must be conditioned with 5 mL of methanol. Methanol is introduced into the cartridge using a Pasteur pipette until reaching the top of the longest end. A 5-mL glass syringe is then rapidly connected to the Sep-Pak C18 cartridge and filled with methanol. It is very important to not dry the C18 phase all through the procedure. The Sep-Pak C18 cartridge is then washed with 10 mL of equilibrium solvent.
3. The dried sample is solubilized in 200 μL of 0.1% (v/v) TFA and loaded directly onto the Sep-Pak C18 cartridge. From this time, the eluate must be immediately collected into a 4-mL screw-capped glass tube. The tube is rinsed twice with 200 μL of 0.1% (v/v) TFA, and both solutions are again directly loaded onto the Sep-Pack C18 cartridge. The glass syringe is then immediately reconnected to the Sep-Pak C18 cartridge and filled with 3 mL of 0.1% (v/v) TFA, which is also collected. The glycans are lyophilized.
4. Peptides and *O*-glycopeptides are eluted *en bloc* by 3 mL of the elution solvent. Prior to the freeze-drying step, the acetonitrile is removed under a stream of nitrogen in a fume hood.

3.5. Release of O-Glycans Using Reductive Elimination

1. The resulting peptides/glycopeptides are dissolved in 200 μL of the sodium hydroxide solution containing 1 M $NaBH_4$, and incubated at 45°C for 16 h.
2. The reaction mixture is left standing at room temperature for 5–10 min.
3. The reaction is stopped by adding glacial acetic acid dropwise until no fizzing is observed (about 3–5 drops).

3.6. Purification of the Released O-Glycans

1. The desalting column consists of a truncated Pasteur pipette, plugged at the tapered end with a small amount of glass wool. A piece of silicone tubing (about 2 cm) is placed at the tapered end, and the flow is blocked using an adjustable clip. The pipette is filled with 5% (v/v) acetic acid and the clip is opened slightly to let the equilibration buffer slowly flow out. While the 5% (v/v) acetic acid is running out, the column is filled with 2 mL of freshly washed Dowex beads (50 × 8, H^+ form) (*see* **Notes 3** and **9**).
2. The sample is loaded onto the column, and the unretained material containing the *O*-glycans and borate salts is immediately

collected into a 4-mL screw-capped glass tube. The column is washed using 4 mL of 5% (v/v) acetic acid solution and the eluates (two fractions of 2 mL) are then lyophilized.

3. Borate salts are removed by repeated evaporation with methanol containing 5% (v/v) acetic acid under a stream of nitrogen in a fume hood. For this purpose, the sample is solubilized in 500 μL of methanol/acetic acid (95:5, v/v) and dried. This procedure is repeated five times and the sample lyophilized.

3.7. Permethylation of Glycans

Aliquots of glycans are permethylated according to Ciucanu and Kerek's procedure *(17)*.

1. The tube containing the freeze-dried glycans is placed in a vacuum vessel saturated with an argon atmosphere and 500 μL of DMSO is added.
2. A few pellets of NaOH (5–10) are placed in a dry mortar. The pellets of NaOH are quickly ground into a fine powder using the pestle.
3. About 25 mg of NaOH is added to the sample.
4. Iodomethane (300 μL) is added and the tube is flushed with a stream of argon. After mixing vigorously, the reaction mixture is placed in an ultrasonic bath for 2 h at room temperature.
5. The reaction is then stopped by adding 1 mL of water, mixing vigorously.
6. Chloroform (600 μL) is added and mixed vigorously, and the mixture is allowed to settle into two layers. The lower chloroform phase is transferred into a new glass tube. This chloroform extraction is repeated twice.
7. Six successive washes of the chloroform phase are carried out with 1 volume of water (*see* **Note 10**). The aqueous phases are discarded.
8. The chloroform phase is then dried under a stream of nitrogen in a fume hood.

3.8. Clean-Up Procedure of Permethylated Glycans Using a Sep-Pak C18

1. The dried sample is dissolved in about 200 μL of methanol and purified using a Sep-Pak C18 cartridge.
2. The Sep-Pak C18 cartridge is conditioned with 5 mL of methanol. Methanol is introduced into the cartridge using a Pasteur pipette until reaching the top of the longest end. A 5-mL glass syringe is then rapidly connected to the Sep-Pak C18 cartridge and filled with methanol.
3. The cartridge is washed with 10 mL of water.
4. The sample is loaded directly onto the Sep-Pack. The tube is rinsed twice with water and the solutions are again directly loaded onto the Sep-Pack cartridge.
5. The cartridge is sequentially washed with 15 mL of water and 2 mL of acetonitrile/water (10:90; v/v) mixture.

6. The permethylated glycans are eluted by 3 mL of the acetonitrile/water (80:20; v/v) mixture, partially evaporated under a stream of nitrogen in a fume hood to remove acetonitrile, and freeze-dried.

3.9. MALDI-TOF-MS Analysis of Permethylated Oligosaccharides

1. Permethylated glycans are dissolved in a methanol/water (1:1; v/v) solution to obtain a concentration of 10 pmol/μL.
2. Directly on the MALDI target, 1 μL of the solution is mixed with 1 μL of the DHB matrix solution. After crystallization at room temperature, the spectra are acquired by submitting each spot to multiple laser shots (100–200) and recorded by a TOF analyzer over a range of *m/z* 1,000–5,000 for permethylated *N*-glycans (**Fig. 1**) or *m/z* 300–5,000 for permethylated

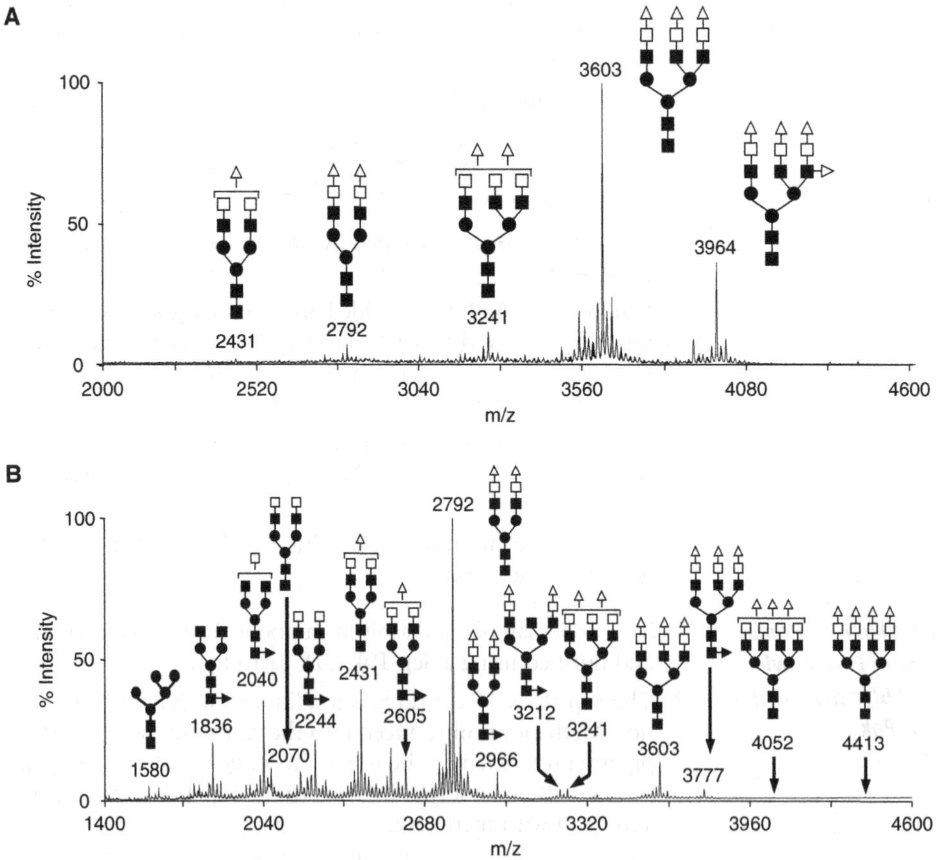

Fig. 1. Positive MALDI-TOF-MS spectra of permethylated *N*-glycans from calf fetuin (**A**) and from the glycoproteins of the normal human serum (**B**). The *N*-glycans were released from tryptic glycopeptides by digestion with PNGase F, separated from peptides and *O*-glycopeptides by Sep-Pak purification, and permethylated. The derivatized glycans were purified by the Sep-Pak. Symbols: *open square* galactose; *filled square* *N*-acetylglucosamine; *crossed square* reduced *N*-acetylgalactosamine; *filled triangle* fucose; *open triangle* *N*-acetyl neuraminic acid; *filled circle* mannose.

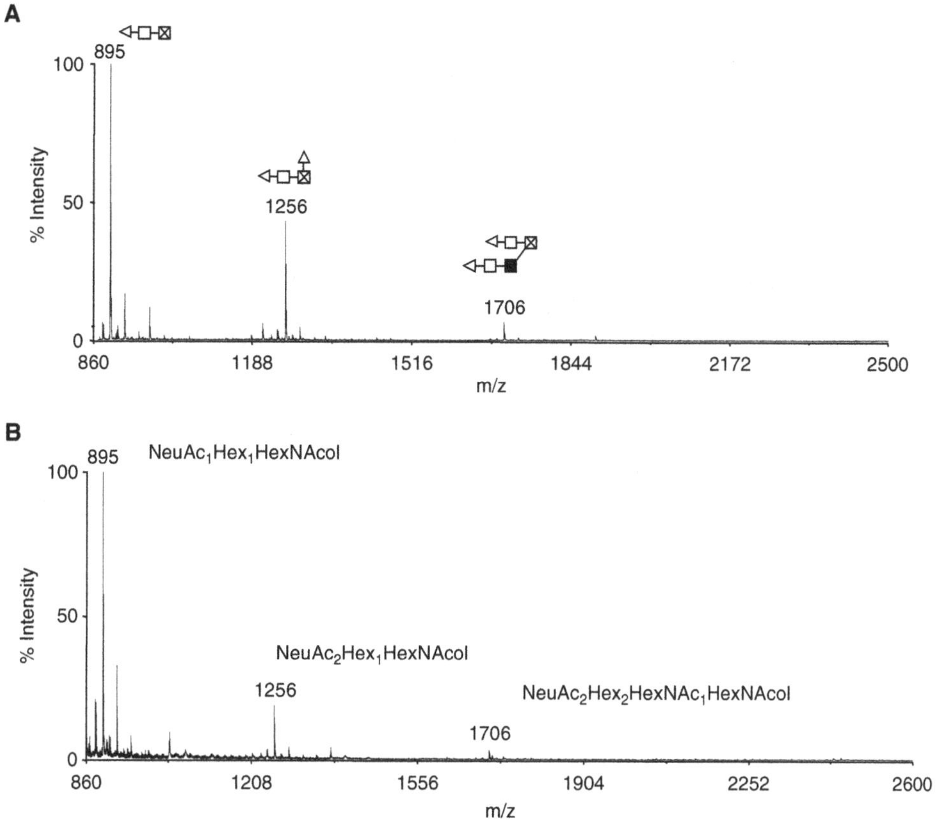

Fig. 2. Positive MALDI-TOF-MS spectra of permethylated *O*-glycans released from the calf fetuin (**A**) and from the glycoproteins of the normal human serum (**B**). *O*-glycans were released by reductive elimination, permethylated, purified by Sep-Pak C18, and analyzed by MALDI-TOF-MS as mixtures. Symbols are as in Fig. 1.

O-glycans (**Fig. 2**). The instrument is operated in the positive ion reflectron mode throughout. An irradiance slightly above the threshold of ion detection is used. Acceleration and reflector voltages are set up as follows: target voltage at 20 kV, first grid at 72% of target voltage, delayed extraction at 550 ns. Spectra are calibrated with an external mixture of isomaltosyl oligosaccharides containing 6–19 glucose units. In order to preclude low-mass ions from saturating the detector, the ion gate was set at 950 for the analysis of permethylated *N*-glycans and at 250 for the analysis of permethylated *O*-glycans.

3.10. Purification of the Underivatized N-Glycans

The confirmation of glycan structures deduced from the MALDI-TOF-MS analysis of permethylated glycans can be obtained using specific exoglycosidases (*see* **Note 11**). Before making exoglycosidase digestions, it is necessary to desalt the *N*-glycans using a carbograph column *(19)*.

1. Prior to use, this column must be conditioned with 5 mL of methanol and washed with 10 mL of equilibrium solvent. Underivatized *N*-glycans are dissolved in 500 μL of equilibrium solution and loaded onto the column.
2. Salts are removed with 15 mL of equilibrium solution.
3. *N*-glycans are recovered by passing 3 mL of elution solvent through the column and directly freeze-drying.

3.11. On-Plate Exoglycosidase Digestions of N-Glycans

1. The exoglycosidase digestions are carried out on the 100-well MALDI sample plate. One microliter of each sample (5–10 pmol) is directly mixed on the plate with 1 μL of the reaction buffer (50 mM ammonium formiate, pH 4.6).
2. One microliter of each exoglycosidase is then added to each spotted sample at the following concentration: α-sialidase, 50 mU; β-galactosidase, 1.25 mU; β-*N*-acetylhexosaminidase, 150 mU; α-fucosidase, 7.7 mU. For each sample, the number of spots depends on the number of exoglycosidases, alone or in combination, to be tested.
3. The MALDI plate is placed at 37°C for 6 h in a crystallization beaker containing water to saturate the atmosphere, thereby avoiding the drying of the drop at the surface of the target. In addition, high-purity water must be regularly added to the spots to prevent air-drying.
4. The enzymatic reactions are stopped by allowing the samples to dry at room temperature.
5. Two microliters of the matrix solution (DHB) is mixed with the dried samples and allowed to dry at room temperature. The MALDI-TOF-MS analysis is performed as described above (*see* **Subheading 3.9**). However, it is important to note that the spectra must be recorded over a range of *m/z* 900–5,000 for underivatized *N*-glycans.

An aliquot (15%) of the *N*-glycans released from 200 μg of calf fetuin (**Fig. 3**) and 20 μL of normal human serum (**Fig. 4**) has been used to illustrate this methodology.

3.12. Assignments of Molecular Ions Observed in MALDI-TOF-MS Spectra of Native and Permethylated Glycans

The interpretation of the MALDI-TOF-MS spectra of permethylated and native glycans are given in **Tables 1** and **2**.

1. $[M + Na]^+$ values are obtained by adding the sum of the residual masses (**Table 1**) and the sum of the sodiated reducing and nonreducing end increments (**Table 2**).
2. For *O*-glycans, it has to be kept in mind that they are reduced with $NaBH_4$ during their release from the peptidic chain. Therefore, $[M + Na]^+$ values for *O*-glycans are obtained by adding the sum of the residual masses (**Table 1**) and the sum of sodiated nonreducing and reduced reducing end increments (**Table 2**).

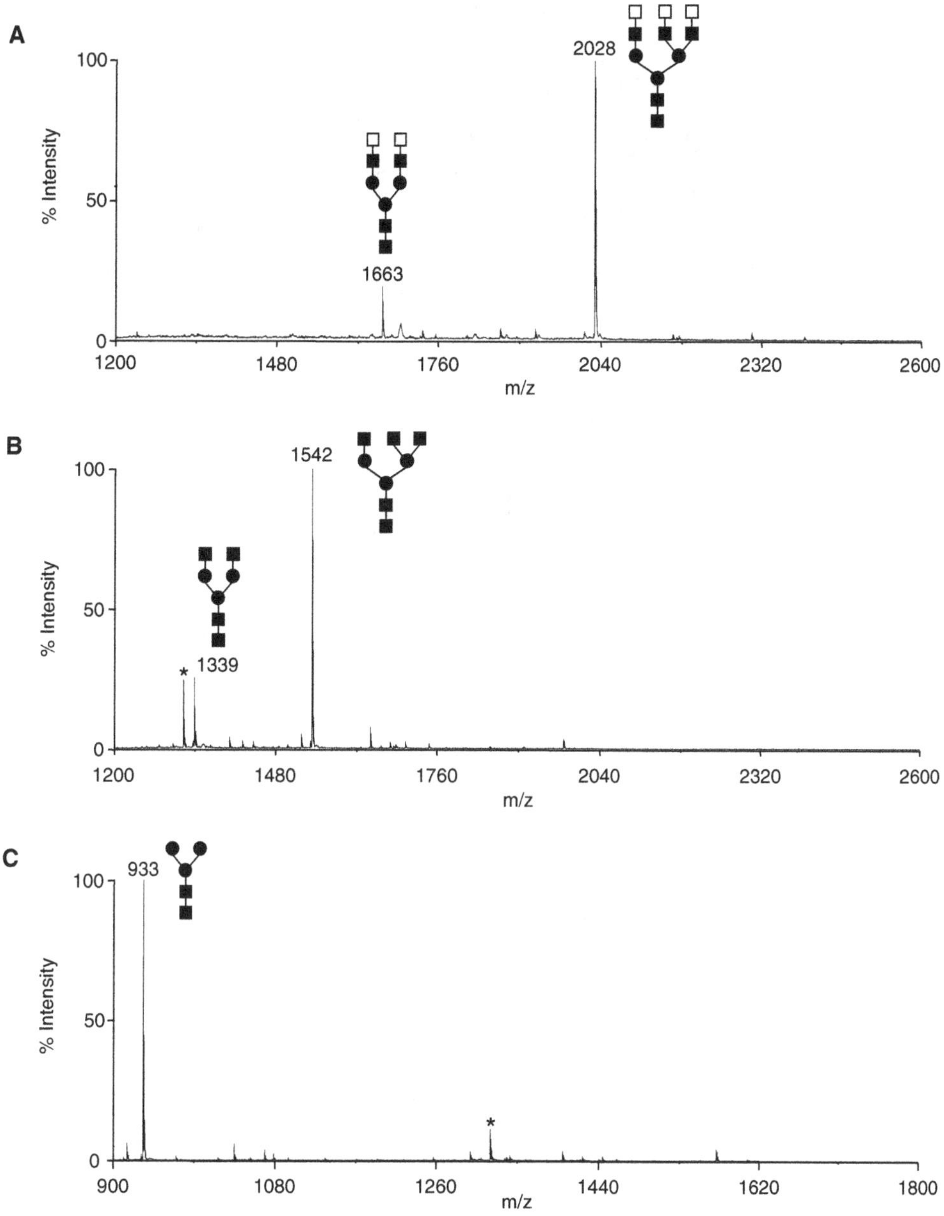

Fig. 3. MALDI-TOF-MS spectra of native *N*-glycans of the calf fetuin digested by α-sialidase (**A**), α-sialidase and β-galactosidase (**B**), and α-sialidase and β-galactosidase and β-*N*-acetylhexosaminidase (**C**). Fetuin was reduced, carboxamidomethylated, and digested with trypsin. Glycans were then released from the resulting peptides/glycopeptides by digestion with PNGase F. Glycans released by PNGase F were separated from the peptides and *O*-glycopeptides using a Sep-Pak cartridge C18 and desalted on a nonporous graphitized carbon solid-phase extraction cartridge. The glycans were then analyzed in positive ion reflective mode after on-target exoglycosidase digestions, as $[M + Na]^+$ pseudomolecular ions. The ions marked with asterisks arise from contaminants. Symbols are as in Fig. 1.

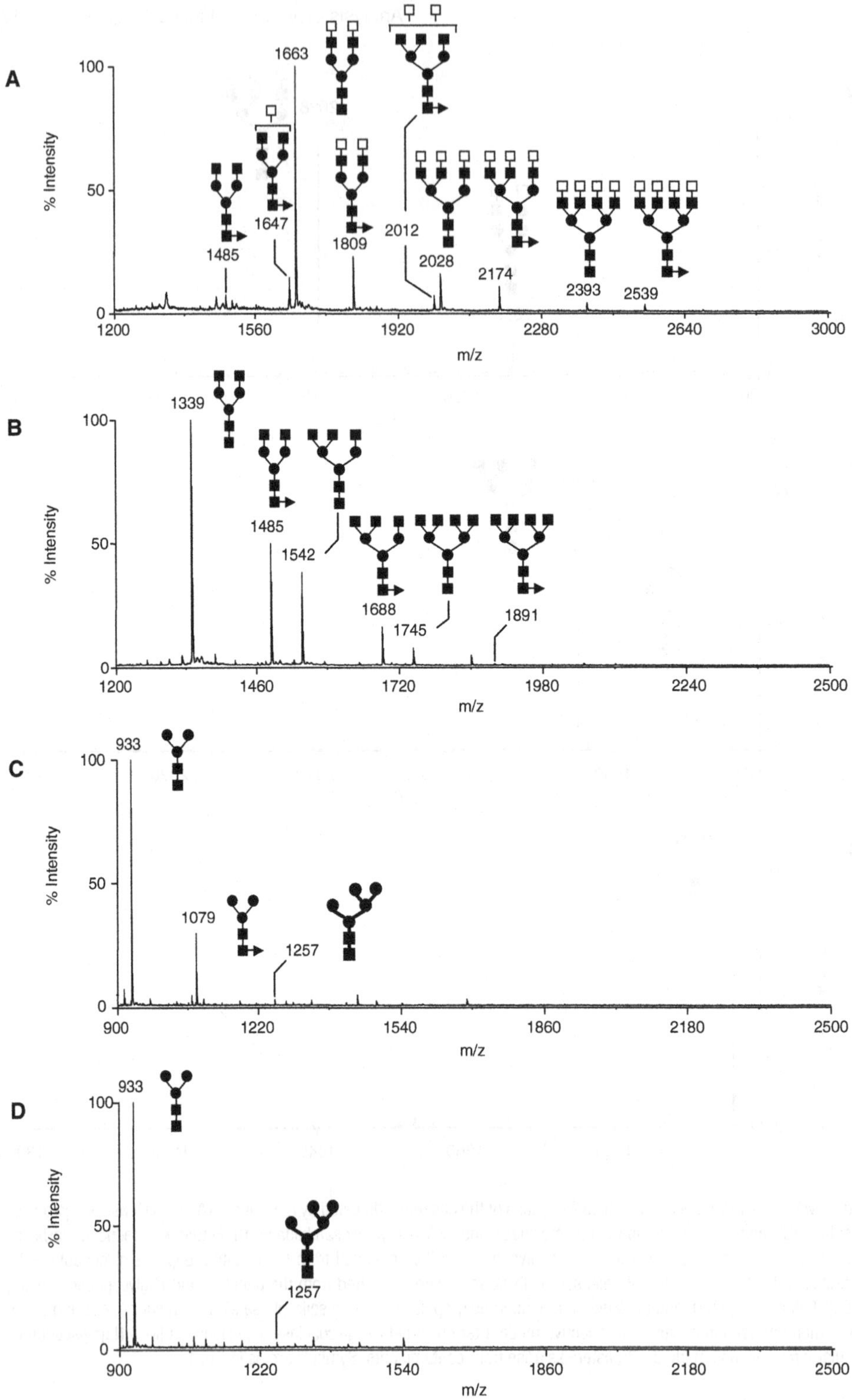

A
100
50
0
% Intensity
1485
1647
1663
1809
2012
2028
2174
2393
2539
1200
1560
1920
2280
2640
3000
m/z
B
100
50
0
% Intensity
1339
1485
1542
1688
1745
1891
1200
1460
1720
1980
2240
2500
m/z
C
100
50
0
% Intensity
933
1079
1257
900
1220
1540
1860
2180
2500
m/z
D
100
50
0
% Intensity
933
1257
900
1220
1540
1860
2180
2500
m/z

Table 2
Masses of nonreducing and reducing end moieties

	Native		Permethylated	
Terminal group	**Accurate mass**	**Average mass**	**Accurate mass**	**Average mass**
Reducing end	18.009	18.014	46.041	46.068
Reduced reducing end	20.025	20.030	62.072	62.111
Sum of masses with the sodium for *N*-glycans	40.998	41.004	69.030	69.058
Sum of masses with the sodium for reduced *O*-glycans	43.014	43.020	85.061	85.100

4. Notes

1. EDTA is added in the dialysis solution to avoid protein degradation by proteases. However, since EDTA cannot be removed during the lyophilization step, it is omitted in the last dialysis bath.
2. Some commercial PNGases F can be stored at – 20°C in presence of glycerol as preservative. However, for the MALDI-TOF-MS studies, glycerol prevents the cocrystallization of the DHB matrix and the sample. For this reason, we use only freeze-dried PNGase F without glycerol.
3. Prior to use, Dowex beads must be stabilized in the H^+ form as follows: Beads are treated twice with 3 volumes of 3 M NaOH for 30 min at room temperature. After extensively washing with water, the beads are treated twice with 3 volumes of a 3 M HCl solution for 30 min at room temperature. Beads are then sequentially washed with water until a neutral pH is obtained, and then washed again three times with

Fig. 4. MALDI-TOF-MS spectra of *N*-glycans released from normal human serum after treatment with α-sialidase (**A**), α-sialidase and β-galactosidase (**B**), α-sialidase and β-galactosidase and β-*N*-acetylhexosaminidase (**C**) and α-sialidase and β-galactosidase and β-*N*-acetylhexosaminidase and α-fucosidase (**D**). Proteins/glycoproteins were reduced, carboxamidomethylated, and digested with trypsin. Glycans were released from the resulting peptides/glycopeptides by digestion with PNGase F. Glycans released by PNGase F were separated from peptides using a C18 Sep-Pak cartridge and desalted on a nonporous graphitized carbon solid-phase extraction cartridge. The glycans were then analyzed in positive ion reflective mode after on-target exoglycosidase digestions, as $[M + Na]^+$ pseudomolecular ions. Symbols are as in Fig. 1.

3 volumes of 5% (v/v) acetic acid. Beads can be stored in the 5% (v/v) acetic acid solution for several months at room temperature. The beads are washed again three times with 3 volumes of 5% (v/v) acetic acid before packing the column.

4. The use of truncated Pasteur pipettes limits the volume of Dowex resin to 2 mL. In these conditions, it is possible to fully remove cationic materials (sodium salts, amino acids, and peptides) obtained from the reductive treatment of glycoproteins up to 2 mg.
5. The mortar and pestle must be placed in an oven. Thirty minutes before permethylation, they are taken out of the oven and allowed to remain at room temperature.
6. Since the PNGase F enzyme does not release any *N*-glycans when glycosylation sites are on the first or last amino acid residue of a peptide, the protease must be carefully chosen. For this reason, when PNGase F digestion is performed on one glycoprotein, the amino acid sequence of this glycoprotein must be known. Trypsin can be replaced by other proteases or by a chemical reagent such as cyanogen bromide in order to generate glycopeptides without glycosylation sites present on the first or last amino acid residue.
7. PNGase F releases practically all protein-bound N-linked carbohydrates except those with fucose attached to the 3 position of the Asn-linked GlcNAc residue. The corresponding glycoproteins are commonly found in plants and in insect cells. Such PNGase F-resistant glycans have been found to be sensitive to PNGase A, an enzyme found in almond emulsin.
8. In spite of their hydrophilic character, *N*-glycans interact very weakly with the hydrophobic C18 phase of the Sep-Pak cartridge. To ensure that they are recovered in the nonretained fractions, it is recommended to add 0.1% (v/v) TFA in the washing buffer to eliminate all interactions between oligosaccharides and the C18 phase.
9. Two milliliters of Dowex 50 × 8 can retain the equivalent of 2 mg of glycoprotein. For higher quantities of glycoprotein, the volume of the column of Dowex 50 × 8 must be proportionally increased.
10. In some cases, the chloroform phase can remain troubled even after six washings with water. It is not necessary to continue to wash the chloroform phase since the Sep-Pak C_{18} cartridge allows full removal of the remaining salts.
11. Before conducting on-plate exoglycosidase digestions, make sure that the quantity of oligosaccharide is sufficient. Therefore, it is necessary to perform the MALDI-TOF-MS analysis using 1 μL of each sample (from 5 to 10 pmol) to check this point.

Acknowledgments

This research was supported by the Centre National de la Recherche Scientifique (Unité Mixte de Recherche CNRS/USTL 8576; Director: Dr Jean-Claude Michalski), the Ministère de la Recherche et de l'Enseignement Supérieur. The Mass Spectrometry facility used in this study was funded by the European Community (FEDER), the Région Nord-Pas de Calais (France), and the Université des Sciences et Technologies de Lille.

References

1. Hart, G.W. (1992) Glycosylation. *Curr Opin Cell Biol.* 4, 1017–1723.
2. Kornfeld, R., and Kornfeld, S. (1985) Assembly of asparagine-linked oligosaccharides. *Annu Rev Biochem.* 54, 631–664.
3. Rudd, P.M., Colominas, C., Royle, L., Murphy, N., Hart, E., Merry, A.H., Hebestreit, H.F., and Dwek, RA. (2001) A high-performance liquid chromatography based strategy for rapid, sensitive sequencing of N-linked oligosaccharide modifications to proteins in sodium dodecyl sulphate polyacrylamide electrophoresis gel bands. *Proteomics* 1, 285–294
4. Takasaki, S., Mizuochi, T., and Kobata, A. (1982) Hydrazinolysis of asparagine-linked sugar chains to produce free oligosaccharides. *Methods Enzymol.* 83, 263–268.
5. Huang, C.C., and Aminoff, D. (1972) Enzymes that destroy blood group specificity. V. The oligosaccharase of *Clostridium perfringens. J. Biol. Chem.* 247, 6737–6742.
6. Ishii-Karakasa I., Iwase, H., Hotta K., Tanaka Y., and Omura S. (1992) Partial purification and characterization of an endo-alpha-N-acetylgalactosaminidase from the culture medium of Streptomyces sp. OH-11242. *Biochem. J.*. 288, 475–482.
7. Chai, W., Feizi, T., Yuen, C.T., and Lawson, A.M. (1997) Nonreductive release of O-linked oligosaccharides from mucin glycoproteins for structure/function assignments as neoglycolipids: application in the detection of novel ligands for E-selectin. *Glycobiology* 7, 861–872.
8. Huang, Y., Mechref, Y., and Novotny, M.V. (2001) Microscale nonreductive release of O-linked glycans for subsequent analysis through MALDI mass spectrometry and capillary electrophoresis. *Anal. Chem.* 73, 6063–6069.
9. Huang, Y., Konse, T., Mechref Y., and Novotny, M.V. (2002) Matrix-assisted laser desorption/ionization mass spectrometry compatible beta-elimination of O-linked oligosaccharides. *Rapid Commun. Mass Spectrom.* 16, 1199–1204.
10. Carlson, D.M. (1968) Structures and immunochemical properties of oligosaccharides isolated from pig submaxillary mucins. *J. Biol. Chem.* 243, 616–626.
11. Harvey, D.J. (2005) Proteomic analysis of glycosylation: structural determination of N- and O-linked glycans by mass spectrometry. *Expert Rev. Proteomics* 2, 87–101.
12. Morelle, W., and Michalski, J.C. (2005) The mass spectrometric analysis of glycoproteins and their glycan content. *Curr. Anal. Chem.* 1, 29–57.
13. Beavis, R.C., and Chait, B. (1990) High-accuracy molecular mass determination of proteins using matrix-assisted laser desorption mass spectrometry. *Anal. Chem.* 62, 1836–1840.
14. Finke, B., Mank, M., Daniel, H., and Stahl, B. (2000) Offline coupling of low-pressure anion-exchange chromatography with MALDI-MS to determine the elution order of human milk oligosaccharides. *Anal. Biochem.* 284, 256–265.
15. Powell, A.K. and Harvey, D.J. (1996) Stabilization of sialic acids in N-linked oligosaccharides and gangliosides for analysis by positive ion matrix-assisted laser desorption/ionization mass spectrometry. *Rapid Commun Mass Spectrom.* 10, 1027–1032.
16. Sekiya, S., Wada, Y., and Tanaka, K. (2005) Derivatization for stabilizing sialic acids in MALDI-MS. *Anal Chem.* 77, 4962–4968.
17. Ciucanu, I., and Kerek, F. (1984) A simple and rapid method for the permethylation of carbohydrates. *Carbohydr.Res.* 131, 209–217.
18. Albersheim, P., Nevins, D.J., English, P.D., and Karr, A. (1967) A method for the analysis of sugars in plant cell wall polysaccharides by gas-liquid chromatography. *Carbohydr. Res.*, 5, 340–345.
19. Packer, N.H., Lawson, M.A., Jardine, D.R., and Redmond, J.W. (1998) A general approach to desalting oligosaccharides released from glycoproteins. *Glycoconj J.* 15, 737–747.

Chapter 2

Infrared Multiphoton Dissociation Mass Spectrometry for Structural Elucidation of Oligosaccharides

Bensheng Li, Hyun Joo An, Jerry L. Hedrick, and Carlito B. Lebrilla

Summary

The structural elucidation of oligosaccharides remains a major challenge. Mass spectrometry provides a rapid and convenient method for structural elucidation on the basis of tandem mass spectrometry. Ions are commonly selected and subjected to collision-induced dissociation (CID) to obtain structural information. However, a disadvantage of CID is the decrease in both the degree and efficiency of dissociation with increasing mass.

In this chapter, we illustrate the use of infrared multiphoton dissociation (IRMPD) to obtain structural information for *O*- and *N*-linked oligosaccharides. The IRMPD and CID behaviors of oligosaccharides are compared.

Key words: Infrared multiphoton dissociation, Oligosaccharide, Collision-induced dissociation, Cross-ring cleavage.

1. Introduction

Glycosylation is an important posttranslational modification of proteins *(1)*. Carbohydrate moieties in glycoproteins are known to mediate a number of essential functions such as protein folding *(4, 5)*, cell–cell and cell–matrix recognition, cellular adhesion, inter- and intracellular interaction, and protection *(6)*. The high complexity of glycans poses a challenge in the determination of carbohydrate structures. The conventional analysis for glycans in glycoconjugates involves a combination of techniques such as NMR, mass spectrometry (MS), chemical derivatization, and monosaccharide composition analysis *(7–10)*. Mass spectrometry

Nicolle H. Packer and Niclas G. Karlsson (eds.), *Methods in Molecular Biology, Glycomics: Methods and Protocols, vol. 534*

DOI: 10.1007/978-1-59745-022-5_2

has become a versatile analytical technique for structural characterization of small organic compounds and biomacromolecules in proteomics, glycomics, and metabolomics because of its high sensitivity, accuracy, and resolution *(11–14)*. Tandem mass spectrometry (MS^n), in which a precursor ion is mass selected, dissociated, and the product ions are analyzed, has been an essential technique for structure analysis of biomolecules. In the structure elucidation of oligosaccharides, collision-induced dissociation (CID) is the most common method to obtain sequence, connectivity, stereochemistry, and even the linkage.

Infrared multiphoton dissociation (IRMPD) is especially well suited for the dissociation of trapped ions in Fourier transform ion cyclotron resonance mass spectrometry (FTICR MS). The resonant absorption of a few IR photons with relatively low energy allows the ions of interest to be fragmented selectively along its lowest-energy dissociation pathways. In an IRMPD event, a precursor ion of interest is mass-selected and trapped in the analyzer cell, dissociated by absorption of IR photons, and the product ions are analyzed for structural information. There are some advantages of IRMPD over other MS^n techniques, such as CID. Most organic molecules are IR active and readily absorb IR photons. In contrast to CID, no collision gas is needed for the dissociation of precursor ions, thus shortening the analysis time. IR photon absorption does not cause translational excitation of ions as CID does, therefore minimizing ion losses due to ejection or ion scattering. Moreover, since both the precursor and product ions absorb IR photons, more extensive fragmentation occurs to yield richer structural information. With IRMPD, ions are fragmented on-axis, yielding more efficient observation of product ions. IRMPD also provides better control of excitation energy with minimal mass discrimination. IRMPD has been extensively applied with FTICR-MS for the structural characterization of intact proteins *(15–18)*, peptides *(19–23)*, oligonucleotides *(24–26)*, and oligosaccharides *(27–29)*.

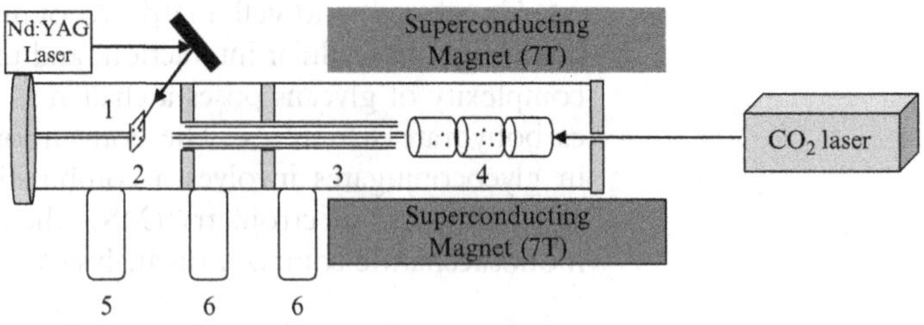

1. MALDI ion source area 2. Multisample plate 3. Quadrupole ion gui
4. FT-ICR cell 5. Turbo pump 6. Cryo pumps

Fig. 1. A schematic assembly of MALDI-FTICR mass spectrometer coupled with a CO_2 infrared laser for IRMPD.

2. Materials

2.1. Release and Purification of Oligosaccharides

1. Materials for release of *O*-linked oligosaccharides, *see* the Chapter 10 "Collision-Induced Dissociation Tandem Mass Spectrometry for Structural Elucidation of Glycans.
2. PNGase F (NEW ENGLAND BioLabs).
3. *N*-linked release solution: 100 mM NH_4HCO_3, pH 7.5.
4. Heating block 37°C.
5. Ethanol.
6. Nonporous graphitized carbon and solutions to clean up oligosaccharides as described in the Chapter 10 "Collision-Induced Dissociation Tandem Mass Spectrometry for Structural Elucidation of Glycans.

2.2. HPLC Separation of Oligosaccharide Components

1. Hypercarb porous graphitized carbon (PGC) column (2.1 × 100 mm, 5 µm, ThermoQ Hypersil Division, Bellfonte, PA).
2. Solvent A (neutral oligosaccharides): H_2O.
3. Solvent B (neutral oligosaccharides): acetonitrile (AcN) containing 0.05% (v/v) trifluoroacetic acid (TFA).
4. Gradient HPLC pump (up to 250 µL/min) with photo diode-array (PDA) detector (206 nm).
5. Fraction collector with 1.5-mL microcentrifuge tubes.

2.3. MALDI-FTICR MS Analyses of Oligosaccharides

1. A commercial MALDI-FT mass spectrometer (IonSpec, Irvine, CA) with an external ion source was used to perform the analysis. The instrument is equipped with a 7.0-T shielded, superconducting magnet and a Nd:YAG laser at 355 nm.
2. MALDI matrix solution: 0.4 M 2,5-dihydroxybenzoic acid (DHB), in 50:50 H_2O/AcN.
3. Positive ion dopant: 0.1 M NaCl in 50:50 H_2O/AcN
4. MALDI stainless steel target plate.

2.4. Infrared Multiphoton Dissociation

1. A continuous-wave, turn-key 25 Watt CO_2 laser; PLX25-s, Parallax Technology, Inc. (Waltham, MA, USA).
2. 2× beam expander from Synrad, Inc. (Mukilteo, WA, USA).

3. Methods

3.1. Release and Purification of Oligosaccharides

1. The release and purification of *O*-linked oligosaccharides from glycoproteins were performed by using a method developed in the laboratory. Briefly, a certain amount of glycoprotein was treated with 1.0 M $NaBH_4$ and 0.1 M NaOH at 42 °C for 24 h,

followed by neutralization with precooled 1.0 M hydrochloric acid to a pH of 2 to get rid of the excess $NaBH_4$ in an ice bath *(30, 31)*.

2. All *N*-linked oligosaccharides were released by incubating the glycoproteins with PNGase F (1 U) in 100 mM NH_4HCO_3 (pH 7.5) for 12 h at 37 °C. The reaction mixture was boiled in water bath for 5 min to stop the reaction, followed by ethanol addition to get a 90% aqueous ethanol solution. Then, reaction mixture was chilled at −20 °C for 30 min and centrifuged. The supernatant was recovered and dried. The residue from the dried supernatant was redissolved in nanopure water for purification.
3. The resulting *O*- or *N*-linked oligosaccharide mixtures were desalted and purified by the solid-phase extraction (SPE) using nonporous graphitized carbon cartridge. The purified oligosaccharides were subjected to HPLC or MALDI-FTICR mass spectrometry.

3.2. HPLC Separation of Oligosaccharide Components

1. Oligosaccharide mixtures purified from SPE were further purified by the separation with HPLC on a Hypercarb porous graphitized carbon (PGC) column.
2. For neutral oligosaccharide separation, the solvent A was H_2O and solvent B was acetonitrile (AcN) containing 0.05% (v/v) trifluoroacetic acid (TFA). The gradient elution system was 5% to 16% of B during the first 44 min, 16–28% of B during the following 12 min (44–56 min), 28–32% of B during the next 13 min (56–69 min), 32% of B during 69–80 min, and finally followed by 5-min of elution with the solvent B down to 5%. The flow rate was set to 250 μL/min for the HPLC separation. The effluents were monitored at 206 nm by a photo diode-array (PDA) detector and collected into 1.5 mL microcentrifuge tubes at 1-min intervals.
3. Each fraction was analyzed with MALDI-FTICR for oligosaccharides after solvent evaporation.

3.3. MALDI-FTICR MS Analyses of Oligosaccharides

1. MALDI sample was prepared by loading 1–6 μL of analyte and 1 μL of matrix solution on a stainless steel target plate. For the positive mode analyses, 1 μL of 0.1 M NaCl in 50:50 H_2O/AcN was applied to the spot to enrich the Na^+ concentration and thus produce the primarily sodiated signals.
2. The plate was placed in ambient air to dry the sample spots before insertion into the ion source.

3.4. Infrared Multiphoton Dissociation

1. The CO_2 laser was installed at the rear of the superconducting magnet to provide infrared photons for IRMPD. The schematic instrument assembly is shown in **Figure 1**.
2. The laser has a working wavelength of 10.6 μm (0.1 eV per photon) and a beam diameter of 6 mm as specified by the

manufacturer. A beam expander was used to expand the laser beam to 12 mm.

3. To perform IRMPD experiments, some modifications were made on the ICR cell and the vacuum chamber. The original trapping plate with the electron filament was removed and replaced with a copper plate with a 13-mm hole in the center. Four copper wires (0.24 gauge) were fixed by screws onto the plate, two horizontally and two vertically, over the hole and set 3.6 mm apart. The existing aluminum vacuum chamber was replaced with a smaller-diameter (101.6 mm OD) tube containing a 70-mm BaF_2 window (Bicron Corp., Newbury, OH, USA).
4. The infrared laser was aimed directly toward the center of the analyzer cell. Ions were subsequently irradiated with an IR laser pulse lasting between 200 and 3,000 ms.

3.5. Interpretation of IRMPD Spectra for O-Linked Glycans

It is not often possible to collect a single oligosaccharide from complex mixtures even with HPLC separation. A fraction from a mixture of glycans obtained from the egg jelly glycoproteins of *X. borealis* may contain a number of oligosaccharides. **Figure 2** shows a typical MALDI-FT mass spectrum of a single HPLC fraction. Four major components were obtained with *m/z* 878.311,

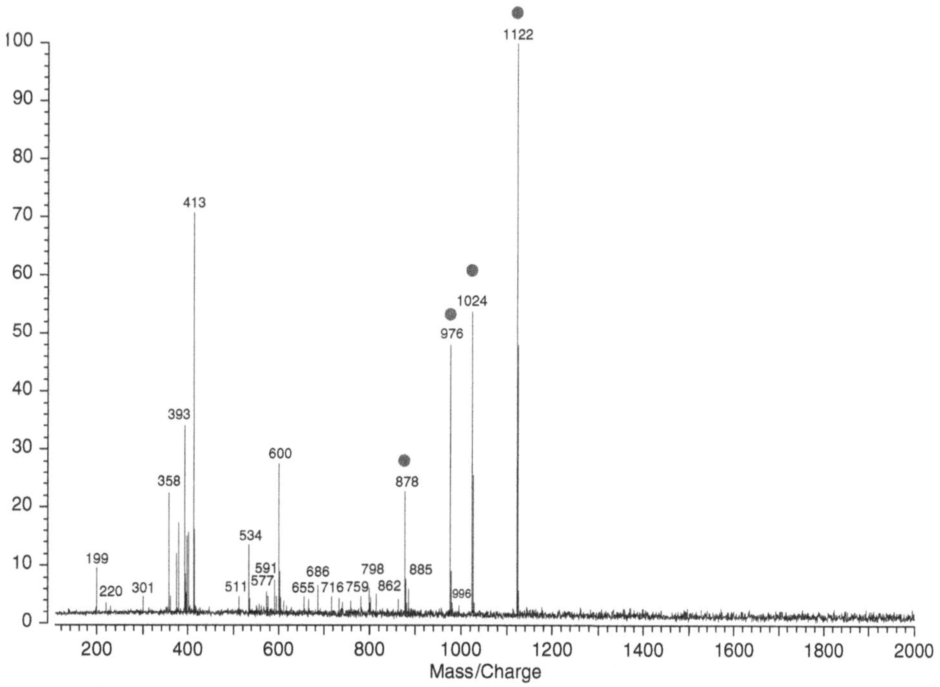

Fig. 2. MALDI mass spectrum (signals are due to sodium-coordinated species) of one HPLC fraction containing several *O*-linked oligosaccharides released from the egg jelly coat of *X. borealis*. *O*-glycans are marked by *filled circles*.

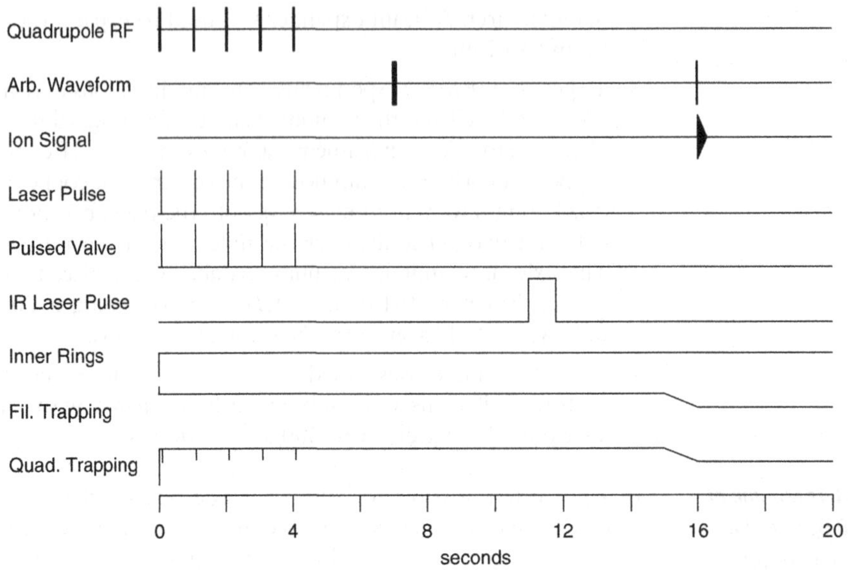

Fig. 3. A typical pulse sequence employed for IRMPD in an FTICR mass spectrometer.

976.360, 1,024.375, and 1,122.421. On the basis of the exact mass, monosaccharide compositions for each component were obtained. The sequence of each component was determined by IRMPD as shown below.

Figure 3 shows a typical pulse sequence for MALDI-IRMPD analysis. The Nd:YAG laser was fired five times at 1,000-ms intervals to get enough ions in the ICR cell. A desired ion is readily selected in the analyzer with the use of an arbitrary waveform generator and a frequency synthesizer. The CO_2 infrared laser was typically fired at 11,000 ms with a duration of 200–3,000 ms depending on the molecular size and structure of the target ion while the trapping plates were elevated to +4 V. The voltages on the trapping plates were then ramped down to + 1.2 V from 15,000 to 16,000 ms for detection.

The resulting spectra are shown in **Fig. 4**. A single component (*m/z* 1,024.375) was isolated from the *O*-glycan mixture of an HPLC fraction (**Fig. 2**) by ejecting all other ions from the ICR cell (**Fig. 4A**). After absorption of IR photons, the selected ions were fragmented into product ions as shown in **Fig. 4B**. The composition was determined based on the exact mass to correspond to three hexoses (Hex), two deoxyhexoses (dHex), and one *N*-acetylhexosamine (HexNAc) (theoretical mass 1,024.369 Da, experimental mass 1,024.375 Da, Δm = 0.006 or 6 ppm).

The primary monosaccharide sequence was elucidated from the IRMPD spectrum. The loss of a dHex (*m/z* 878), a Hex (*m/z* 862) or HexNAc-ol (the reduced HexNAc, *m/z* 801)

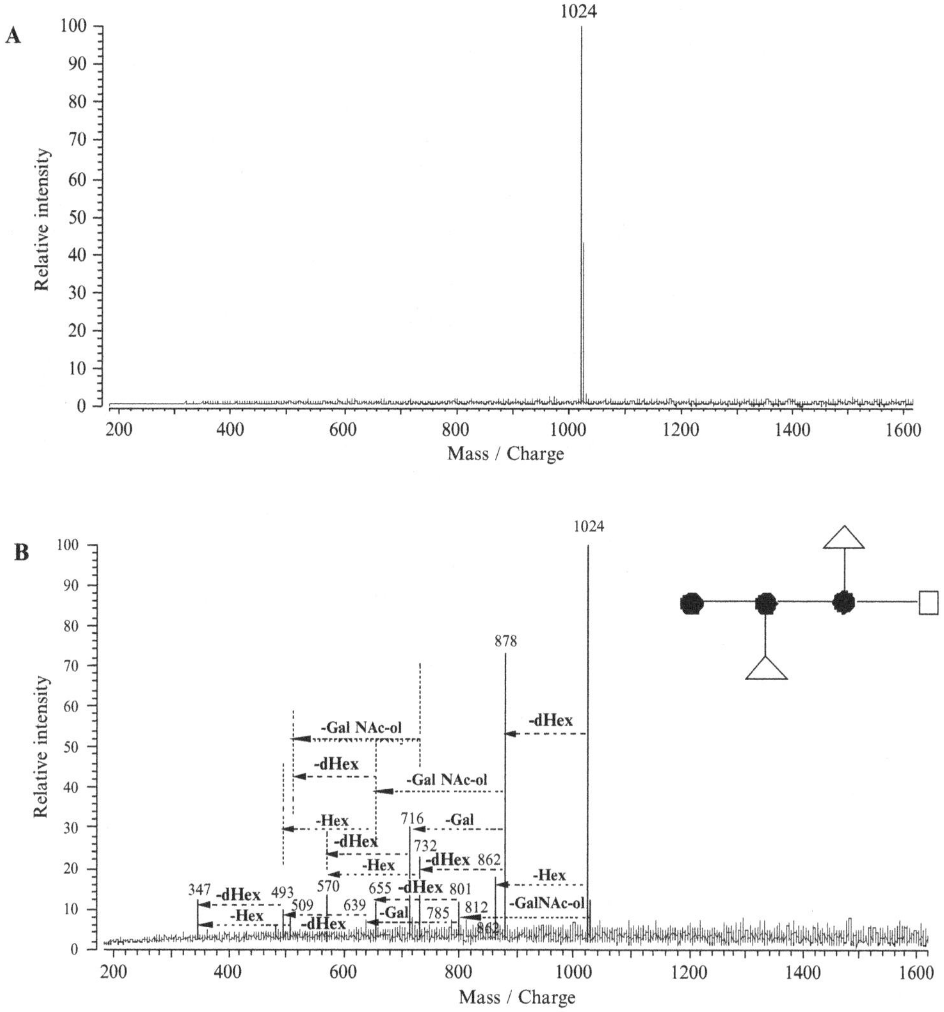

Fig. 4. The sequence of an *O*-linked glycan (*m/z* 1,024, [M + Na]$^+$) released from glycoprotein of *X. borealis* egg jelly was determined primarily by IRMPD. (**A**) The precursor ion (*m/z* 1,024) was selected and isolated from all other ions. (**B**) The ion was fragmented by absorption of IR photons over 1 s with laser power of 12.7 W. *Filled circle* Hex; *open square* GalNAc-ol; *open triangle* dHex.

indicates that the three residues are in terminal positions. The HexNAc-ol is the reducing end and is putatively a GalNAc-ol. The other two residues are in the nonreducing ends indicating at least one branched point. The ion with *m/z* 862 can lose one dHex (*m/z* 716) or one GalNAc-ol (*m/z* 639), but no further loss of one Hex was observed (*m/z* 700 not present) suggesting that at least one dHex is attached to a second Hex that is internal. The ion with *m/z* 639 loses only one dHex (*m/z* 493) with no further loss of the second Hex indicating that the other dHex must be connected to a third internal Hex. The primary sequence of the precursor ion (*m/z* 1,024) can be deduced as being Hex–Hex(dHex)–Hex(dHex)–GalNAc-ol. The remainder of the fragment ions are internal fragments that are consistent with the proposed structure (inset of **Fig. 4B**).

Another oligosaccharide component in the HPLC fraction with a mass of *m/z* 1,122.421 (*see* **Fig. 2**) was also subjected to IRMPD. On the basis of the exact mass, the ion consists of two hexoses (Hex), one deoxyhexose (dHex), and three *N*-acetyl hexosamines (HexNAc) (theoretical mass 1,122.417 Da, experimental mass 1,122.421 Da, m = 0.004 or 4 ppm). The IRMPD spectrum of isolated *m/z* 1,122 is shown in **Fig. 5**. The quasimolecular ion $[M + Na]^+$ loses a single dHex (*m/z* 976) followed by the loss of one Hex (*m/z* 814) indicating that the previously lost dHex is attached to a Hex residue at the nonreducing end. There is no Hex loss (*m/z* 960) from the quasimolecular ion suggesting that there is no Hex at the nonreducing end. The ion *m/z* 814 loses either one Hex (*m/z* 652), one GalNAc-ol (*m/z* 591), or one HexNAc (*m/z* 611). It is noted that the loss of GalNAc-ol or the loss of HexNAc from the same precursor ion *m/z* 814 occurs by two separate pathways, namely, the former follows the loss of the Hex connecting to reducing end and the latter is preceded

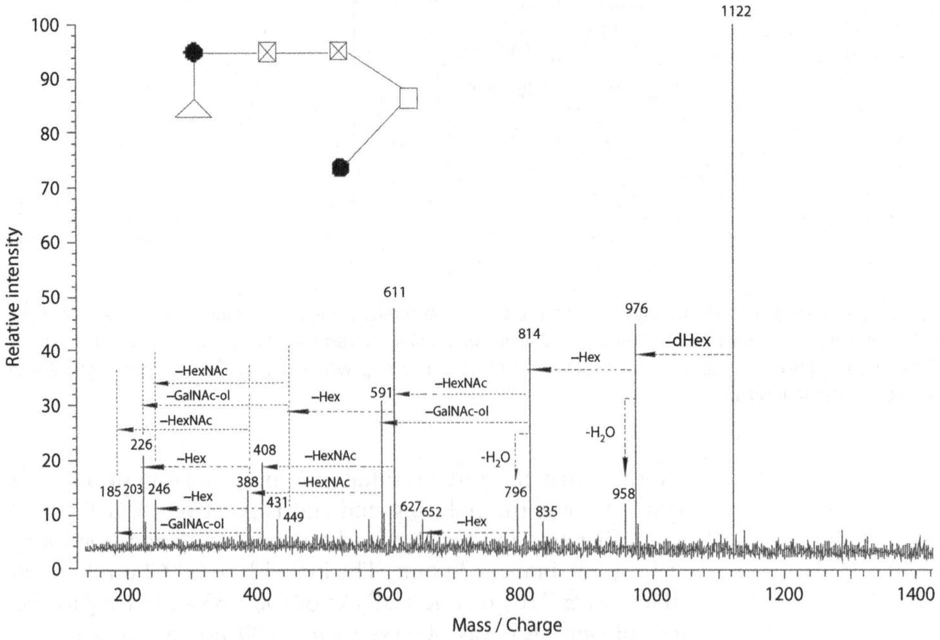

Fig. 5. IRMPD spectrum of an *O*-linked oligosaccharide with *m/z* 1,122 ($[M + Na]^+$, see Fig. 2). The IR laser was fired for 1 s with laser power setting at 12.7 W. The primary component sequence of the *O*-linked glycan was determined based on the spectrum and shown in the *inset*. *Filled circle* Hex; *open square* GalNAc-ol; *open triangle* dHex; *crossed square* HexNAc.

by the loss of the other internal Hex. There is another HexNAc loss (*m/z* 408) from the ion *m/z* 611, which comes directly from *m/z* 814 through the loss of one HexNAc, thus indicating that the two HexNAc are connected to each other. On the basis of the above information, the primary sequence of the ion *m/z* 1,122 is determined as shown in the inset of **Fig. 5**.

The sequence is further confirmed by the additional fragments. The ion *m/z* 611 loses one Hex (*m/z* 449) corresponding to HexNAc-GalNAc-ol, or one HexNAc to yield Hex-GalNAc-ol (*m/z* 408). The ion *m/z* 449 loses either one HexNAc to yield a GalNAc-ol (*m/z* 246) or one GalNAc-ol to produce a HexNAc (*m/z* 226). The *m/z* 408 ion is fragmented into two ions to yield either a Hex (*m/z* 185) or a GalNAc-ol (*m/z* 246). In addition, the ion *m/z* 591, Hex–HexNAc–HexNAc, loses one HexNAc to yield Hex–HexNAc (*m/z* 388), which then loses one Hex to yield a HexNAc (*m/z* 226) (*see* **Note 1**).

3.6. Interpretation of IRMPD Spectra for N-Linked Glycans

A common problem with CID for large molecules is the lack of fragments so that MS^n ($n > 2$) is required to obtain the complete sequence. IRMPD requires only a single MS/MS event to obtain the complete sequence resulting in greater sensitivity (better signal-to-noise).

As an example, a high-mannose type *N*-linked glycan consisting of two GlcNAc and nine mannoses (Man9) was examined *(32)*. The IRMPD spectrum of the quasimolecular ion (*m/z* 1,905.634, $[M + Na]^+$) is shown in **Fig. 6B**. The extensive fragmentation was observed in a single MS/MS event. In the contrast, the CID of this compound (**Fig. 6A**) did not produce the complete sequence even after MS^3 at which point the signal was lost. For *N*-linked glycans, the putative structure can be deduced based on the composition.

Often the goal of tandem MS is to confirm the putative structure and obtain linkage information. The CID and IRMPD of Man9 are compared in **Fig. 6**. Large fragment ions were obtained corresponding to the B_5 and B_4 fragments (*m/z* 1,685 and 1,481, respectively) in the CID. Subsequent mannose losses from B_4 ion were clearly discernable (*m/z* 1,319, 1,157, 995, 833, 671, and 509), but no series of mannose losses from B_5 were observed. In addition, a number of cross-ring cleavages were observed. These ions originated from the internal cleavage of the major branching mannose ($^{0,3}A_4$, *m/z* 923) and the subsequent mannose losses from the $^{0,3}A_4$ fragment ($^{0,3}A_4/Y_{5\alpha}$, *m/z* 761; $^{0,3}A_4/Y_{5\alpha}$, *m/z* 599; $^{0,3}A_4/Y_{4\alpha',5\alpha}$, *m/z* 437, respectively). A minor fragment corresponding to $^{0,4}A_4/Y_{5\alpha'}$ was also observed (*m/z* 731) with an accompanying mannose loss ($^{0,4}A_4/Y_{5\alpha}$, *m/z* 569). The cross-ring cleavages were limited, but provided information regarding the first branched residue and the linkages of the antennae.

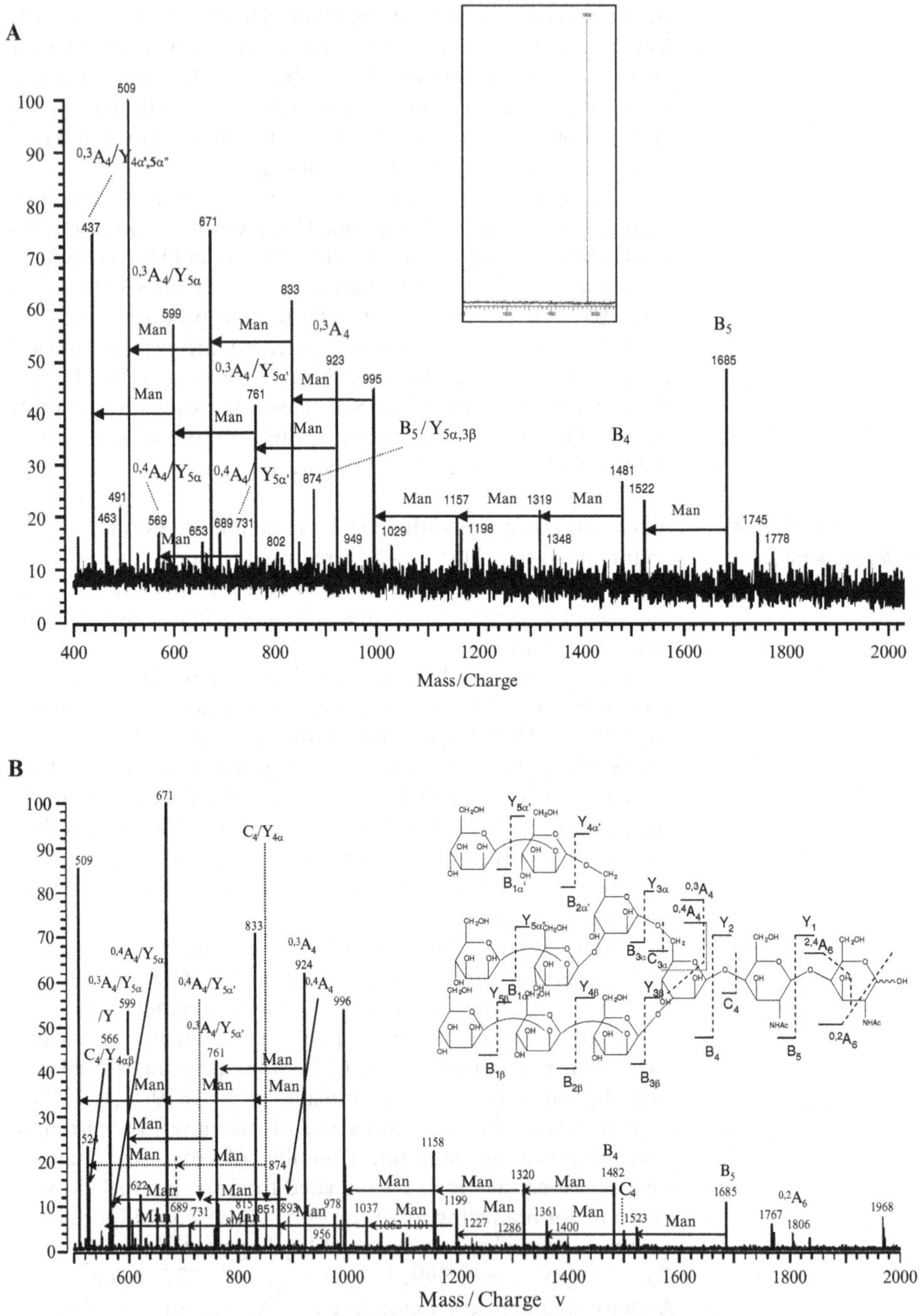

Fig. 6. (**A**) CID and (**B**) IRMPD spectra of *N*-linked oligosaccharide Man9. The spectra show cross-ring cleavages with some branching information as well as B and Y fragments. The inset of (**B**) shows the structure of Man9 with labeled fragmentations.

The IRMPD spectrum of Man9 (**Fig. 6B**) showed significantly more structural information and better signal-to-noise ratio. A fragment ion, $^{0,2}A_6$, corresponding to the cross-ring cleavage of the reducing end, was the largest fragment ion detected (*m/z* 1,806). The B_5 and B_4 ions (*m/z* 1,685 and 1,482, respectively) were readily observed along with fragments corresponding to subsequent mannose losses, respectively. The extensive fragmentation represents a major difference between the CID and IRMPD spectra. The mannose losses from the B_5 ion were incomplete in the CID spectrum (**Fig. 6A**), while both B_4 and B_5 showed the subsequent mannose losses in the IRMPD spectrum. The results demonstrate that with IRMPD essentially every glycosidic bond cleavage is represented while ion loss is minimized.

Another ion that was not observed in the CID spectrum is the C_4 fragment (*m/z* 1,499) and subsequent losses of mannose (C_4/Y_n). These ions were present in IRMPD spectrum at *m/z* 1,337, 1,175, 1,013, 851, 689, and 527. Some internal fragments were also present in the IRMPD spectrum, but not in the CID spectrum. These fragments corresponded to products of cross-ring cleavages occurring at the core branching mannose. These ions were present at *m/z* 893, 731, and 569 and corresponded to the $^{0,4}A_4$ ion and those with subsequent mannose losses, respectively. Another series of cross-ring cleavages representing the $^{0,3}A_4$ fragment from the same mannose core with subsequent mannose losses (*m/z* 923, 761, and 599) were observed and were more intense than the $^{0,4}A_4$ series. The fragmentation patterns observed in the IRMPD spectrum provided substantial information for the structural elucidation of $GlcNAc_2Man_9$ compared to the CID (*see* **Note 2**).

4. Notes

1. The primary sequence of *O*-linked glycans could be determined based solely on IRMPD analysis. However, their linkages and anomeric characters could not be determined. A method for obtaining this information as well as the identity of the residue involving the use of exoglycosidases with IRMPD or CID has been published previously *(30)*.
2. Owing to the symmetric nature of the antennae for Man9, there are several peaks in the CID (**Fig. 6A**) and IRMPD (**Fig. 6B**) that could be attributed to several cleavages. For simplicity, only a single assignment is provided for each cleavage.

Acknowledgements

The authors gratefully acknowledge the financial support from the National Institutes of Health (R01 GM049077).

References

1. Apweiler, R., Hermjakob, H. and Sharon, N. (1999) On the frequency of protein glycosylation, as deduced from analysis of the SWISS-PROT database. *Biochim. Biophys. Acta* 1473, 4–8.
3. Moens, S. and Vanderleyden J. (1997) Glycoproteins in prokaryotes. *Arch. Microbiol.* 168, 169–175.
4. Cyster, J. G., Shotton, D. M. and Williams, A. F. (1991) The dimensions of the T lymphocyte glycoprotein leukosialin and identification of linear protein epitopes that can be modified by glycosylation. *EMBO J.* 10, 893–902.
5. Helenius, A. and Aebi, M. (2001) Intracellular Functions of N-Linked Glycans. *Science* 291, 2364–2369.
6. Varki, A. (1993) Biological roles of oligosaccharides: all of the theories are correct. *Glycobiology* 3, 97–130.
7. Carpita, N.C., Shea, E.M., Niermann, C.J. and McGinnis, G.D., Eds. (1989) *Analysis of Carbohydrates by GLC and MS;* CRC: Boca Raton, FL, pp 151–216.
8. Hellerqvist, C. G. (1990) Linkage analysis using lindberg method. *Methods Enzymol.* 193, 554–573.
9. Mukerjea, R., Kim, D. and Robyt, J. F. (1996) Simplified and improved methylation analysis of saccharides, using a modified procedure and thin-layer chromatography. *Carbohydr. Res.* 292, 11–20.
10. Solouki, T., Reinhold, B. B., Costello, C. E., O'Malley, M., Guan, S. and Marshall, A. G. (1998) Electrospray ionization and matrix-assisted laser desorption/ionization Fourier transform ion cyclotron resonance mass spectrometry of permethylated oligosaccharides. *Anal. Chem.* 70, 857–864.
11. Song, A., Zhang, J., Lebrilla, C. B. and Lam, K. S. (2004) Solid-phase synthesis and spectral properties of 2-Alkylthio-6H-pyrano[2,3-f]benzimidazole-6-ones: A Combinatorial approach for 2-Alkylthioimidazocoumarins. *J. Comb. Chem.* 6(4), 604–610.
12. Skop, A. R., Liu, H., Yates, J. R., Meyer, B. J. and Heald, R. (2004) Functional proteomic analysis of the mammalian midbody reveals conserved cell division components, *Science* 305, 61–66.
13. An, H. J., Miyamoto, S., Lancaster, K. S., Kirmiz, C., Li, B., Lam, K. S., Leiserowitz, G. S. and Lebrilla, C. B. (2006) Global profiling of glycans in serum for the diagnosis of potential biomarkers for ovarian cancer. *J. Proteome Res.* 5, 1626–1635 (Web release at 05-27-2006).
14. Des Rosiers, C., Lloyd, S., Comte, B. and Chatham, J. C. (2004) A critical perspective of the use of 13C-isotopomer analysis by GCMS and NMR as applied to cardiac metabolism. *Metab. Eng.* 6, 44–58.
15. Little, D. P., Speir, J. P., Senko, M. W., O'Connor, P. B. and McLafferty, F. W. (1994) Infrared multiphoton dissociation of large multiply charged ions for biomolecule sequencing. *Anal. Chem.* 66, 2809–2815.
16. Mortz, E., O'Connor, P. B., Roepstorff, P., Kelleher, N. L., Wood, T. D., McLafferty, F. W. and Mann, M. (1996) Sequence tag identification of intact proteins by matching tandem mass spectral data against sequence data bases. *Proc. Natl. Acad. Sci. U.S.A.* 93, 8264–8267.
17. Li, W., Hendrickson, C. L., Emmett, M. R. and Marshall, A. G. (1999) Identification of intact proteins in mixtures by alternated capillary liquid chromatography electrospray ionization and LC ESI infrared multiphoton dissociation fourier transform ion cyclotron resonance mass spectrometry. *Anal. Chem.* 71, 4397–4402.
18. Meng, F., Cargile, B. J., Patrie, S. M., Johnson, J. R., McLoughlin, S. M. and Kelleher, N. L. (2002) Processing complex mixtures of intact proteins for direct analysis by mass spectrometry. *Anal. Chem.* 74, 2923–2929.
19. Jockusch, R. A., Paech, K. and Williams, E. R. (2000) Energetics from slow infrared multiphoton dissociation of biomolecules. *J. Phys. Chem. A* 104, 3188–3196.
20. Masselon, C., Anderson, G. A., Harkewicz, R., Bruce, J. E., Pasa-Tolic, L. and Smith, R. D. (2000) Accurate mass multiplexed tandem mass spectrometry for high-throughput polypeptide identification from mixtures. *Anal. Chem.* 72, 1918–1924.

21. Flora, J. W. and Muddiman, D. C. (2001) Selective, sensitive, and rapid phosphopeptide identification in enzymatic digests using ESI-FTICR-MS with infrared multiphoton dissociation. *Anal. Chem.* 73, 3305–3311.

22. Hakansson, K., Cooper, H. J., Emmett, M. R., Costello, C. E., Marshall, A. G. and Nilsson, C. L. (2001) Electron capture dissociation and infrared multiphoton dissociation MS/MS of an N-glycosylated tryptic peptide to yield complementary sequence information. *Anal. Chem.* 73, 4530–4536.

23. Li, L., Masselon, C. D., Anderson, G. A., Pasa-Tolic, L., Lee, S. W., Shen, Y., Zhao, R., Lipton, M. S., Conrads, T. P., Tolic, N. and Smith, R. D. (2001) High-throughput peptide identification from protein digests using data-dependent multiplexed tandem FTICR mass spectrometry coupled with capillary liquid chromatography. *Anal. Chem.* 73, 3312–3322.

24. Little, D. P. and McLafferty, F. W. (1995) Sequencing 50-mer DNAs using electrospray tandem mass spectrometry and complementary fragmentation methods. *J. Am. Chem. Soc.* 117, 6783–6784.

25. Little, D. P., Aaserud, D. J., Valaskovic, G. A. and McLafferty, F. W. (1996) Sequence information from 42–108-mer DNAs (complete for a 50-mer) by tandem mass spectrometry *J. Am. Chem. Soc.* 118, 9352–9359.

26. Little, D. P. and McLafferty, F. W. (1996) Infrared photodissociation of non-covalent adducts of electrosprayed nucleotide ions. *J. Am. Soc. Mass Spectrom.* 7, 209–210.

27. Xie, Y. and Lebrilla, C. B. (2003) Infrared multiphoton dissociation of alkali metal-coordinated oligosaccharides. *Anal. Chem.* 75, 1590–1598.

28. Zhang, J., Schubothe, K., Li, B., Russell, S. and Lebrilla, C. B. (2005) Infrared multiphoton dissociation of *O*-linked mucin-type oligosaccharides. *Anal. Chem.* 77, 208–214.

29. Goldberg, D., Bern, M., Li, B. and Lebrilla, C. B. (2006) Automatic determination of *O*-glycan structure from fragmentation spectra. *J. Proteome Research* 5, 1429–1434.

30. Zhang, J., Lindsay, L. L., Hedrick, J. L. and Lebrilla, C. B. (2004) Strategy for profiling and structure elucidation of mucin-type oligosaccharides by mass spectrometry. *Anal. Chem.* 76, 5990–6001.

31. Tseng, K., Hedrick, J. L. and Lebrilla, C. B. (1999) Catalog-library approach for the rapid and sensitive structural elucidation of oligosaccharides. *Anal. Chem.* 71, 3747–3754.

32. Lancaster, K. S., An, H. J., Li, B. and Lebrilla, B. C. (2006) Interrogation of *N*-linked oligosaccharides using infrared multi-photon dissociation in FT-ICR mass spectrometry. *Anal. Chem.* 78 (14), 4990–4997.

Chapter 3

Mass Spectrometry of N-Linked Glycans

Parastoo Azadi and Christian Heiss

Summary

The analysis of the *N*-glycan portions of glycoproteins has become important for the detection of various diseases involving altered glycosylation patterns, including cancer and congenital disorders of glycosylation, and in the quality control of therapeutic recombinant glycoproteins. Here in this chapter we give detailed analysis procedures necessary for the complete structural elucidation of any *N*-glycan mixture found in naturally occurring and recombinant glycoproteins. The protocols include monosaccharide composition analysis on the intact glycoprotein, the enzymatic release of *N*-glycans from glycoprotein and their separation from *O*-glycopeptides and peptides, and the structural characterization of the glycans by exoglycosidase digestions, methylation ("linkage") analysis, and mass spectrometry/tandem mass spectrometry of permethylated glycans using matrix-assisted laser desorption and electrospray ionization methods.

Key words: *N*-Glycan, Peptide *N*-glycosidase, Methylation analysis, Glycoprotein, Exoglycosidase, Mass spectrometry, MALDI, ESI, Permethylation.

1. Introduction

The function of glycoproteins has been found to be strongly influenced by the structure of their glycan moieties. Embryonic development is characterized by profound changes in glycosylation, primarily associated with cell adhesion. Several types of hereditary diseases have their cause in abnormal glycosylation patterns. Altered glycosylation is also a characteristic feature of cancer cells *(1–3)*. Recombinant glycoproteins for use as drugs require precise glycosylation patterns for their efficacy and safety *(4)*. For these reasons, analytical techniques that are able to reliably and simply elucidate glycan structure are essential. The often minute quantities of glycoprotein available make it necessary that such techniques be extremely sensitive.

Nicolle H. Packer and Niclas G. Karlsson (eds.), *Methods in Molecular Biology, Glycomics: Methods and Protocols, vol. 534*

DOI: 10.1007/978-1-59745-022-5_3

Mass spectrometry, because of its potential for superior sensitivity and high information output, has been used extensively for glycan structure determination *(5, 6)*. Modern soft ionization techniques such as the matrix-assisted laser desorption ionization, fast atom bombardment, and electrospray ionization have made it possible to obtain molecular ions with minimal fragmentation. Conversely, controlled fragmentation of ions by MS/MS allows sequence analysis of oligosaccharides by sequentially removing monosaccharides from either end of the carbohydrate.

Structure determination by mass spectrometry and Gas Chromatography-Mass Spectrometry (GC-MS) requires that the glycans be released from the protein. While this is difficult for *O*-glycans, *N*-glycans are easily cleaved from the protein by non-specific peptide *N*-glycosidases *(7)*. However, it is often necessary to digest the protein with a protease to provide sterical access for the glycosidase and in cases where a glycoprotein contains several disulfide bonds these need to be reduced and carboxyamidomethylated before protease treatment.

Derivatization of *N*-glycans by permethylation increases the volatility of the relatively high molecular weight *N*-glycans and improves sensitivity in mass spectrometry.

Mass spectrometry of the permethylated oligosaccharides does not usually provide information about the linkage positions between the individual monosaccharides. If previous data on the *N*-glycan structures of related cell lines are available, the linkages can be confirmed by sequential treatment of the released *N*-glycans with several linkage-specific exoglycosidases *(8, 9)*. Taking into account the mass spectrometry results, the glycans are digested with exoglycosidase enzymes, sequentially cleaving monosaccharide residues from the nonreducing end, and, after permethylation, re-analyzed by MS to observe any corresponding losses in mass. In this way, the linkages can be determined according to the known specificities of the enzymes. However, most exoglycosidases are not completely specific for their preferred substrates and may also cleave other linkages. Also, this methodology is not suitable to detect unusual linkages. Therefore it is advisable, if sufficient material is available, to confirm the linkages by GC-MS analysis of the partially methylated alditol acetates derived from the glycans ("glycosyl linkage analysis" or "methylation analysis") *(10, 11)*. For this purpose, a portion of the permethylated glycan is hydrolyzed into partially methylated monosaccharides, which are reduced with sodium borodeuteride and acetylated. In the resulting partially methylated alditol acetates the positions that were involved in a glycosidic linkage in the glycan are substituted with acetyl groups and those that were unsubstituted in the glycan are now methylated. GC-MS analysis enables separation and quantitation by GC and identification by characteristic fragment ions in MS.

2. Materials

2.1. Release of N-Glycans

1. *Ammonium bicarbonate (AmBic) (Sigma, St. Louis, MO).* Prepare 50 mM in water (*see* **Note 1**).
2. *Dithiothreitol (DTT) (Sigma).* 25 mM in AmBic.
3. *Iodoacetamide (IAM) (Sigma).* 90 mM in AmBic.
4. *Trypsin (Sigma).* 10 mg/mL in AmBic.
5. *Peptide N-glycosidase F (PNGase F) (New England Biolabs, Ipswich, MA).* 500 U/μL supplied in: 50 mM NaCl, 20 mM Tris–HCl (pH 7.5 at 25°C), and 5 mM Na_2EDTA.
6. *PNGase F buffer.* 0.1 M sodium phosphate (Sigma) pH 7.5.
7. Five percent acetic acid (Baker, Phillipsburg, NJ) (5% AcOH).
8. Twenty percent 2-propanol (Fisher, Fair Lawn, NJ) in 5% AcOH.
9. Forty percent 2-propanol in 5% AcOH.
10. C18-Sep-Pak cartridges (Waters, Milford, MA) (*see* **Note 2**).

2.2. Exoglycosidase Digestions

1. *Sodium phosphate buffer (Baker).* 0.5 M in water, pH 5.
2. α-2→3 Neuraminidase from *Salmonella typhimurium* (New England Biolabs).
3. Neuraminidase from *Arthrobacter ureafaciens* (Sigma).
4. β-1→3 Galactosidase from *Xanthomonas manihotis* (New England Biolabs).
5. β-1→4 Galactosidase (Sigma).
6. α-1→(2,3,4)-Fucosidase solution from *Xanthomonas* sp. (Sigma).

2.3. Base Preparation and Permethylation

1. Dimethyl sulfoxide (DMSO) (Aldrich, Milwaukee, WI) (*see* **Note 3**).
2. Iodomethane (Sigma).
3. Dichloromethane (Aldrich).
4. Fifty percent sodium hydroxide solution (Fisher).
5. Anhydrous methanol (Aldrich).

2.4. Linkage Analysis

1. *Trifluoroacetic acid (TFA) (Pierce, Rockford, IL).* Prepare a 2 M solution in water.
2. 2-Propanol (2-PrOH) (Baker).
3. *Ammonium hydroxide (NH_4OH) (Fisher).* Prepare a 1 M solution in water.

4. *Sodium borodeuteride (Aldrich)*. Prepare a 10 mg/mL solution in 1 M NH_4OH (*see* **Note 4**).
5. Glacial acetic acid (HOAc) (Fisher).
6. *Sodium carbonate (Baker)*. Prepare a 0.20 M solution in water.
7. Dichloromethane (Aldrich).
8. Acetic anhydride (Aldrich).

2.5. MALDI-TOF Mass Spectrometry

1. *Matrix*. Prepare a 100 mM solution of 2,5-dihydroxybenzoic acid (Aldrich) in 1:1 acetonitrile – water (*see* **Note 5**).
2. *Calibrant*. Prepare a 1 mg/mL solution of maltooligosaccharides (Sigma) in water.

2.6. Ion Trap Mass Spectrometry with Electrospray Ionization

1. *Infusion buffer*. Prepare a 1 mM solution of sodium hydroxide (Baker) in 1:1 methanol – water.

3. Methods (Fig. 1)

3.1. Release of N-Glycans

1. Dissolve 200–1,000 μg glycoprotein in 100 μL AmBic (*see* **Note 6**).
2. Add 25 μL DTT solution and incubate for 45 min at 50°C.
3. Add 25 μL IAM solution and incubate for 45 min at 25°C in the dark.
4. Add 1 μL trypsin solution; incubate for 5–6 h at 37°C. Then add an additional 1 μL of trypsin and continue digestion overnight.
5. Stop the trypsin digestion by heating reaction to 100°C for 5 min; then cool in ice.
6. Centrifuge for 5 min at maximum speed and collect the supernatant; add equal volume of water, mix, and spin down again (5 min); combine the supernatant fractions and dry down in a speed-vac.
7. Reconstitute in 5% AcOH and apply the mixture to a C18 Sep-Pak.
8. Wash with 10 mL 5% AcOH.

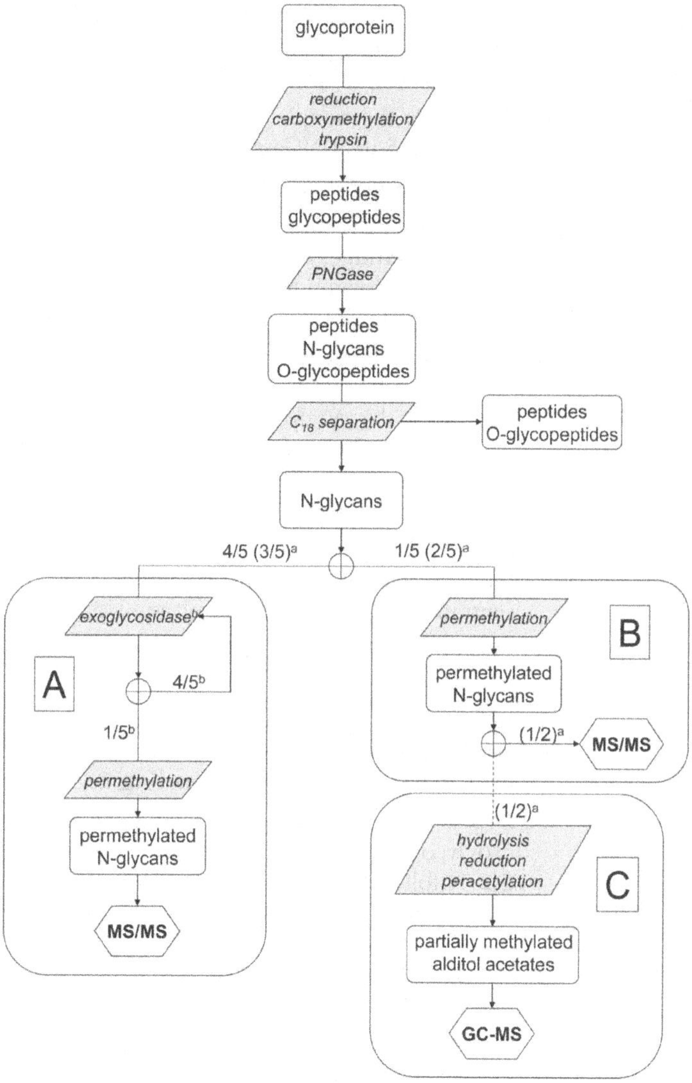

Fig. 1. *N*-Glycan analysis strategy. The entries in rounded rectangles represent the flow and chemical state of the analyte. The entries in parallelograms represent chemical or enzymatic treatment steps. The entries in hexagons represent instrumental analysis methods. [a] If more than 500 μg of glycoprotein are available, methylation analysis (panel C) can be performed in addition to MS/MS (panel B), in order to confirm the linkages obtained by sequential exoglycosidase digestions (panel A). In this case, use only three fifths for exoglycosidase treatments, otherwise use four fifths. [b] The *N*-glycans are subjected to sequential digestions with various exoglycosidases. After each digestion one fifth of the initial sample amount (i.e. the portion that was taken for exoglycosidase treatments) is separated, permethylated, and analyzed by MALDI-TOF mass spectrometry.

9. Elute the glycopeptides with 3 mL 20% 2-PrOH in 5% AcOH and 3 mL 40% 2-PrOH in 5% AcOH; collect the eluate in tubes and dry in the speed vac.
10. Redissolve the sample in 25 μL PNGase F buffer, 25 μL water, and 5 μL PNGase F; incubate at 37°C for 16–18 h.
11. Dry the reaction mixture in a speed vac concentrator.
12. Reconstitute the sample in 5% AcOH and apply to a C18 Sep-Pak cartridge; elute the *N*-glycans with 5% AcOH, the O-linked glycopeptides with 40% 2-PrOH, and the peptides with 70% 2-PrOH.
13. Dry down the fractions on a speed-vac concentrator.

3.2. Exoglycosidase Digestions

1. Redissolve the *N*-glycan sample in 100 μL water and set aside 20 μL of the resulting solution (40 μL if linkage is to be done) for permethylation and MS analysis (*see* **Subheadings 3.3** and **3.4**).
2. Add the appropriate volume of 0.5 M sodium phosphate to the remainder of the *N*-glycan solution to make its final concentration 50 mM in sodium phosphate.
3. Add 1 μL α-2→3-neuraminidase and incubate at 37°C overnight.
4. Remove 20 μL (10 μL), dry down, permethylate (*see* **Subheadings 3.3** and **3.4**), and analyze by MALDI-TOF mass spectrometry (*see* **Subheading 3.7**).
5. Digest the remainder sequentially with the appropriate exoglycosidases. Generally, the order given in **Subheading 2.2** is followed, but may be modified based on the mass spectral data.

3.3. Base Preparation

1. Combine 100 μL 50% NaOH and 200 μL anhydrous MeOH (*see* **Note** 7). Vortex briefly.
2. Add 4 mL dry DMSO and vortex.
3. Centrifuge briefly and discard supernatant (*see* **Note 8**).
4. Repeat **steps 2** and **3** about five times to remove all water from the gelatinous NaOH pellet.
5. Once the pellet is clear, add 2 mL dry DMSO and vortex.

3.4. Permethylation

1. Add 200 μL of dry DMSO to the sample.
2. Purge the headspace of the tube with dry nitrogen gas.
3. Sonicate the sample until it is dissolved.

4. Add 230 μL base to the sample and sonicate.
5. Add 100 μL of iodomethane to the sample.
6. Purge the headspace of the tube with dry nitrogen.
7. Sonicate the sample for 10 min.
8. Add 2 mL water and mix well. The sample will turn cloudy.
9. Remove iodomethane by sparging the mixture with nitrogen (*see* **Note 9**).
10. Add 2 mL dichloromethane and mix well (*see* **Note 10**).
11. Centrifuge and remove the aqueous (top) layer.
12. Add 2 ml of water and mix. Centrifuge briefly.
13. Remove the aqueous (top) layer.
14. Repeat **steps 12** and **13** twice.
15. Remove the organic (bottom) layer into another tube. Dry this down using a stream of dry air (*see* **Note 11**).

3.5. Linkage Analysis

1. Add 4 μg of *myo*-inositol to one fifth of the permethylated sample.
2. Add 230 μL 2 M TFA to the sample. Cap tightly and incubate on a heating block for 2 h at 121°C.
3. Remove the sample and let it cool to room temperature. Evaporate the TFA with dry air.
4. Add ten drops of 2-PrOH and dry down
5. Repeat **step 4** twice.
6. Add 150 μL of a 10 mg/mL solution of sodium borodeuteride in 1 M ammonium hydroxide to each sample. This must incubate at room temperature for at least 1 h (*see* **Note 12**)
7. Neutralize the borodeuteride solution with 3–5 drops of glacial acetic acid.
8. Add five drops MeOH and dry down.
9. Add ten drops of 9:1 MeOH:AcOH and dry down
10. Repeat **step 9**.
11. Add ten drops of MeOH and dry down.
12. Repeat **step 11** twice. You should have a crusty, white residue on the sides of the tube.
13. Add 250 μL of acetic anhydride to the tube and vortex to dissolve the sample. Add 230 μL conc. TFA to the sample. Incubate at 50°C for 10 min.

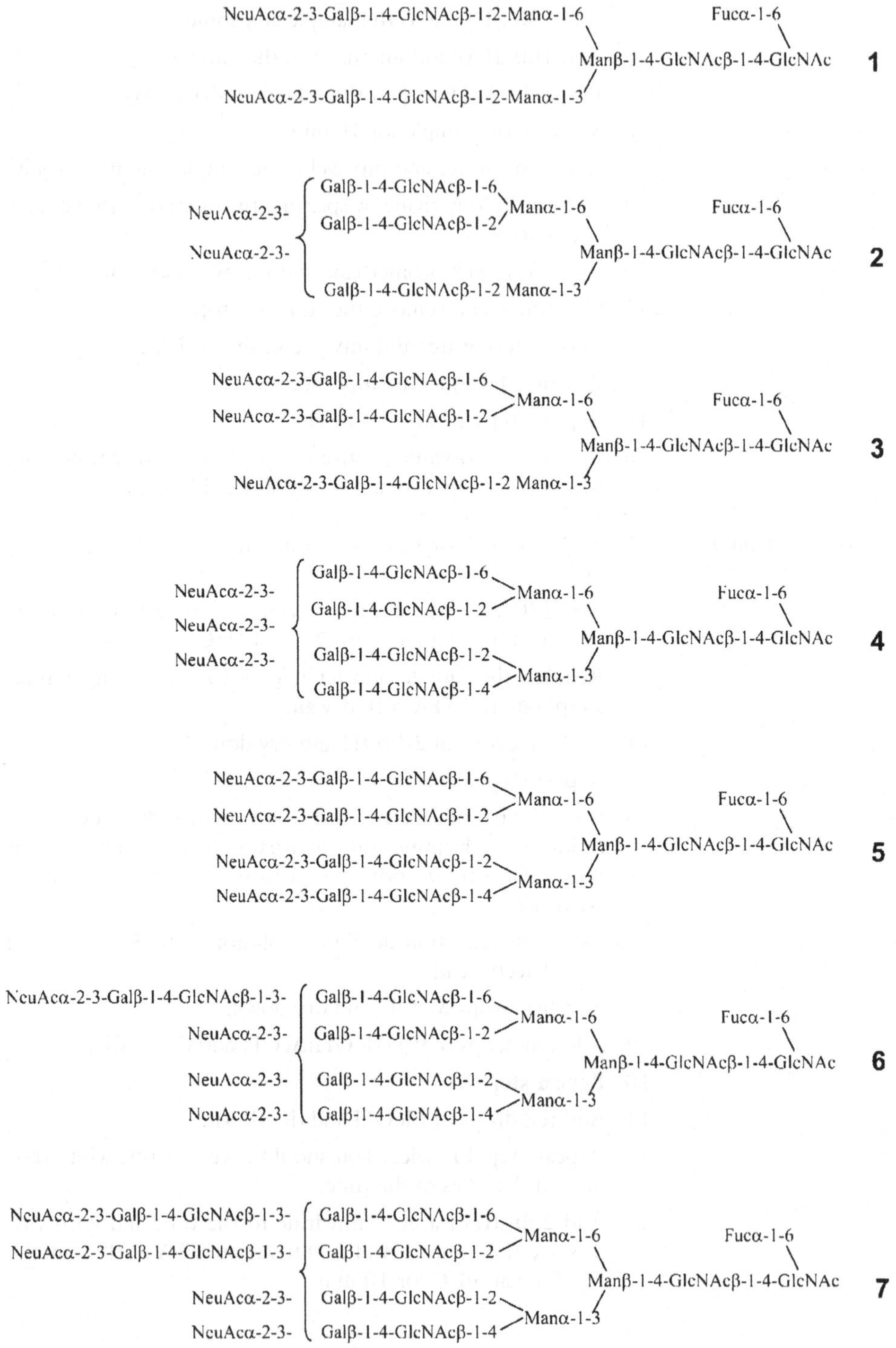

Fig. 2. *N*-linked glycan structures found in EPO.

14. Remove from heat and allow it to cool to room temperature. Add approximately 2 mL of 2-PrOH to the sample and dry down.
15. Add approximately 2 mL of 0.20 M sodium carbonate to the sample. Vortex.
16. Add approximately 2 mL of methylene chloride to the sample and mix gently.
17. Follow **steps 12–15** in the permethylation procedure (*see* **Subheading 3.4**).
18. Dissolve the dried sample in 10–15 drops of methylene chloride and inject 1 μL onto the GC-MS.

3.6. Gas Chromatography-Mass Spectrometry (Fig. 2)

Any commercially available GC-MS instrument can be used for the analysis of partially methylated alditol acetates. We use a Hewlett Packard 5890 GC interfaced to a 5970 mass selective detector.

Perform the separation on a 30 m Supelco 2330 bonded phase fused silica capillary column using the following temperature program: After 2 min at the initial temperature of 80°C, ramp to 170°C at a rate of 30°/min, then ramp to 235°C at rate of 4°/min and hold at 235°C for 20 min. Both inlet and outlet should be at 250°C.

3.7. MALDI-TOF Mass Spectrometry (Figs. 3 and 4)

MALDI-TOF MS is performed in the positive ion mode using 2,5-dihyroxybenzoic acid as matrix. The instrument is calibrated with a mixture of maltooligosaccharides. We use a Voyager V-DE Mass Spectrometer (Applied Biosystems). The sample is dissolved in 20 μL methanol, and 5 μL of this solution is mixed with 5 μL matrix solution (*see* **Note 12**). Likewise, 5 μL of the calibrant solution is mixed with matrix, and 1 μL of both sample-matrix and calibrant-matrix mixtures are spotted on the sample plate and dried under vacuum or by a stream of air.

3.8. Ion Trap Mass Spectrometry with Electrospray Ionization (Figs. 5–8)

Mass analysis is performed on an ESI-MS^n mass spectrometer by direct infusion of permethylated glycans dissolved in 1 mM NaOH in 50% methanol (~10–100 μL). We use an Advantage-LCQ mass spectrometer (Thermo Finnigan) in the positive ion mode. The sample is introduced via a syringe pump at a constant flow rate of 0.4 μL/min. Mass spectra are obtained with a capillary temperature of 210°C.

MS/MS spectra were obtained with 28% normalized collision energy.

All the above methods, when used together, will give the detailed structures of the *N*-linked glycans attached to the glycoprotein (**Fig. 9**)

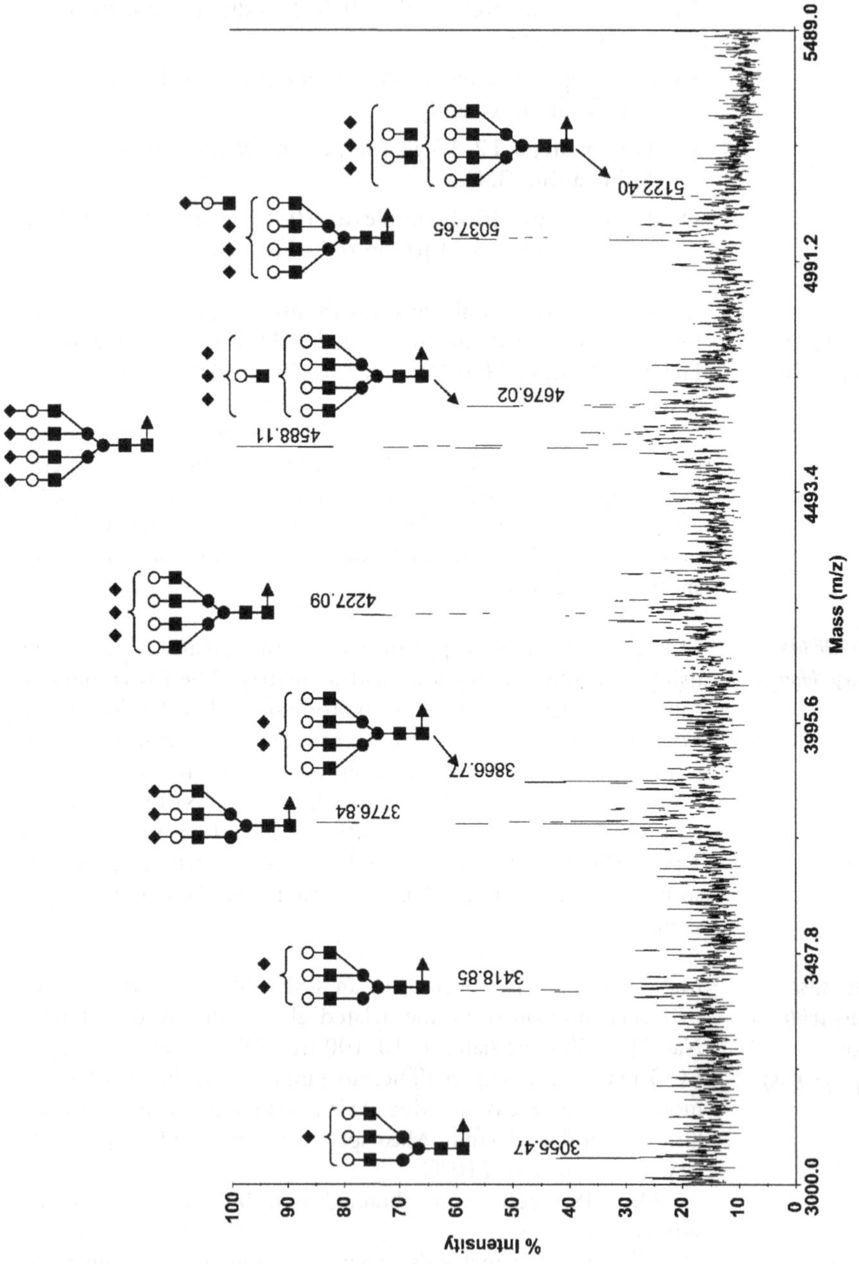

Fig. 3. MALDI-TOF mass spectrum and structural assignment of EPO *N*-glycans, which have been released by PNGase F treatment from the disulfide-reduced, carboxymethylated, and trypsin-digested glycoprotein. *Rectangles* GlcNAc; *filled circles* Man; *empty circles* Gal; *diamonds* NeuAc; *triangles* Fuc. We did not determine which of the antennae are terminated with NeuAc in those structures in which not all antennae are terminated by NeuAc, The spectrum was obtained on an Applied Biosciences 4700 Proteomics Analyzer in the positive ion mode using 2,5-dihydroxybenzoic acid as matrix.

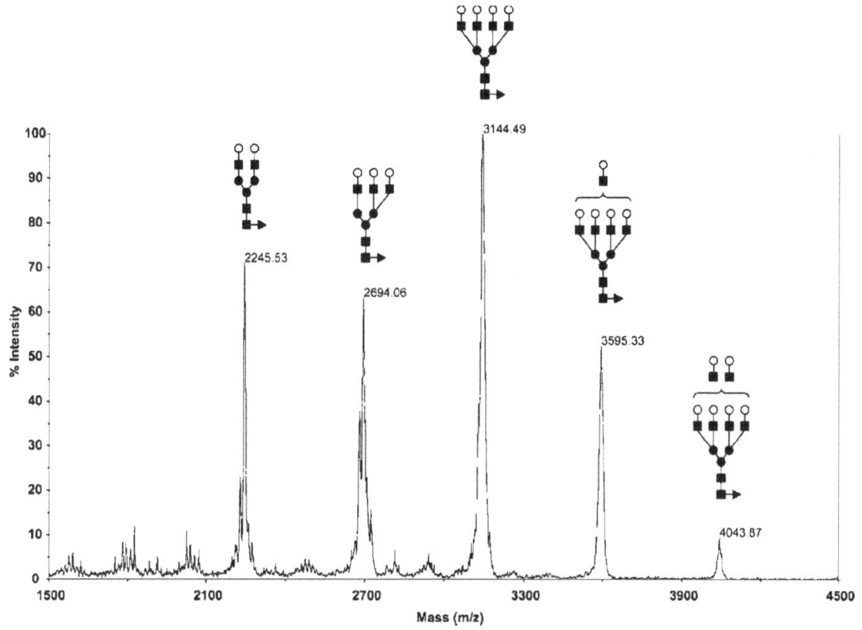

Fig. 4. MALDI-TOF mass spectrum and structural assignment of 2,3-sialidase treated, permethylated *N*-glycans from EPO. *See* **Fig.3** legend for structural symbols. The spectrum was acquired on an Applied Biosystems Voyager instrument in the positive ion mode using 2,5-dihydroxybenzoic acid as matrix.

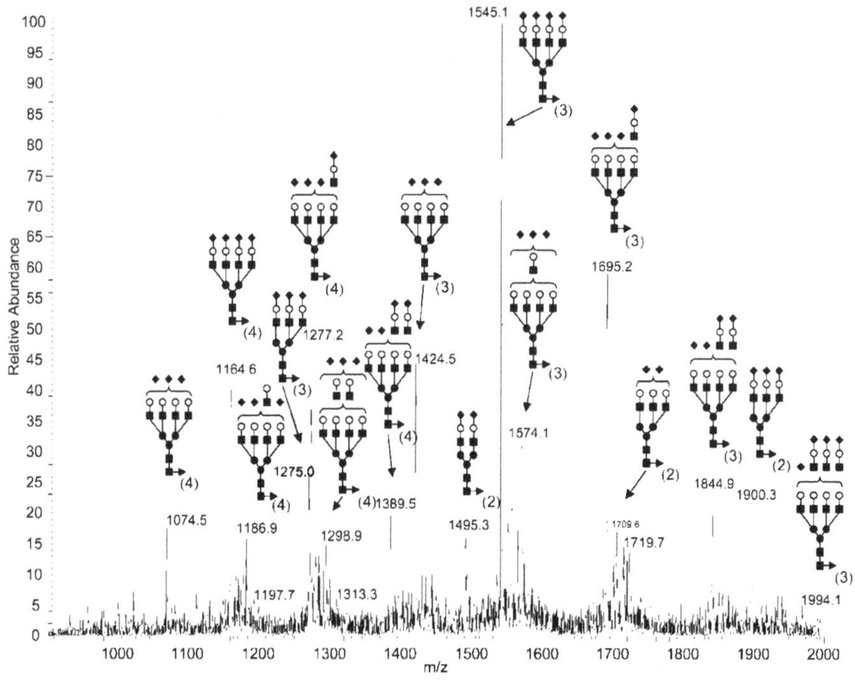

Fig. 5. ESI-ion trap mass spectrum of released, permethylated *N*-glycans from EPO. See **Fig.3** legend for structural symbols. The numbers in parentheses refer to the number of positive charges in each species. The spectrum was acquired on a Thermo Finnigan Advantage-LCQ mass spectrometer in the positive ion mode using direct infusion of a solution of the sample in 1 mM NaOH in 50% methanol at a flow rate of 0.4 μL/min.

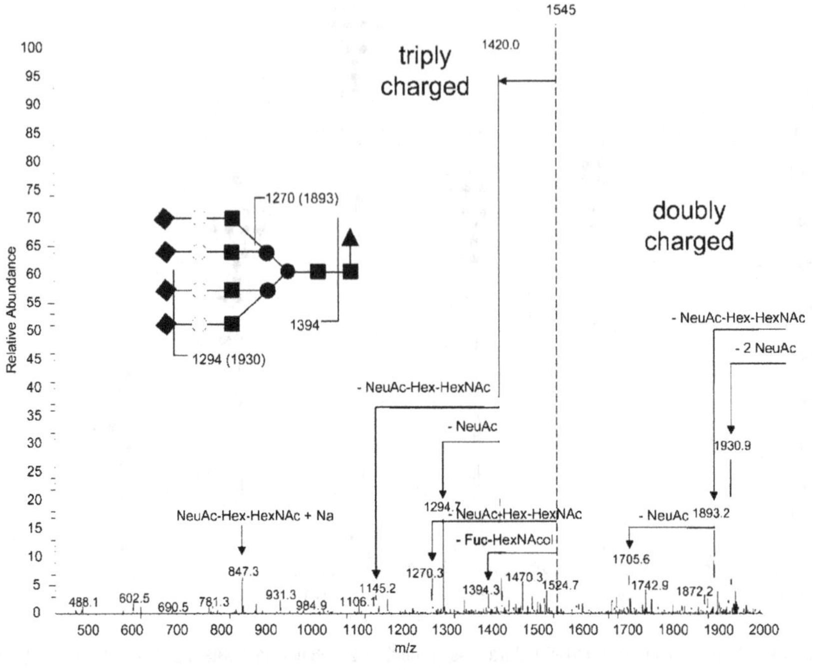

Fig. 6. ESI-MS/MS spectrum of the major ion $m/z = 1,545$ (tetrasialo-teraantennary, triply charged) in the MS of permethylated *N*-glycans from EPO. A normalized collision energy of 28% was used. The parent ion is indicated by a *dashed vertical line*. The main fragments are those resulting from loss of NeuAc residues. The ion at $m/z = 1,394$ corresponds to loss of Fuc-GlcNAcol and shows that the Fuc residue is attached to the reducing end GlcNAc. The ions at $m/z = 1,893$, 1,270, and 1,145 arise from loss of the NeuAc–Gal–GlcNAc trisaccharides.

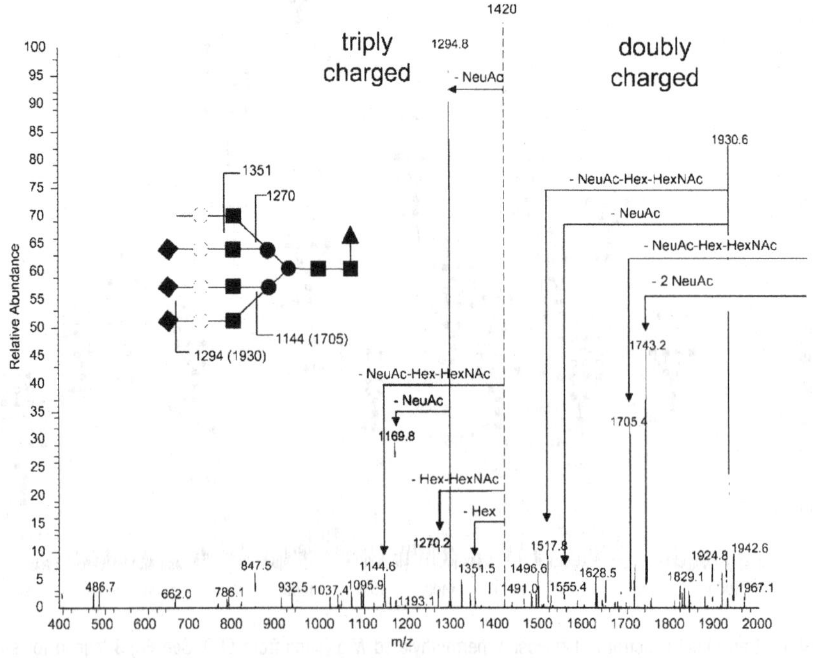

Fig. 7. ESI-MS^3 spectrum of the major ion $m/z = 1,420$ (trisialo-teraantennary, triply charged) in the MS/MS of permethylated *N*-glycans from EPO. The parent ion is indicated by a *dashed vertical line*. The main fragments are those resulting from loss of NeuAc residues. The ion at $m/z = 1,351$ corresponds to loss of Gal and the ion at $m/z = 1,270$ is due to loss of Gal-GlcNAc. The ions at $m/z = 1,144$ and 1,705 arise from loss of the NeuAc–Gal–GlcNAc trisaccharides.

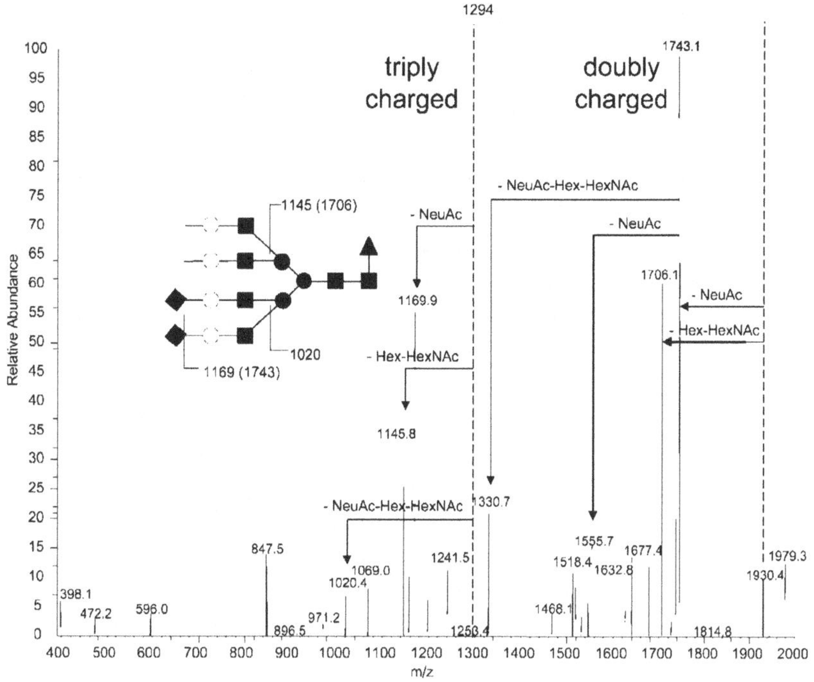

Fig. 8. ESI-MS[4] spectrum of the major ion $m/z = 1{,}294$ (disialo-tetraantennary, triply charged) in the MS[3] spectrum. Losses of NeuAc, Gal–GlcNAc, and NeuAc–Gal–GlcNAc can be seen, further confirming the structural identity of the parent ion.

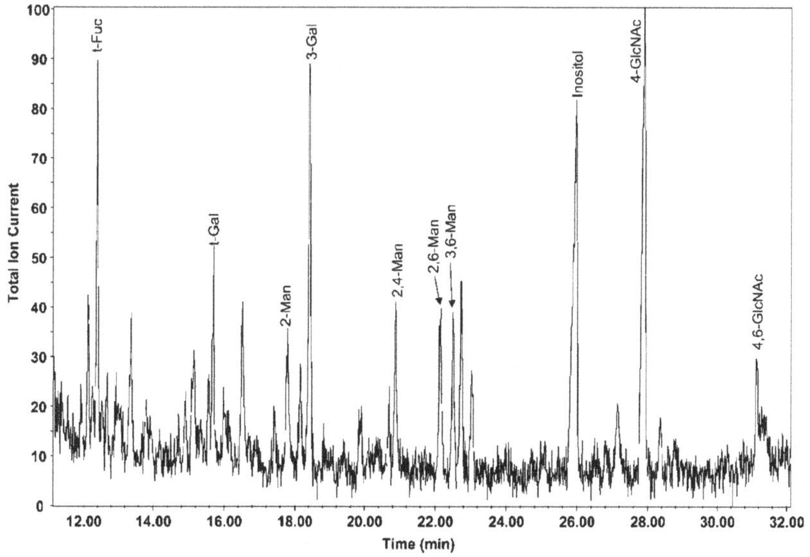

Fig. 9. Methylation analysis by GC-MS of EPO *N*-glycans. The presence of 3-Gal shows that NeuAc is 2–3 linked, and the presence of 4,6-GlcNAc shows that Fuc is 1–6 linked. The chromatogram was acquired on a Hewlett Packard 5890 GC interfaced to a 5970 mass selective detector. The capillary column was a 30 m Supelco 2330 bonded phase fused silica capillary column using the following temperature program: After 2 min at the initial temperature of 80°C, ramp to 170°C at a rate of 30°/min, then ramp to 235°C at a rate of 4°/min and hold at 235°C for 20 min.

4. Notes

1. Unless stated otherwise, deionized water with a resistivity of 18.2 MΩ-cm should be used for all aqueous solutions.
2. Prepare the Sep-Pak cartridges for use by washing first with10 mL MeOH and then with 20 mL 5% acetic acid.
3. Dimethylsulfoxide is stored over 4A molecular sieves that have been activated by drying at 150–300°C for several hours, followed by cooling in a desiccator.
4. This solution has to be prepared immediately before use.
5. This solution can be stored at –20°C for several months.
6. This protocol has been optimized for 1 mg of glycoprotein, although amounts of up to 3 mg can be processed. If a larger amount of glycoprotein is to be digested, adjust the volume of reagents accordingly.
7. Use a plastic pipette for the NaOH solution and a glass syringe for the MeOH.
8. During this base preparation, a significant quantity of sodium carbonate precipitates on the wall of the tube; this must be removed as thoroughly as possible with a Pasteur pipette.
9. Place the tip of the nitrogen line into the sample solution at the bottom of the tube. Use a very low flow of nitrogen. The sample will become clear as the iodomethane is bubbled off. Make certain that all of the iodomethane is gone by moving the pipette around in the solution. Some bubbles will stick to the walls of the tube.
10. Do not allow the solution to come into contact with the cap. At this point, this will introduce contaminants into the sample.
11. If several samples are to be analyzed at once, it is advisable to use an air-manifold similar to the one described in York et al. *(10)*.
12. The sample can be left overnight after addition of sodium borodeuteride.

References

1. Varki, A., Cummings, R., Esko, J., Freeze, H., Hart, G. & Marth, J., eds. (1999) Essentials of Glycobiology (Cold Spring Harbor Laboratory Press, New York, NY).
2. Freeze, H. H. & Aebi, M. (2005) Altered glycan structures: the molecular basis of congenital disorders of glycosylation. *Current Opinion in Structural Biology* 15, 490–498.
3. Ohtsubo, K. & Marth, J. D. (2006). Glycosylation in Cellular Mechanisms of Health and Disease. *Cell* 126, 855–867.
4. Sethuramana, N. & Stadheim, T. A. (2006) Challenges in therapeutic glycoprotein

production. *Current Opinion in Biotechnology* 17, 341–346.

5. Dell, A. & Morris, H. R. (2001) Glycoprotein Structure Determination by Mass Spectrometry. *Science* 291, 2351–2356
6. Medzihradszky, K. F. (2005) Characterization of Protein N-Glycosylation. *Methods in Enzymology* 405, 116–138
7. Maley, F., Trimble, R. B., Tarentino, A. L. Thomas, H. & Plummer, J. (1989) Characterization of glycoproteins and their associated oligosaccharides through the use of endoglycosidases. *Analytical Biochemistry* 180, 195–204
8. Kobata, A. (1979) Use of endo- and exoglycosidases for structural studies of glycoconjugates. *Analytical Biochemistry* 100, 1–14
9. Prime, S., Dearnley, J., Ventom, A. M., Parekh, R. B. & Edge, C. J. (1996) Oligosaccharide sequencing based on exo- and endoglycosidase digestion and liquid chromatographic analysis of the products. *Journal of Chromatography A.* 720, 263–274
10. York, W. S., Darvill, A. G., McNeil, M., Stevenson, T. T. & Albersheim, P. (1985) Isolation and characterization of plant cell walls and cell wall components. *Methods in Enzymology* 118, 3–40
11. Anumula, K. R. & Taylor, P. B. (1992) A comprehensive procedure for preparation of partially methylated alditol acetates from glycoprotein carbohydrates. *Anal Biochem* 203, 101–108

Chapter 4

Solid-Phase Permethylation for Glycomic Analysis

Yehia Mechref, Pilsoo Kang, and Milos V. Novotny

Summary

This chapter discusses in detail a miniaturized version of the widely used permethylation technique which permits quantitative derivatization of oligosaccharides derived from minute quantities of glycoprotein. The approach involves packing of sodium hydroxide powder or beads in a microcolumn format, including spin columns, fused silica capillaries (500 μm i.d.) and plastic tubes (1 mm i.d.). The derivatization proceeds effectively in less than a minute time scale and it is applicable to glycans derived from femtomole quantities of glycoproteins. Prior to mass spectrometry (MS), methyl iodide is added to analytes suspended in dimethyl sulfoxide solution containing traces of water. The reaction mixture is then immediately infused through the microreactor. The packed sodium hydroxide powder or beads inside the microcolumns minimize oxidative degradation and peeling reactions which are otherwise commonly associated with the conventional permethylation technique. In addition, this solid-phase permethylation approach eliminates the need for excessive sample clean-up. As demonstrated below, picomole amounts of various types of glycans derived from model glycoproteins as well as real samples, including linear and branched, sialylated and neutral glycans were shown to become rapidly and efficiently permethylated through this approach.

Key words: Permethylation, Oligosaccharides, Glycans derived from glycoproteins, Sodium hydroxide-packed capillaries, Sodium hydroxide-packed spin columns, Mass spectrometry.

1. Introduction

While matrix-assisted laser desorption-ionization/mass spectrometry (MALDI/MS) allows the structural analyses of numerous glycans can in their native forms, it is more advantageous to convert such compounds to their methylated derivatives. The advantages include an easier determination of branching, interglycosidic linkages and the determination of configurational and

Nicolle H. Packer and Niclas G. Karlsson (eds.), *Methods in Molecular Biology, Glycomics: Methods and Protocols, vol. 534*

DOI: 10.1007/978-1-59745-022-5_4

conformational isomers. In addition, permethylation also stabilizes the sialic acid residues in acidic oligosaccharides, thus allowing a simultaneous analysis of both sialylated and neutral glycans present in the same sample. Moreover, methylated sugars ionize easier, relative to their native counterparts. For a number of years, permethylation has also been shown to effectively yield detailed structural analyses in conjunction with electrospray ionization (ESI) and collision-induced dissociation *(1–7)*.

Permethylation has customarily employed one of the two successful methodologies. The first, originally described by Hakomori *(8)*, utilizes the dimethyl sulfoxide anion ($DMSO^-$, commonly referred to as "dimsyl anion") to remove protons from the sample analyte molecules prior to their replacement with methyl groups. The second, that is currently a more widespread approach introduced in 1984 by Ciucanu and Kerek *(9)*, is based on the addition of methyl iodide to dimethylsulfoxide containing powdered sodium hydroxide and a trace of water. The popularity of the latter procedure stems from its speed, experimental simplicity, "cleaner" products of reaction, and its effectiveness in replacing protons at both oxygen and nitrogen sites in oligosaccharides. This methylation procedure has now been used for derivatization of carbohydrates *(9–11)*, other polyols *(12)*, and fatty acids including, their hydroxy derivatives *(13)*. However, it has gained a particular popularity with complex oligosaccharides and glycolipids. More recently, the yield of this procedure was substantially improved by the addition of traces of water to the reaction mixture, thus enhancing reaction yields and minimizing side-products *(14)*.

Although these permethylation procedures have now been used successfully in various MS structural studies of neutral and sialylated glycans, they are not satisfactory at the range of low picomole to femtomole quantities which are usually the amounts at which glycoproteins are present in biological samples. The problems are primarily due to oxidative degradation and peeling reactions, resulting from the high pH attained upon dissolving the sodium hydroxide powder which is a common step proceeding liquid-liquid extraction. Such side reactions are particularly visible with small sample amounts. These limitations have been largely overcome through the development of solid-phase permethylation protocol *(15)*.

Solid-phase permethylation permits a simultaneous methylation of neutral and acidic oligosaccharides, using microspin columns or microreactors (capillary or larger tubes) packed with sodium hydroxide powder or beads. This procedure facilitates a highly quantitative conversion of minute quantities of glycoprotein-derived glycans. It is further compatible with the microscale glycan release methods *(16–19)* commonly employed in glycoprotein analysis.

2. Materials

2.1. Sodium Hydroxide Reactors

1. Sodium hydroxide, 20–40 mesh beads, 97% (Aldrich, Milwaukee, WI).
2. Dry dimethylsulfoxide, 2,5-dihydroxybenzoic acid (DHB) and acetonitrile (Aldrich, Milwaukee, WI).
3. Chloroform and iodomethane (EM Science, Gibbstown, NJ).
4. Fused silica capillaries, 500 μm i.d.×750 μm o.d. (Polymicro Technologies, Phoenix, AZ).
5. Micor-spin columns (Harvard Apparatus, Holliston, MA).
6. Peak tubing, 1 mm i.d., nuts and ferrules (Upchurch Scientific, Oak Harbor, WA).

2.2. Release of Glycans from Glycoproteins

1. Pancreatic bovine ribonuclease B, fetuin from fetal calf serum, human α_1-acid glycoprotein (Sigma Co., St. Louis, MO).
2. Non-small cell lung cancer cell pellet acquired from the National Cancer Institute panel of 60 human tumor cell lines.
3. *CHAP-based lysis buffer.* 150 mM sodium chloride, 0.5% 3-[3-(cholamidopropyl)dimethylammonio]-1-propanesulfonate (CHAPS), 20 mM Tris–HCl (pH 7.5), 2.5 mM sodium pyrophosphate, 1 mM ethylenediamine tetraacetic acid (EDTA) and 1 mM ethylene glycol bis(2-aminoethyl ether)tetraacetic acid (EGTA).
4. PNGase F (2 mU/μL) in a buffer of 52 mM Tris–HCl, 36 mM sodium chloride, 9 mM sodium acetate, and 5 mM EDTA (pH 7.5).
5. Borane–ammonia complex (Sigma Co., St. Louis, MO).
6. 28% aqueous ammonium hydroxide (Sigma Co., St. Louis, MO).

2.3. MALDI Matrix

1. Suspend 10 mg of 2,5-dihydroxybenzoic acid (DHB) in a 1 mL of 50/50 water/methanol solution in a clean microtube.
2. Shake the microtube thoroughly.
3. Centrifuge for 10 min prior to use.

3. Methods

An optimum chemical derivatization of analytes should modify quantitatively the required functional groups, yielding preferably a single, stable reaction product. Moreover, such an analyte modification should be fast, simple, and applicable to minute sample quantities. The last requirement is particularly crucial for the glycans derived from trace quantities of glycoproteins. In general, solid-phase derivatization has been explicitly recognized as an effective approach at the small sample scale *(20)*. Two different versions of the solid-phase treatment, a capillary reactor derivatization and a spin column derivatization (**Fig. 1**), are discussed below.

3.1. Preparation of Sodium Hydroxide Capillary Reactors

1. Crush carefully sodium hydroxide beads to produce a fine powder (*see* **Note 1**).
2. Suspend the crushed sodium hydroxide powder in acetonitrile immediately to protect the packing material from moisture (*see* **Note 2**).
3. Pressure-pack the powdered sodium hydroxide suspended in acetonitrile inside 500 μm i.d. fused silica capillaries or 1 mm i.d. peak tubing. The packing is achieved through the use of a packing bomb at 750 psi, using a fritted union to hold the packed material at the reactor's end. Another union must be securely attached to the inlet of the reactor upon the completion of packing.

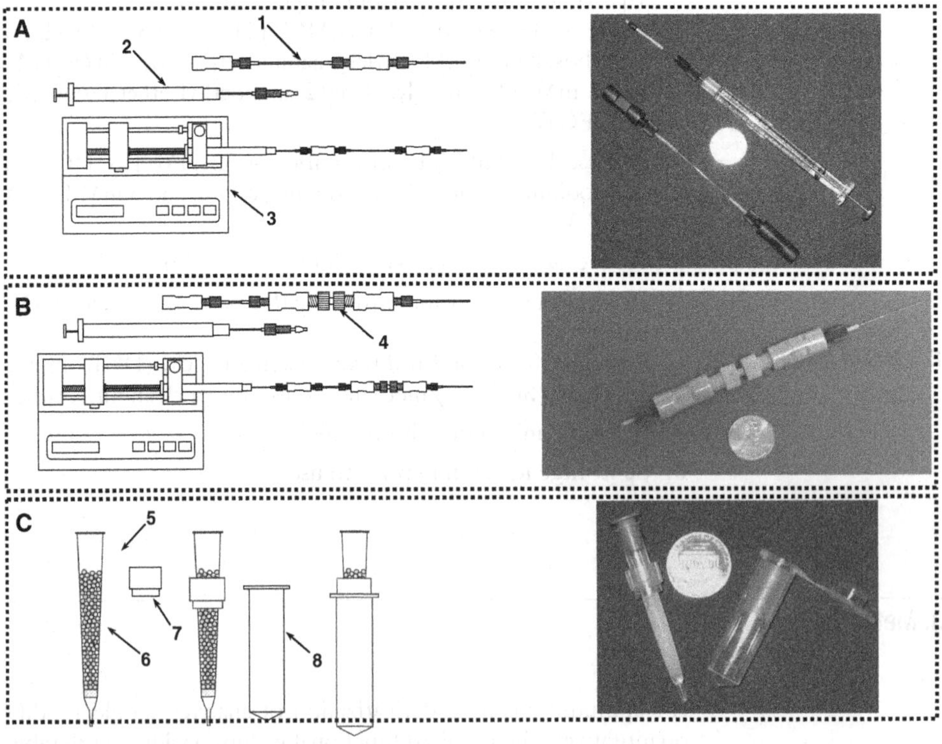

Fig. 1. Schematic illustration of the solid-phase permethylation set-up for (**A**) 500 μm i.d. fused silica capillary, (**B**) sodium hydroxide 1 mm i.d. peak column, and (**C**) micro-spin column. Key: (1) 500 μm i.d. fused silica capillary; (2) syringe; (3) syringe pump; (4) sodium hydroxide column; (5) spin column; (6) sodium hydroxide beads; (7) column holder; (8) micro-centrifuge tube.

4. After packing the capillary with sodium hydroxide using acetonitrile, assemble the capillary set-up, as illustrated in **Fig. 1a,** for fused silica and peak tubing, respectively.
5. A 100-μL Hamilton syringe, and a syringe pump (KD Scientific, Inc., Holliston, MA) can be employed for introducing a sample solution into the capillary.
6. Infuse DMSO through the sodium hydroxide reactor to replace acetonitrile and condition the reactor prior to the application of sample (*see* **Notes 3** and **4**).

3.2. Preparation of Sodium Hydroxide Micro-spin Column Reactors

1. Suspend sodium hydroxide beads in acetonitrile to prevent the absorption of atmospheric moisture (*see* **Note 1**).
2. Pack the suspended beads in the spin column to about 3-cm depth.
3. Centrifuge the spin column at 100 × g for 2 min (**Fig. 1C**).
4. Wash the sodium hydroxide-packed spin column several times with DMSO to replace acetonitrile and condition the column prior to sample application (*see* **Note 5**).

3.3. Cell Lysis and Sample Preparation

1. Suspend cell pellet (10–100 × 10^6 cells) in a 2-mL aliquot of CHAPS-based lysis buffer (*see* **Subheading 2.2**).
2. Incubate the cell pellets in the lysis buffer at 4°C for 1 h.
3. Centrifuge at 100,000 × *g* for 90 min.
4. Remove the supernatants containing cytosol proteins.
5. Resolubilize the remaining pellet, containing the membrane proteins, in a 2-mL aliquot of CHAPS-based lysis buffer.
6. Sonicate the solution for 30 min.
7. Reduce both cytosol and membrane proteins through the addition of a 100-μL aliquot of 10 mM DTT.
8. Incubate the reaction mixture at 56°C for 45 min.
9. Alkylate the reduced proteins by adding a 100-μL aliquot of 55 mM iodoacetamide.
10. Incubate the reaction mixture in the dark at room temperature for 30 min.

3.4. Release of N-Glycans from Glycoproteins

1. Prepare a 0.1 mg/mL solution of glycoproteins by suspending the appropriate amount of glycoproteins in 10 mM sodium phosphate buffer, pH 7.0, including 1% 2-mercaptoethanol.
2. Transfer a 1–5 μL aliquot of the glycoprotein stock solution to a 0.6-μL Eppendorf tube.
3. Incubate the glycoprotein solution at 95°C for 10 min to thermally denature the glycoprotein.
4. Cool the tube to room temperature.
5. Add a 0.1-μL aliquot of PNGase F (0.2 mU).

6. Incubate the reaction mixture at 37°C overnight to ensure complete release of *N*-glycans (*see* **Note 6**).

3.5. Purification of Released Glycans

The released *N*-glycans are purified through a solid-phase extraction using both C18 and activated charcoal cartridges to remove peptides and O-linked glycopeptides and to trap and wash released *N*-glycans, respectively (*see* **Note** 7).

1. Equilibrate Sep-Pak Classic C18 cartridge (Waters Corp.) using ethanol and water.
2. Dilute samples to a 1 mL volume with water and apply to the conditioned cartridge.
3. Wash cartridges with 1 mL of water (*see* **Note 8**).
4. Wash the C18 cartridges with 5 mL of 85% aqueous acetonitrile solution containing 0.1% formic acid to collect peptides and O-linked glycopeptides.
5. Vacuum dry the collected peptides and O-linked glycopeptides which will be then subjected to β-elimination to release *O*-glycans (*see* **Subheading 3.6**).
6. Apply the C18-eluents to a microspin column filled with activated charcoal.
7. Wash, first, the microspin column with 5 mL of 50% acetonitrile aqueous solution to remove salt and other contaminants.
8. Apply the glycan samples to the microspin column and wash first using H_2O and then using 0.1% trifluoroacetic acid in 5% acetonitrile/H_2O (v/v).
9. Elute glycans, using 0.25 mL (4×) of 50% (v/v) aqueous acetonitrile solution containing 0.1% trifluoroacetic acid.
10. Dry the collected samples using an Eppendorf Vacufuge.

3.6. Release of O-Glycans Through β-Elimination

1. Subject the peptides and O-linked glycopeptides, eluted from the C18 cartridges as described in **Subheading 3.5**, to *β*-elimination for releasing O-linked glycans.
2. Resuspend the dried sample in a 15 μL aliquot of borane–ammonia complex solution, prepared by dissolving 5 mg/mL of borane–ammonia complex in 28% aqueous ammonium hydroxide.
3. Allow the reaction to proceed at 45°C for 18 h.
4. Remove the ammonium hydroxide from the samples using an Eppendorf Vacufuge (*see* **Note 9**).
5. Reconstitute the samples in 40 μL of water.
6. Vortex the sample thoroughly.
7. Remove water using an Eppendorf Vacufuge.
8. Repeat **steps 6** and 7 three times to ensure a complete removal of ammonium hydroxide

9. Purify the released *O*-glycans using C18 and activated charcoal cartridges and spin columns, respectively, as described in **Subheading 3.5**.

3.7. Solid-Phase Permethylation of Glycans by Capillary Sodium Hydroxide Reactor

1. Reconstitute released and purified glycans (*see* **Note 10**) in a 50-μL aliquot of DMSO, a 22-μL aliquot of iodomethane and a trace of water (0.3 μL).
2. Infuse the reaction mixture immediately through the sodium hydroxide capillary column, prepared as described in **Subheading 3.1**, at a 2 μL/min flow rate.
3. Collect and pass through solution in a clean 1.6-mL Eppendorf microtube.
4. Wash the column with 100 μL of acetonitrile or DMSO two times at 10 μL/min flow rate (*see* **Note 11**).
5. Add the wash to the same Eppendorf microtube.
6. Add 200 μL chloroform and 200 μL water to Eppendorf microtube.
7. Shake the Eppendorf microtube thoroughly and discard the aqueous layer (upper layer).
8. Add 200 μL water again and shake.
9. Discard the aqueous layer and repeat **step 8** at least three times.
10. Dry the chloroform solution using an Eppendorf Vacufuge.
11. Resuspend the dried sample in 5 μL of 50% aqueous methanol solution containing 1 mM sodium acetate.

3.8. Solid-Phase Permethylation of Glycans by Sodium Hydroxide Micro-spin Column Reactors

1. Reconstitute the released and purified glycans in a 50-μL aliquot of DMSO, a 22-μL aliquot of iodomethane and a trace of water (0.3 μL).
2. Apply the reaction mixture immediately over the sodium hydroxide micro-spin column reactor, prepared as described in **Subheading 3.2**.
3. Allow the reaction mixture to pass through either by gravity flow or by centrifugation at 400 × g.
4. Re-apply and pass through over the reactor five times (*see* **Note 12**)
5. Wash the reactor with 100 μL acetonitrile or DMSO two times, either by gravity flow or centrifugation at 400 × g.
6. Add 200 μL chloroform and 200 μL water to sample microtube.
7. Shake thoroughly and discard the aqueous layer (upper layer).
8. Add 200 μL water again and shake.
9. Discard the aqueous layer and repeat the addition of water, shaking and discarding three times.

10. Dry the chloroform using an Eppendorf Vacufuge.
11. Resuspend the dried sample in 5 μL of 50% aqueous methanol solution containing 1 mM sodium acetate.

3.9. MALDI Analysis of Permethylated Glycans

1. Spot 0.5 μL of permethylated sample on a stainless steel MALDI plate.
2. Add a 0.5 μL aliquot of DHB matrix solution, prepared as described in **Subheading 2.3**, on top of the sample spot.
3. Dry the spot under vacuum prior to MS analysis (*see* **Note 13**).
4. Acquire mass spectra using a MALDI mass spectrometer such as Applied Biosystems 4700 Proteomics Analyzer (Applied Biosystems), which is equipped with a 200 Hz Nd:YAG laser (355 nm). MALDI mass spectra of methylated *N*-glycans derived from model glycoproteins such as ribonuclease B, fetuin, and α_1-acid glycoproteins are depicted in **Fig. 2A–C**, respectively, while MALDI mass spectrum of *N*-glycans released from a mixture of these model glycoproteins is illustrated in **Fig. 3**. Methylated *N*- and *O*-glycans derived from the cell extracts are shown in **Fig. 4A, B**, respectively.

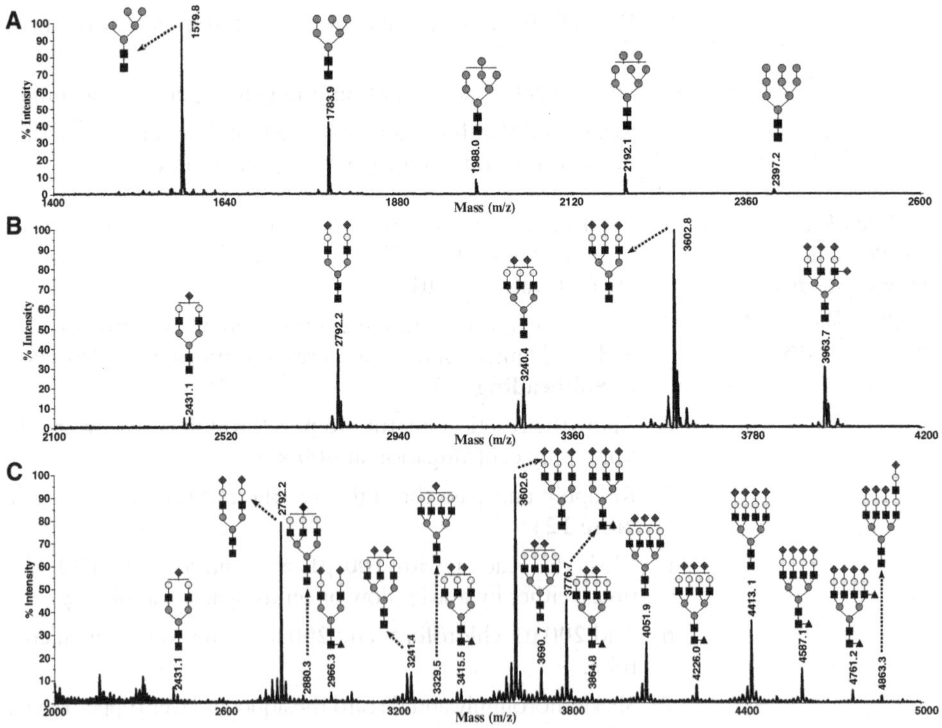

Fig. 2. MALDI mass spectra of permethylated *N*-glycans derived from **(A)** 0.1 μg ribonuclease B, **(B)** 0.1 μg fetuin, and **(C)** 0.1 μg α_1-acid glycoprotein by using 500 μm i.d. fused silica capillary sodium hydroxide reactor. Symbols: *filled square*, *N*-acetylglucosamine; *filled circle*, manose; *empty circle*, galactose; *filled triangle*, fucose; filled diamond, *N*-acetylneuraminic acid.

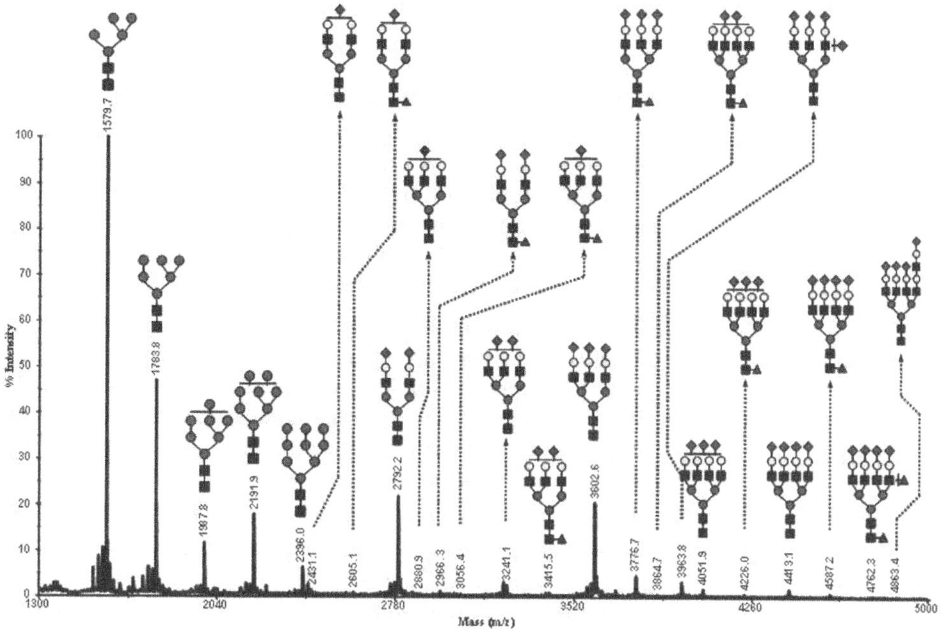

Fig. 3. MALDI mass spectrum of permethylated *N*-glycans derived from 0.5 μg mixture of ribonuclease B, fetuin, and α_1-acid glycoprotein using 500 μm i.d. fused silica capillary sodium hydroxide reactor. Symbols as in **Fig. 2**.

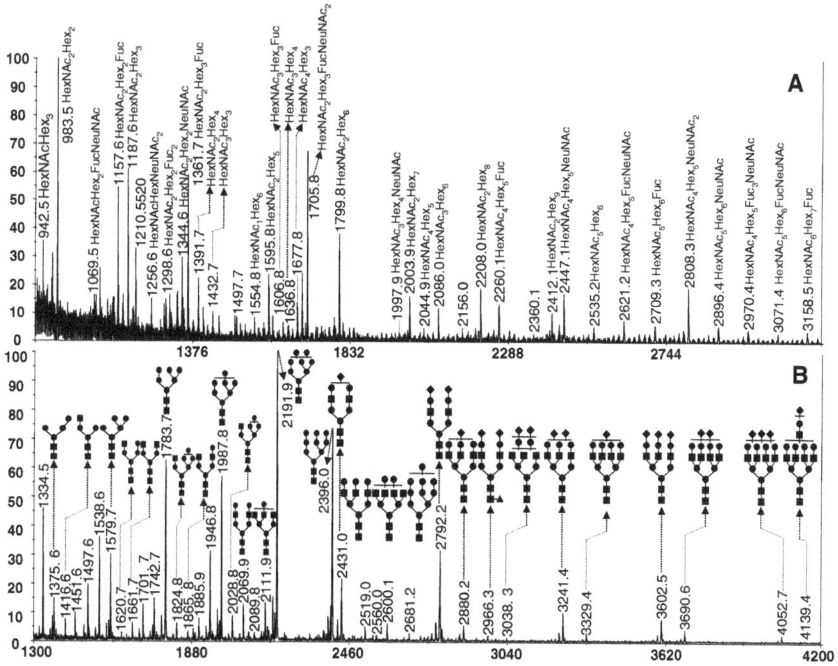

Fig. 4. MALDI mass spectra of permethylated (**a**) *O*-glycans and (**b**) *N*-glycans derived from non-small cell lung cancer cell pellet. Symbols as in **Fig. 2**.

4. Notes

1. Sodium hydroxide is one of the most common hygroscopic materials which absorb moisture from the atmosphere. It is very corrosive and very harmful by ingestion, a skin contact or through inhalation of dust. It causes severe burns, or may cause serious permanent eye damage. Therefore, it should be handled with care. Crushing of the beads to create the powder should be completed rapidly and under a stream of nitrogen.
2. The use of acetonitrile in this step is recommended; however, DMSO could be used. Note that DMSO is an irritant to eyes, respiratory system and skin. It is a combustible liquid, and is readily absorbed through skin. Accordingly, packing sodium hydroxide with acetonitrile is more convenient and requires lower packing pressures due to its low solvent pressure resistance.
3. Although the use of acetonitrile for packing is recommended, the reactors have to be conditioned thoroughly with DMSO prior to methylation. This is needed to achieve highly efficient methylation. Methylation in acetonitrile is attainable; however, at a much lower yield and efficiency.
4. Due to the hygroscopic nature of sodium hydroxide (*see* **Note 1**), acetonitrile or DMSO have to be constantly infused through the reactors to prevent clogging and extend the lifetime of reactors. Under these conditions, the reactors can be easily used over few weeks without any loss in activity.
5. Micro-spin columns can be used several times if they are capped securely after filling them with DMSO and acetonitrile.
6. Although overnight incubation is recommended for the efficient enzymatic release of *N*-glycans, efficient release is attained in a shorter incubation time (ca. 2 h) when the reaction volume is reduced to few microliters *(19)*.
7. The use of C18 cartridges prior to the solid-phase extraction of glycans using the activated charcoal medium is necessary. This is due to the fact that peptides also interact with activated charcoal, and would thus compete with glycans for interactions if not effectively removed. Moreover, the use of C18 cartridge is needed to retrieve O-linked glycopeptides for a further treatment to release *O*-glycans.
8. Quantitative recovery of glycans is attained as a result of an exhaustive washing of the activated charcoal medium. The eluent contains the released *N*-glycans, while peptides and O-linked glycopeptides are retained on the C18 cartridges.
9. Effective evaporation of ammonium hydroxide is a critical step to permit efficient trapping of glycans on the activated charcoal microspin column. Trapping efficiency of activated charcoal is drastically reduced in the presence of traces of ammonium hydroxide.

10. Purification of the released glycans through the solid-phase extraction with activated charcoal media is crucial to methylation reaction. The efficiency of methylation is adversely influenced by the presence of salts or other impurities.
11. Methylated glycans can be eluted from the reactors through washing with either DMSO or acetonitrile; however, the latter is easier and more convenient to handle.
12. The reaction mixture needs to be reapplied through the reactor several times to ensure sufficient contact time, thus ensuring efficient methylation.
13. Since the nature of the matrix and the method of sample preparation are critical to obtain strong signals, ensuring a proper crystallization of the DHB matrix is essential. When a mixture of acetonitrile or methanol and water is used, DHB typically crystallizes at room temperature and atmospheric pressure as long needle-shaped crystals that originate at the periphery of the spot and project toward the center. An amorphous mixture of the analytes, contaminants and salts are present in the central region of the spot. Such crystallization does not ensure strong signal; therefore, vacuum drying is needed to ensure a homogenous thin and even film of the matrix and analyte crystals, allowing efficient ionization and strong signal intensities.

Acknowledgments

This work was supported by grants No. GM24349 from the National Institute of General Medical Sciences, U.S. Department of Health and Human Services, and RR018942 from the National Institute of Health through the National Center for Research Resources (NCRR) for the National Center for Glycomics and Glycoproteomics. PK is the recipient of a fellowship from Merck Research laboratory.

References

1. Muhlecker, W., Gulati, S., McQuillen, D. P., Ram, S., Rice, P. A., andReinhold, V. N. (1999) An essential saccharide binding domain for the MAb 2C7 established for Neisseria gonorrhoeae LOS by ES-MS and MS^n. *Glycobiology* 9, 157–171.
2. Reinhold, V. N., and Sheeley, D. M. (1998) Detailed characterization of carbohydrate linkage and sequence in an ion trap mass spectrometer: glycosphingolipids. *Anal. Biochem.* 259, 28–33.
3. Sheeley, D. M., and Reinhold, V. N. (1998) Structural characterization of carbohydrate sequence, linkage, and branching in a quadrupole ion trap mass spectrometer: Neutral oligosaccharides and N-linked glycans. *Anal. Chem.* 70, 3053–3059.
4. Viseux, N., de Hoffmann, E., and Domon, B. (1997) Structural analysis of permethylated oligosaccharides by electrospray tandem mass spectrometry. *Anal. Chem.* 69, 3193–3198.

5. Viseux, N., de Hoffmann, E., and Domon, B. (1998) Structural assignment of permethylated oligosaccharide subunits using sequential tandem mass spectrometry. *Anal. Chem.* 70, 4951–4959.
6. Weiskopf,A.S.,Vouros,P.,andHarvey,D.J.(1997) Characterization of oligosaccharide composition and structure by quadrupole ion trap mass spectrometry. *Rapid Commun. Mass Spectrom.* 11, 1493–1504.
7. Weiskopf, A. S., Vouros, P., and Harvey, D. J. (1998) Electrospray ionization-ion trap mass spectrometry for structural analysis of complex N-linked glycoprotein oligosaccharides. *Anal. Chem.* 70, 4441–4447.
8. Hakomori, S.-I. (1964) Rapid permethylation of glycolipids and polysaccharides, catalyzed by methylsulfinyl carbanion in dimethyl sulfoxide. *J. Biochem.* 55, 205–208.
9. Ciucanu, I., and Kerek, F. (1984) A simple and rapid method for the permethylation of carbohydrates. *Carbohydr. Res.* 131, 209–217.
10. Ciucanu, I., and Konig, W. (1994) Immobilization of peralkylated β-cyclodextrin on silica gel for high-performance liquid chromatography. *J. Chromatogr. A* 685, 166–171.
11. Ciucanu, I., and Luca, C. (1990) Avoidance of degradation during the methylation of uronic acids. *Carbohydr. Res.* 206, 731–734.
12. Ciucanu, I., and Gabris, P. (1987) Peralkylation of pentaerythritol for gas chromatographic-mass spectrometric analysis. *Chromatographia* 23, 574–578.
13. Ciucanu, I., and Kerek, F. (1984) Rapid and simultaneous methylation of fatty and hydroxy fatty acids for gas-liquid chromatographic analysis. *J. Chromatogr.* 286, 179–185.
14. Ciucanu, I., and Costello, C. E. (2003) Elimination of oxidative degradation during the per-O-methylation of carbohydrates. *J. Am. Chem. Soc.* 125, 16213–16219.
15. Kang, P., Mechref, Y., Klouckova, I., and Novotny, M.V. (2005) Solid-phase permethylation of glycans for mass-spectrometric analysis. *Rapid Commun. Mass Spectrom.* 19, 3421–3428.
16. Huang,Y.,Konse,T.,Mechref,Y.,andNovotny, M. V. (2002) MALDI/MS-compatible β-elimination ofO-linked oligosaccharides. *Rapid Commun. Mass Spectrom.* 16, 1199–1204.
17. Huang, Y., Mechref, Y., and Novotny, M. V. (2001) Microscale nonreductive release of O-linked glycans for subsequent analysis through MALDI mass spectrometry and capillary electrophoresis. *Anal. Chem.* 73, 6063–6069.
18. Mechref, Y., and Novotny, M. V. (1998) Mass spectrometric mapping and sequencing of N-linked oligosaccharides derived from submicrogram amounts of glycoproteins. *Anal. Chem.* 70, 455–463.
19. Palm, A., and Novotny, M. V. (2005) A monolithic PNGase F enzyme microreactor enabling glycan mass mapping of glycoproteins by mass spectrometry. *Rapid Commun. Mass Spectrom.* 19, 1730–1738.
20. Rosenfeld, J. M. (1999) Solid-phase analytical derivatization: enhancement of sensitivity and selectivity of analysis. *J. Chromatogr A* 843, 19–27.

Chapter 5

Method for Investigation of Oligosaccharides Using Phenylhydrazine Derivatization

Erika Lattová and Hélène Perreault

Summary

The reaction of carbohydrates with phenylhydrazine provides corresponding phenylhydrazone derivatives, yielding increased sensitivity for mass spectrometry (MS) and ultraviolet (UV) detection during high-performance liquid chromatography (HPLC) separation. After a simple derivatization procedure, 1 h of incubation at 37–70°C, the samples can be analyzed by mass spectrometry techniques or with a combination of MS with HPLC using electrospray (ESI). Another useful aspect of this tagging is shown in the direct determination of glycosylation sites in proteins. Reaction conditions have enabled visualization of glycans next to their "mother peptides", that are otherwise usually not detectable because of suppression by abundant molecular ions of deglycosylated peptides.

We have shown that oligosaccharides as phenylhydrazone derivatives under MALDI-tandem mass spectrometry (MS/MS) and post-source decay (PSD) conditions provide very useful data for the structural elucidation of saccharides, including assignment of dominant isomers. Sialylated oligosaccharides also produce B, C fragment ions and cross-ring cleavage fragments comprising sialic acid.

Key words: Saccharides, Glycopeptides, Phenylhydrazone, High-performance liquid chromatography, HPLC/MS, Matrix-assisted laser desorption/ionization.

1. Introduction

It is undoubtedly true that carbohydrates, either in free state or as constituents of glycoproteins, play various roles in cell systems *(1)*. Therefore the growing interest in understanding their functions has stimulated the development of methods for glycoprotein analysis. Mass spectrometry techniques such as electrospray ionization (ESI) and especially matrix-assisted laser desorption/ionization (MALDI) belong to the most sensitive methods for

Nicolle H. Packer and Niclas G. Karlsson (eds.), *Methods in Molecular Biology, Glycomics: Methods and Protocols, vol. 534*

DOI: 10.1007/978-1-59745-022-5_5

the characterization of carbohydrate compounds. In order to facilitate their detection, it is common practice to introduce a chromophore at the reducing termini of oligosaccharides. In this respect, a large number of derivatization procedures for mono- and oligosaccharides have been developed. Among derivatization agents frequently used are e.g. 2-aminopyridine *(2)*, 4-aminobenzoic acid *(3)* and, 2-aminobenzamide *(4)*. Most reactions of saccharides with amino compounds are followed by reduction, providing open ring structures at the former reducing ends. These so called reduced derivatives are stable and produce intense signals in mass spectra. Some studies have shown that unreduced alkylamine derivatives of carbohydrates exist as closed-ring glycosylamines *(5)* and provide improved structural information on linkage and anomeric configuration than the above mentioned reduced derivatives *(6)*.

Reaction with phenylhydrazine *(7)* also does not include a cyanoborohydride step to reduce the **>C=N–** bond. On the samples of small oligosaccharides examined – fucosyllactose, *N*-acetyllactosamine and sialylyllactose, we did not observe the cleavage of the glycosidic bonds, losses of acetyl groups or of sialic acid moieties. *N*-glycans, enzymatically released from hen ovalbumin by PNGase and analyzed as their corresponding phenylhydrazones by the combination of on-line HPLC/electrospray ionization mass spectrometry, were separated at different extents depending on the mobile phase used and the flow rate *(8)*. In some cases, oligosaccharides of the same molecular weight were separated into individual isomeric glycan structures using reversed-phase conditions. Some of these were confirmed by MALDI-MS/MS and –PSD *(9)*. Perhaps the most important feature of the method is that the reaction does not require any additional reagent, besides phenylhydrazine. Thus, no salts are produced upon reaction and derivatized mixtures can be analyzed directly after incubation, without sample cleanup *(10)*. This feature of the method has enabled mass spectrometric investigation of detached oligosaccharides in the presence of their mother deglycosylated peptides *(11)*. Tagged oligosaccharides were thus analyzed on one target spot together with their corresponding deglycosylated peptides.

It should be pointed out that glycans labeled as unreduced PHN derivatives show very good fragmentation patterns under tandem MALDI-MS/MS and -PSD conditions. Abundant *m/z* fragment ions at higher *m/z* values correspond to only B and C cleavages. B ions formed by loss of the chitobiose core undergo easier losses of residues from 3-linked mannose than from 6-linked mannose. Also, we observed that the abundance of ions corresponding to the cleavage of bisecting GlcNAc depends on the number of residues attached to the 6-antenna – three GlcNac residues attached directly to 6-mannose caused the most extensive loss of bisecting GlcNAc moiety *(9)*. This fact has enabled us to distinguish isomeric structures for glycans of the same molecular weights directly from MALDI-MS/MS spectra without previous separation. With regard to the difference of other derivatization agents used for labeling saccharides, we noted that with PHN, the cross-ring cleavage ions are of relatively high abundance, sufficient to provide linkage information. MALDI-MS/MS spectra of sialylated oligosaccharides found in ovalbumine and IgG from human plasma contain fragment ions with sialic acids that have not been observed before in the spectra of the native or other derivatives of sialylated saccharides.

2. Materials

2.1. General Procedure for PHN Derivatization of Saccharide In-Tube

1. Reducing oligosaccharides (up to 5 μmol).
2. HPLC-grade deionized distilled water (e.g. Milli-Q-plus TOC water purification system Millipore, Bedford, MA).
3. Phenylhydrazine (e.g. 97% Sigma, St. Louis, MO).
4. Ethyl acetate (HPLC grade, e.g. Fischer Scientific, Fair Lawn, NJ).
5. Heating block or oven (75°C).

2.2. On-Target MALDI Derivatization

1. Reducing oligosaccharides dissolved in deionized water (1μL; 1–30 pmol).
2. Saturated 2,5-dihydroxybenzic acid (DHB) (e.g. Sigma) in acetonitrile/water 1:1.
3. *Phenylhydrazine reagent.* (3 μL) in deionized water (10 μL) and acetonitrile (3 μL), mixed properly.

2.3. Reversed-Phase HPLC of PHN-Saccharides

1. Phenylhydrazine derivatives of saccharides.
2. C8 column (Zorbax 300-SB C8 (4.6 × 150 mm) column (Chromatographic Specialties, Brockville, ON, Canada)).
3. *Solvent A.* 0.1 M acetic acid in 10:90 acetonitrile–water.
4. *Solvent B.* 0.1 M acetic acid in 25:75 acetonitrile–water.
5. HPLC-gradient pump (up to 0.5 mL/min).
6. UV-detector (245 nm).

2.4. Reversed-Phase HPLC/ESI-MSAnalysis for Glycans of Glycoprotein

1. Glycoprotein (e.g. hen ovalbumin as standard).
2. Peptide-*N*-glycosidaseF (PNGaseF, Prozyme, San Leandro, CA). The Glyko™ deglycosylation kit is supplied with a 5× incubation buffer kit. Alternatively, use a 100 mM solution of ammonium bicarbonate.

3. Heating facilities (water bath ~100°C, 75°C and 37°C incubation facilities).
4. Cooling facility (–20°C).
5. Cold Ethanol (–20°C).
6. Phenylhydrazine.
7. HPLC-gradient pump (up to 0.5 mL/min).
8. C18 column (e.g. Vydac 218-TP54 Protein & Peptide C_{18} (Hesperia, CA, USA)).
9. *Solvent A*. deionized water.
10. *Solvent B*. 0.05 M acetic acid in acetonitrile.
11. ESI-Mass spectrometer (e.g. Z-Spray™ ESI source and Quattro-LC triple quadrupole MS (Micromass, UK)).
12. Flow splitter to feed 1/10 of the HPLC flow to ESI (T-junction).

2.5. C_{18} Zip-Tip Purification of Sample Derivatized on the Target

1. C_{18} Zip-Tips (Millipore).
2. Syringe for loading sample into Zip-Tip.
3. Micropipette.
4. *Matrix solution (system A)*. 30 mg of DHB in 10% acetonitrile (1,000 μL).
5. *System B*. Acetonitrile/water = 1:1.
6. Acetonitrile.
7. Phenylhydrazine derivatized oligosaccharides (**Subheading 3.4**).

2.6. Determination of Oligosaccharides from Glycopeptides with Direct Determination of Glycosylation Sites

1. Glycoprotein (0.5 mg).
2. 25 mM ammonium bicarbonate.
3. Trypsin (e.g. Sequencing grade Trypsin from Promega, Madison, USA).
4. Heating block and/or oven (37°C and 70°C).
5. *Solvent A*. 5% acetonitrile.
6. *Solvent B*. 90% acetonitrile in 0.1% TFA.
7. HPLC gradient pump with UV detector e.g. with the analytical column Vydac 218 TP54 Protein &Peptide C18 (Separation Group, Hesperia, CA, USA), chromatograph equipped with a Rheodyne injector (5 μL loop); flow rate 0.5 mL/min; fractions collected e.g. manually.
8. Matrix solution 30 mg of DHB in 30% acetonitrile (1,000 μL).
9. HPLC-grade deionized distilled water (e.g. Milli-Q-plus TOC water purification system Millipore, Bedford, MA).
10. Peptide-*N*-glycosidaseF (PNGaseF, Prozyme, San Leandro, CA). The Glyko™ deglycosylation kit is supplied with a

5× incubation buffer kit. Alternatively, use 100 mM solution of ammonium bicarbonate.

11. Phenylhydrazine.

3. Methods

Derivatization with PHN is done with no additional reagent; to achieve full conversion to tagged carbohydrates, it is important to provide sufficient mixing of the reaction sample during incubation. In the case of on-target derivatization it is also recommended to mix the sample analyzed with PHN solution prior to loading onto the target spot with dried DHB matrix.

3.1. General Procedure for PHN Derivatization of Saccharides In-Tube

1. Dissolve the saccharides (up to 5 μmol) in distilled water (~30 μL).
2. Add phenylhydrazine reagent (1 μL) to the tube containing dissolved saccharide.
3. Leave mixture to incubate at 70°C temperature approximately for 1 h. During this time, mix sample periodically.
4. After incubation, the sample can be analyzed directly (depending on the purity of the sample before derivatization); or purified using common techniques, as e.g. HPLC, cartridges, or home-packed microcolumns. Nonreacted phenylhydrazine can be removed by extraction with ethylacetate: add additional water to the reaction mixture (100 μL) and ethylacetate (2 × 30 μL). Mix by vortexing and centrifuge briefly (~5 s) to separate ethylacetate from the water portion. Pipette out the upper ethylacetate layer and pour into the waste container.
5. Dilute or evaporate the aqueous portion to desired concentration.

3.2. On-Target MALDI Derivatization

1. Deposit saturated matrix solution of DHB in acetonitrile/water = 1:1 (0.7 μL) and leave to air dry.
2. Prepare reagent by dissolving phenylhydrazine (3 μL) in deionized water (10 μL) and mix properly.
3. To the sample of oligosaccharide dissolved in deionized water (1 μL; 1–30 pmol) add diluted phenylhydrazine solution (0.5 μL) and load ~1 μL of this mixture onto the target with dried DHB.
4. Leave to react at a temperature of 37°C (~30 min) or at room temperature until the spot is dried.

3.3. Reversed-Phase HPLC of PHN-Saccharides

1. Mix solution of saccharides with phenylhydrazine and follow the procedure described in the **Subheading 3.1**. After incubation evaporate solution to get volume to ~10 μL.
2. Inject aliquot of derivatized sample into the HPLC and allow for separation. An example of HPLC separation of PHN derivatized saccharides is shown in **Fig. 1**.

3.4. Reversed-Phase HPLC/ESI-MS Analysis for Glycans of Glycoprotein

1. Detach oligosaccharides from glycoprotein e.g. enzymatically using enzyme PNGaseF: dissolve glycoprotein (200–500 μg) in deionized water (36 μL) in a small microcentrifuge tube. Add Buffer 5× (9 μL) supplied with the Glyko™ deglycosylation kit or 100 mM solution of ammonium bicarbonate (9 μL). Leave mixture to boil in a water bath for 5–10 min to denature glycoprotein. After cooling, add PNGaseF (2 μL), mix by vortexing and centrifuge briefly (1–2 s). Keep mixture at a temperature 37°C (18–20 h). After incubation, add cold ethanol (150 μL) to the reaction mixture and leave for at least 4 h. in the freezer to precipitate the protein. . Then centrifuge the tube (~15 min), pipette out the supernatant containing the oligosaccharides into another microcentrifuge tube and evaporate the ethanol.
2. Dissolve the white residue in deionized water (30 μL), add phenylhydrazine (1 μL) and leave to react at a temperature

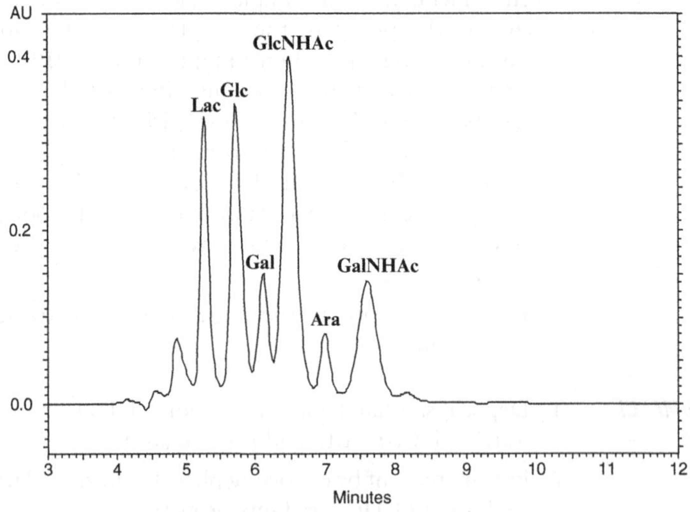

Fig. 1. HPLC-UV chromatogram of PHN derivatized mixture of lactose (Lac), glucose (Glc), galactose (Gal), *N*-acetylglucosamine (GlcNAc), arabinose (Ara) and *N*-acetylgalactosamine(GalNAc). The derivatized sample, containing 20 nmol of each PHN saccharide, was injected into a System Gold HPLC chromatograph equipped with a system Gold 166 UV Detector and 32 Karat Software (Beckman-Coulter Canada, Missisauga, ON, Canada) and separated on a Zorbax 300-SB C8 (4.6 × 150 mm) column (Chromatographic Specialties, Brockville, ON, Canada). Elution was carried out at a flow-rate 0.5 mL/min with solvents A (0.1 M acetic acid in 10:90 acetonitrile–water) and B (0.1 M acetic acid in 25:75 acetonitrile–water). The proportion of B was linearly increased from 5 to 30% over 6 min, then kept at 50% until the end of the run. UV detection was performed at 245 nm.

of 75°C for 1 h with periodical mixing. Evaporate reaction mixture to a volume of approximately 10 μL.

3. Inject a 5 μL aliquot in the HPLC instrument of which the column effluent is split and introduced directly into the ESI source of the mass spectrometer.

An example of on-line HPLC/ESI-MS analysis of oligosaccharides detached from hen ovalbumin is shown in **Fig. 2**.

3.5. C_{18} Zip-Tip Purification of Sample Derivatized on the Target

Purification of derivatized mixtures on C_{18} carriers (commercially available Zip-Tips) can help to achieve separation of isomers and to acquire MS/MS or PSD spectra from single spots. This allows confirmation of the existence of structures of the same molecular weights.

1. Prepare matrix solution by dissolving 30 mg of DHB in e.g. acetonitrile/water = 1:9 (system A) and acetonitrile/water = 1:1 (B).
2. Wash the pipette tip with 100% acetonitrile (2 × 50 μL) and then with deionized water.
3. With a syringe, slowly withdraw 5 μL of derivatized mixture (prepared according to procedure described above in **Subheading 3.4**). Insert the syringe into the Zip-Tip pipette tip and carefully load sample onto the wet cartridge.
4. Attach the Zip-Tip column to the micropipette and by pressing the plunger, allow sample to permeate into the cartridge.
5. Load 5 μL of solution A on the top of the Zip-Tip cartridge and by pressing the plunger, allow solution to wash the column. Repeat again with system A (2 × 5μL) and then with solvent B (2 × 5μL). Always deposit solvent eluting solvent with sample as single spots (~0.7 μL) onto the target.

Figure 3 shows an example where Zip-Tip cleaning of derivatized sample was used for the separation of oligosaccharide isomers.

3.6. Determination of Oligosaccharides from Glycopeptides with Direct Determination of Glycosylation Sites

1. *Tryptic digestion.* Dissolve an intact glycoprotein (0.5 mg) in 25 mM ammonium bicarbonate (200 μL); add trypsin (5 μg) and digest at 37°C for 20 h. After the incubation, evaporate the sample to a volume ~20 μL, store in the freezer or use directly for HPLC separation.
2. Inject evaporated tryptic digested sample (10 μL) into HPLC, with 5% acetonitrile in water as solvent A, and 90% acetonitrile in 0.1% TFA as solvent B. Collect fractions, including those with peaks of very low intensities in UV.
3. Separately dilute all evaporated fractions in 10 μL of deionized water. Deposit approximately 0.7 μL of each fraction on the target with dried DHB matrix.
4. To each tube with confirmed presence of glycopeptides in water solution (~10 μL) add incubation buffer (2.5 μL) with

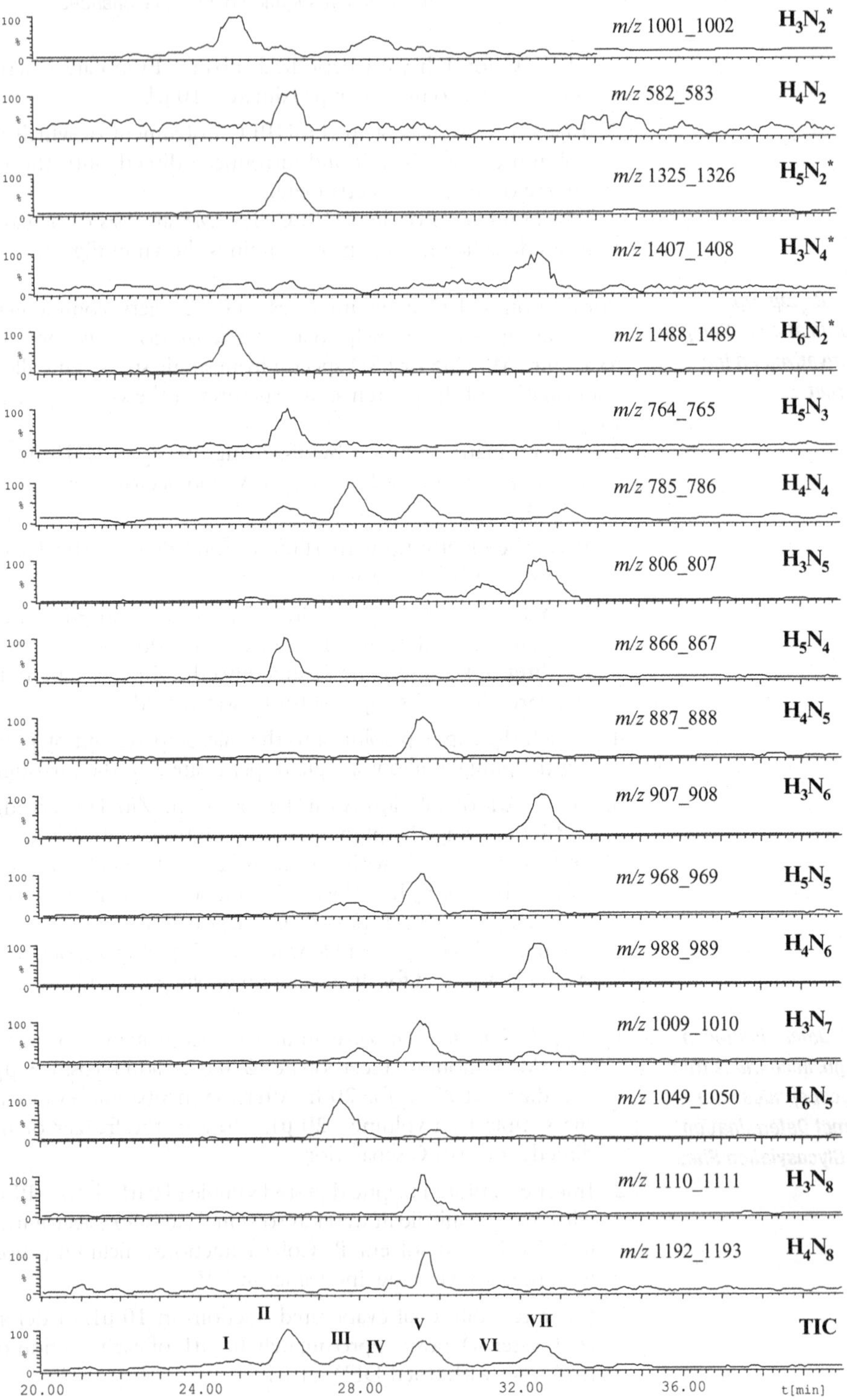

m/z 1001_1002 H3N2*
m/z 582_583 H4N2
m/z 1325_1326 H5N2*
m/z 1407_1408 H3N4*
m/z 1488_1489 H6N2*
m/z 764_765 H5N3
m/z 785_786 H4N4
m/z 806_807 H3N5
m/z 866_867 H5N4
m/z 887_888 H4N5
m/z 907_908 H3N6
m/z 968_969 H5N5
m/z 988_989 H4N6
m/z 1009_1010 H3N7
m/z 1049_1050 H6N5
m/z 1110_1111 H3N8
m/z 1192_1193 H4N8
TIC
I
II
III
IV
V
VI
VII
100
%
0
20.00
24.00
28.00
32.00
36.00
t[min]

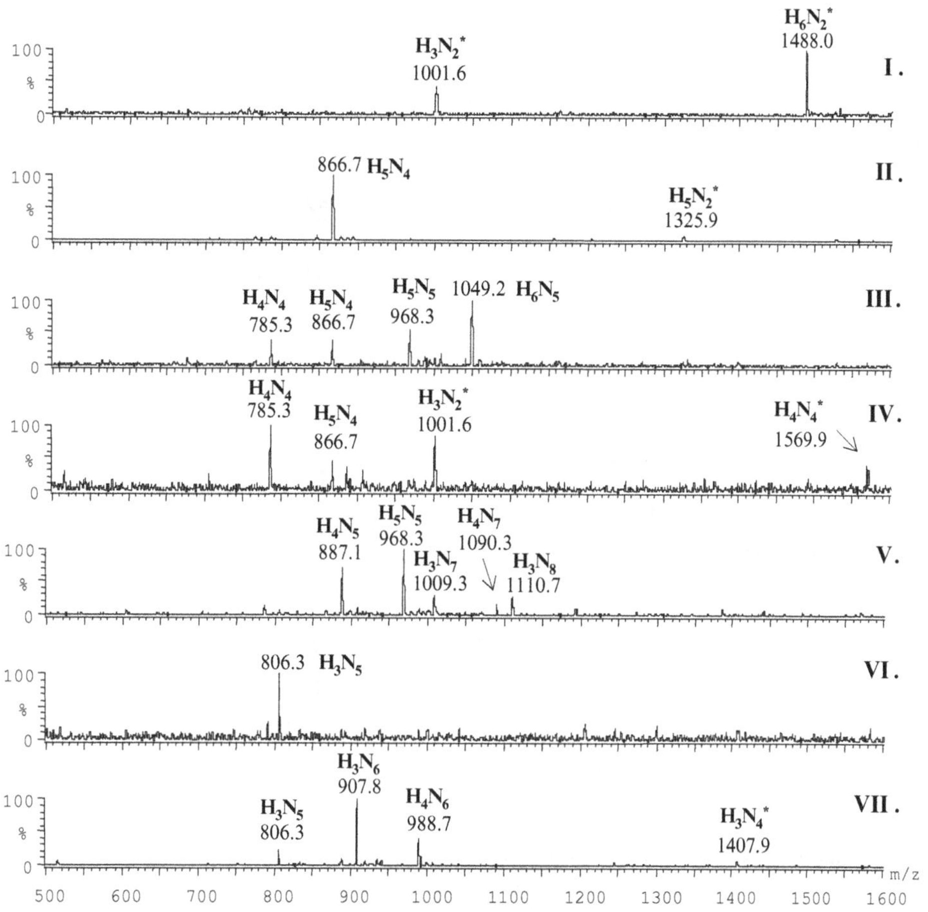

Fig. 2. (**a**) HPLC/MS selected and total ion chromatograms obtained for PHN-glycans of hen ovalbumin (Grade VII). HPLC analysis was performed on a Hewlett-Packard 1100 (Agilent Technologies, Mississauga, ON), with a Vydac 218-TP54 Protein & Peptide C_{18} (Hesperia, CA, USA) as the analytical column. The sample was separated at a flow rate of 0.5 mL/min with variable concentration of acetic acid during separation. The proportion of B (0.05 M acetic acid in acetonitrile) was increased from 0 to 10% over 30 min and then to 15% over the next 30 min. As eluent A, deionized water was used. The column effluent was split and introduced directly in the Z-Spray™ ESI source of a Quattro-LC triple quadrupole MS (Micromass, UK). (**b**) ESI mass spectra acquired from peaks shown in TIC chromatogram (labeled with roman numbers, (**Fig. 2**). Electrospray produced mainly $[M + 2H]^{2+}$, and some PHN-oligosaccharides formed $[M + H]^{+}(*)$ ions.

PNGaseF (0.3 μL). After mixing and centrifugation for a short time leave tubes at 37°C for 18–20 h.

5. After incubation, it is possible to detect only peaks corresponding to deglycosylated peptides. Detached oligosaccharides are suppressed, but they can be visualized following derivatization:
   (a) To the tube of PNGase treated sample add deionized water (5 μL), phenylhydrazine (0.3 μL), acetonitrile (2 μL) and leave to incubate at ~37°C for 30 min (mix periodically

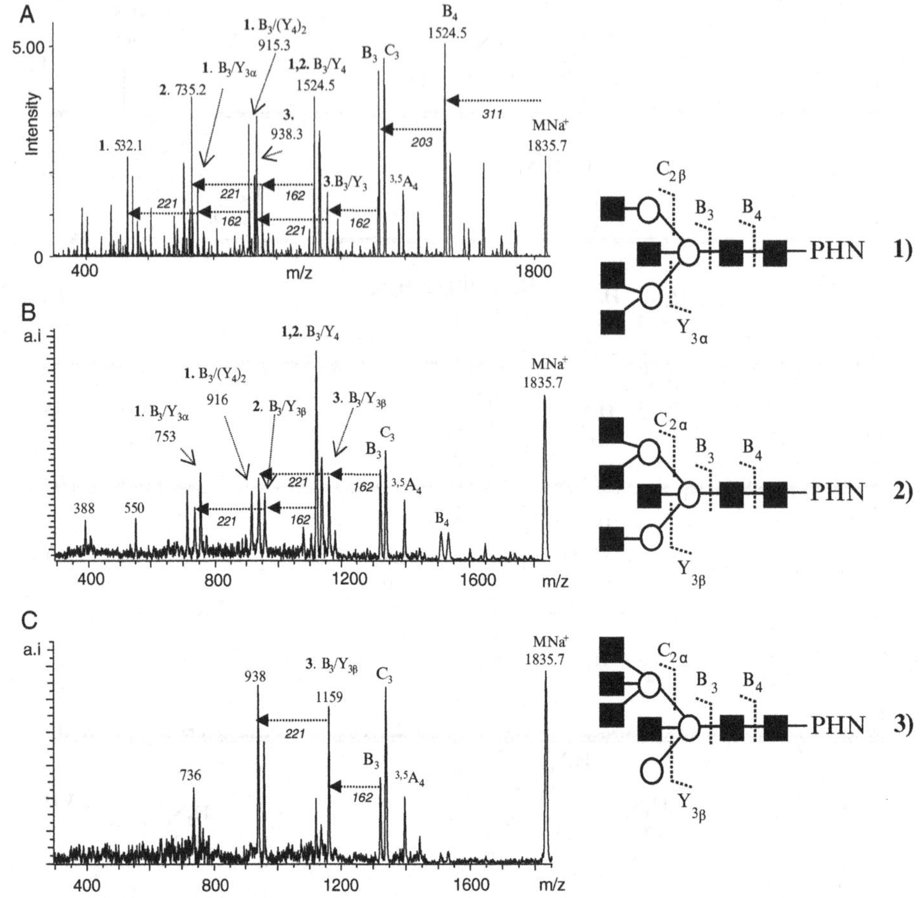

Fig. 3. MALDI CID-MS/MS (**a**) and MALDI-PSD (b, c) spectra of [M + Na]$^+$ ions of complex PHN-glycans recorded at *m/z* 1,835.6 from hen ovalbumin (Grade V). Spectrum (**a**) was acquired from the whole derivatized sample by using a prototype quadrupole-quadrupole-TOF mass spectrometer and corresponds to three possible isomers. PSD-spectra (**b, c**) were recorded on a Biflex IV spectrometer (Bruker Daltonics, Billerica, MA) from spots obtained by on-target purification with C_{18}- ZipTip. Dominant Isomers 1 and 2 present in spectrum (**a**) and partially in spectrum (**b**), were separated out from the Isomer 3 detected in the spectrum (**c**).

during incubation by vortexing). After derivatization, evaporate to approx. 5 μL and use for MS analysis.

(b) In case of on-target derivatization, mix deglycosylated sample (1 μL) with a phenylhydrazine solution (0.5 μL; 2 μL PHN in 10 μL water). Immediately after mixing, spot 1 μL of this mixture onto dried DHB on the target and leave to react at 37°C until the spot is dried.

An example of determination of oligosaccharides next to deglycosylated peptides is shown in **Fig. 4**.

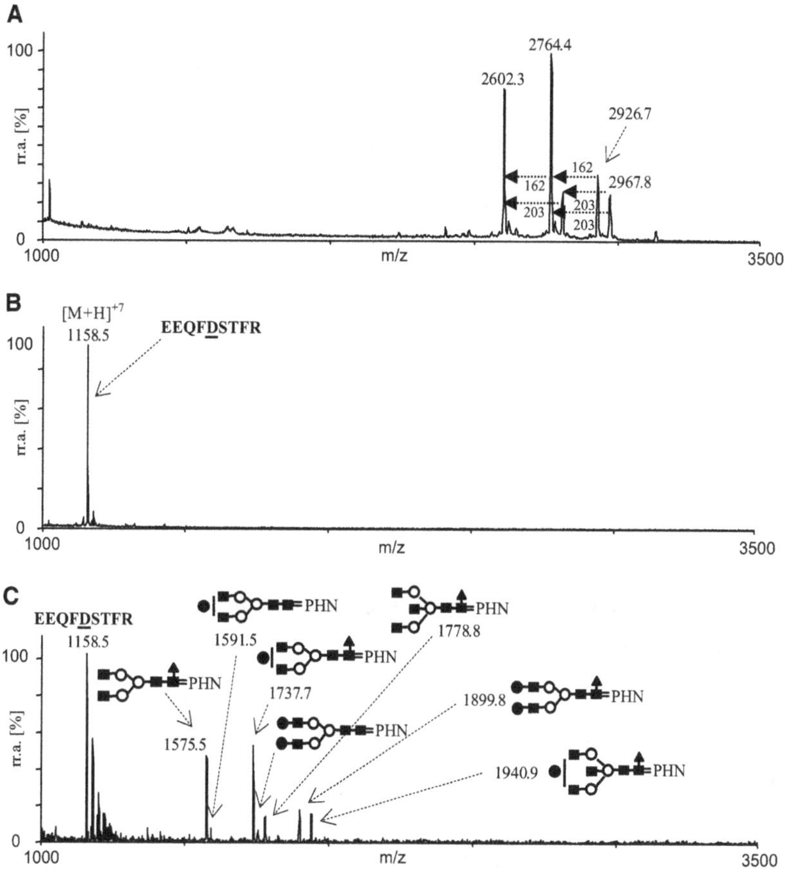

Fig. 4. Positive MALDI-MS spectra acquired from an HPLC fraction of polyclonal IgG, obtained at elution time 28 min (elution was carried out at a flow rate 1 mL/min): (**a**) after trypsin digestion, all ions are $[M + H]^+$; (**b**) after enzymatic deglycosylation with PNGaseF; (**c**) on target derivatized with PHN, oligosaccharide ions are $[M + Na]^+$.

4. Notes

1. PHN reagent is highly reactive; for a better result it is recommended that the PHN solution always be prepared immediately before reaction with the saccharide sample.
2. PHN is an oily liquid. Thorough mixing of the prepared reagent solution before pipetting it out (especially for on-target derivatization) and all reaction mixtures during derivatization is a key step for maximal conversion of carbohydrates to their corresponding phenylhydrazones.
3. Avoid excess amounts of phenylhydrazine, especially in the case of on-target reaction; it can prevent crystallization of the sample.

4. We recommend analyzing the samples immediately after derivatization. Otherwise, derivatized samples should be stored under argon at a temperature ~ –20°C.
5. For Zip-Tip separation and on-target deposition, derivatized samples can be fractionated on home-made reverse phase micro-columns constructed from standard Geloader™ pipette tip packed with 15 μL of C18 packing material.
6. We found this method suitable for characterization of all types of saccharides with free reducing ends. Acidic oligosaccharides (with sialic/uronic acids) can be successfully derivatized without apparent cleavage during derivatization (e.g. **Fig. 5**).
7. Under mass spectrometric conditions, neutral PHN-oligosaccharides generally yield better sensitivity in the positive mode and by contrast, acidic carbohydrates accomplish better detection in the negative mode.

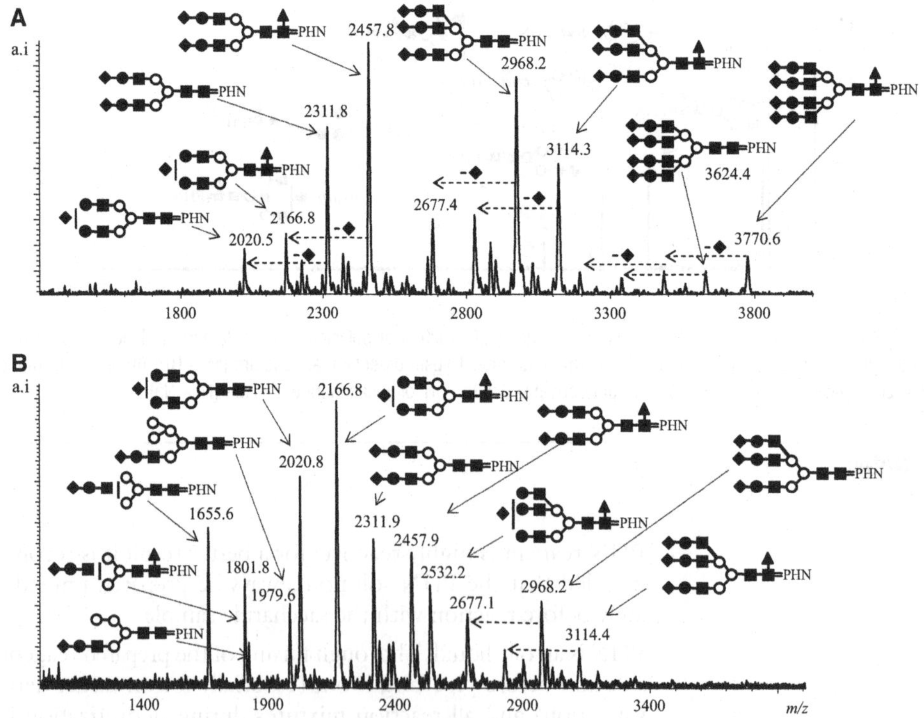

Fig. 5. Negative MALDI-MS spectra of PHN-sialylated oligosaccharides recorded from HPLC fractions (**A**) with elution time 3 min and (**B**) elution time 4 min. Oligosaccharides were released from total human integrin (~50 μg) by PNGaseF. Sample was fractionated on HPLC on reversed-phase column (elution was carried out at a flow rate 0.5 mL/min with solvents A composed of 10% acetonitrile in water and B composed of 50% acetonitrile; the proportion of B was linearly increased from 5 to 30% over 6 min, then kept at 50% until the end of the run). Individual manually collected fractions (~100 μL) were evaporated (~5 μL) and each of them derivatized on-target. All oligosaccharide ions are $[M-H]^-$.

8. Derivatization with PHN produces phenylhydrazones of saccharides, thus augmenting their mass by 90 units. Generally, mass spectra of derivatized oligosaccharides, native carbohydrates and glycopeptides produce peaks with characteristic differences of 146 (Fuc), 162 (Hex), 203 (HexNAc), 291 (NeuAc) or 307 (NeuGly) units.

Acknowledgements

Acknowledgements are made to Professors Kenneth G. Standing and Werner Ens for the use of the Qq-TOF mass spectrometer. This work was supported by grants from the Natural Sciences and Engineering Research Council of Canada (NSERC), from the Canadian Foundation for Innovation (CFI) and the Canada Research Chairs Program (CRC).

References

1. Helenius, A., Aebi, M. (2001) Intracellular function of N-linked glycans. *Science* 291, 2364–2369.
2. Okamoto, M., Takahashi, K., Doi, T., Takimoto, Y. (1997) High-sensitivity detection and postsource decay of 2-aminopyridine-derivatized oligosaccharides with matrix-assisted laser desorption/ionization mass spectrometry. *Anal. Chem.* 69, 2919–2926.
3. Takao, T., Tambara, Y., Nakamura, A., Yoshino, K. I., Fukuda, H., Fukuda, M., Shimonishi, Y. (1996) Sensitive analysis of oligosaccharides derivatised with 4-Aminobenzoic acid 2-(Diethylamino)Ethyl Ester by matrix-assisted laser desorption/ionization mass spectrometry. *Rapid Commun. Mass Spectrom.* 10, 637–640.
4. Bigge, J., Patel, T.P., Bruce, J.A., Goulding, P.N., Charles, S.M., Parekh, R.B. (1995) Non-selective and efficient fluorescent labeling of glycans using 2-aminobenzamide and anthranilic acid. *Anal. Biochem.* 230, 229–238.
5. Ojala, W. H., Ostman, J. M., Ojala, C. R. (2000) Schiff bases or glycosylamines: crystal and molecular structures of four derivatives of d-mannose. *Carbohydr. Res.* 326, 104–112.
6. Zaia, J., Li, X. Q., Chan, S. Y., Costello, C. (2003) Tandem mass spectrometric strategies for determination of sulfation position and uronic acid epimerization in chondroitin sulfate oligosaccharides. *J. Am. Soc. Mass Spectrom.* 14, 1270–1281.
7. Lattová, E., Perreault, H. (2003) Labelling saccharides with phenylhydrazine for electrospray and matrix-assisted laser desorption-ionization mass spectrometry. *J. Chromatogr. B* 793, 167–179.
8. Lattová, E., Perreault, H. (2003) Profiling of N-linked oligosaccharides using phenylhydrazine derivatization and mass spectrometry. *J. Chromatogr. A*, 1016, 71–87.
9. Lattová, E., Krokhin, O., Perreault, H. (2004) Matrix-assisted laser desorption/ionization mass spectrometry and post source decay fragmentation study of phenylhydrazones of N-linked oligosaccharides from ovalbumin. *J. Am. Soc. Mass Spectrom.* 15, 725–735.
10. Lattová, E., Krokhin, O., Perreault, H. (2005) Influence of labeling group on ionization and fragmentation of carbohydrates in mass spectrometry. *J. Am. Soc. Mass Spectrom.* 15, 725–735.
11. Lattová, E., Kapková, P., Krokhin, O., Perreault, H. (2006) A method for investigation of oligosaccharides from glycopeptides: direct determination of glycosylation sites in proteins. *Anal. Chem.* 78, 2977–2984, DOI 10 1021/ac060278V

Chapter 6

Two-Dimensional HPLC Separation with Reverse-Phase-Nano-LC-MS/MS for the Characterization of Glycan Pools After Labeling with 2-Aminobenzamide

Manfred Wuhrer, Carolien A.M. Koeleman, and André M. Deelder

Summary

N-Glycans, *O*-glycans, glycolipid glycan chains, and other oligosaccharide pools released from various natural sources often represent very complex mixtures. Various tandem mass spectrometric techniques may be applied to glycan pools, resulting in valuable structural information. For a very detailed analysis of these pools, however, the following approach may be preferable: In a first step, reducing-end oligosaccharides are labeled with the aromatic tag 2-aminobenzamide (2-AB). The 2-AB glycans are separated by analytical-scale, normal-phase (NP)-HPLC (first dimension) and peak-fractionated using fluorescence detection. Peak fractions are analyzed by nano-LC-ESI-IT-MS/MS using a conventional reverse-phase (RP) nanocolumn (second dimension). Chromatography may be monitored by measuring the UV absorbance of the AB tag. Tandem mass spectrometry may be performed on deprotonated species (negative-ion mode), on proton adducts, as well as on sodium adducts (positive-ion mode). This approach has the following particular advantages: (1) the combination of the two HPLC dimensions usually separates isobaric species from each other, thereby allowing the tandem mass spectrometric characterization of individual glycan structures; (2) the AB-mass tag helps with the unambiguous assignment of fragment ions; (3) the second-dimension RP-nano-LC-MS/MS analyses can be performed in many mass spectrometric laboratories, as only standard equipment is needed. Alternatively, for a less in-depth characterization of complex glycan pools, the RP-nano-LC-MS/MS technique may be used as a stand-alone technique. In conclusion, the presented methods allow the detailed mass spectrometric characterization of complex *N*-glycan pools released from various biological sources.

Key words: LC-MS, Mass spectrometry, Normal phase, Reverse phase.

1. Introduction

Oligosaccharides exhibit a variety of biological activities *(1)*. They act as ligands for cell surface receptors in cellular communication *(2, 3)*. Cell surface glycans have, therefore, a crucial role in

Nicolle H. Packer and Niclas G. Karlsson (eds.), *Methods in Molecular Biology, Glycomics: Methods and Protocols, vol. 534*

DOI: 10.1007/978-1-59745-022-5_6

the cellular differentiation and development of multicellular organisms *(4)*. Hence, many types of cancers are associated with changes of cell surface glycosylation *(5)*. Moreover, glycans often form the outmost layer of pathogens (bacteria, parasites, viruses), which gets into contact with the host, and are targets of the humoral as well as cellular immune system *(6–8)*.

Many mass spectrometric (MS) techniques are available for analyzing and monitoring the glycosylation status of biological samples or systems, which may be multicellular organisms, organs, particular cell types, body fluids, or a (group of) purified glycoconjugate(s) (glycoproteins or glycolipids) *(9–12)*. Glycans may be released from glycoconjugates by chemical or enzymatic methods, purified, optionally derivatized, and analyzed by mass spectrometry. In order to obtain a detailed picture of a glycan pool, chromatographic and electrophoretic separation methods, i.e., liquid chromatography (LC) *(13)* and capillary electrophoresis (CE) *(14)*, are often combined with MS techniques (LS-MS and CE-MS, respectively) by online electrospray ionization. LC analysis of oligosaccharides with online electrospray MS at the nano or capillary scale may be performed with a porous graphitized carbon stationary phase *(13, 15)* or various normal phases *(13, 16, 17)*, or by high-pH anion exchange chromatography with online desalting *(18)*. Alternatively, when derivatization of the reducing end is performed with an aromatic tag, oligosaccharides may be analyzed by RP-nano-LC-MS/MS *(13)*, as described here. For this approach, glycans are labeled with 2-AB by reductive amination and purified via a cartridge (*see* **Fig. 1**). RP-nano-LC-MS/MS analysis may then be performed on standard-type LC-MS systems as used in many mass spectrometry laboratories working in the fields of proteomics and metabolomics. Remarkably, no changes of the hardware and only minor changes of the running solvents may be required, which is a major advantage of RP-nano-LC compared to the other stationary phases mentioned above.

In the case of very complex glycan pools from, e.g., pathogens or other nonmammalian sources, the RP-nano-LC system may not sufficiently resolve the mixture for a detailed analysis. In these cases, RP-nano-LC-MS/MS can be combined as a second separation dimension with a first-dimension preparative NP-HPLC of 2-AB-labeled glycans with fluorescence detection and peak fractionation (*see* **Fig. 1**) *(19)*. The latter NP-HPLC method was introduced by Dr Pauline M. Rudd and others from the Oxford Glycobiology Institute, who combined exoglycosidase arrays with repetitive NP-HPLC runs and a reference database with standardized retention properties for structural analysis of 2-AB-labeled glycans *(20–24)*. The two-dimensional HPLC separation system with MS/MS detection, which is used by us *(19)* and described in the present paper, allows the detailed mass

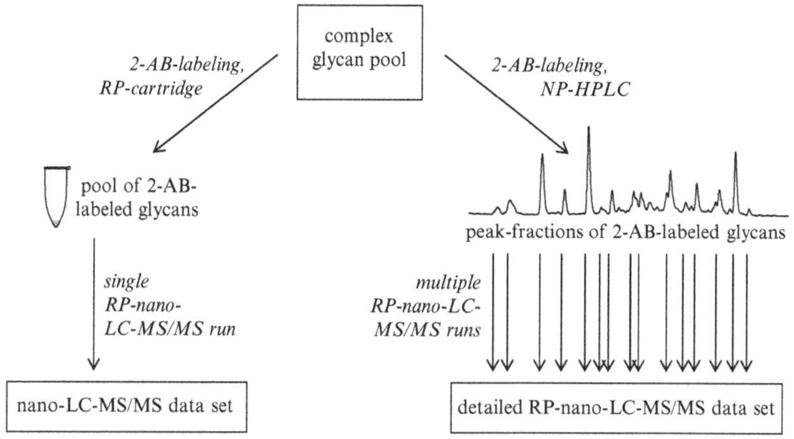

Fig. 1. RP-nano-LC-MS/MS with or without NP-HPLC.

spectrometric characterization of very complex biological samples and exhibits the following features:

- Low femtomole sensitivity of fluorescence detection in the first-dimension NP-HPLC separation.
- Low femtomole sensitivity of the second-dimension online nano-ESI mass spectrometric detection, both in MS and in MS/MS mode.
- Possibility of RP-nano-LC followed by registering UV absorption of the 2-AB tag, allowing facile quantification.
- Routine separation of isobaric structures in both first and second dimension, resulting in their differentiation and characterization by MS and MS/MS experiments.
- The possibility of MS analysis of the 2-AB-labeled glycans in positive-ion mode (sodium adducts or proton adducts) as well as in negative-ion mode.
- When required, further mass spectrometric characterization of specific parts of the glycan by MS^3 experiments using ion trap MS.

2. Materials

2.1. 2.-Aminobenzamide Labeling

1. Trifluoroacetic acid (TFA; Fluka 73645, for protein sequence analysis).
2. Deionized water (Milli-Q; Millipore) (*see* **Note 1**).

3. *Dextran hydrolysate*: Dextran T10 (formerly available from AmershamBiosciences, but discontinued; now available from Pharmacosmos A/S, Holbaek, Denmark) is dissolved at 10 mg/mL in 0.1 M trifluoroacetic acid in a screw-capped glass vial sealed with a Teflon-lined cap (add 995 μL water and 7.8 μL trifluoroacetic acid to 10 mg of dextran T10). Incubate for 1 h at 100°C. Transfer an aliquot of 1 mg to an Eppendorf vial. Dry by lyophilization and use for 2-AB labeling (100 μL of labeling reagent). Add 1 mL water and store at –20°C.
4. Dimethylsulfoxide (DMSO; 99.9% purity, Sigma D8779).
5. Sodium cyanoborohydride ($NaCNBH_3$, Fluka 71435).
6. 2-Aminobenzamide (2-AB, anthranilamide; 98 + % purity, Sigma, A8,980-4).
7. Oligosaccharide standards (e.g. maltopentaose, maltohexaose and maltoheptaose; Sigma).

2.2. Sample Clean-Up with RP Cartridge

1. C18 Sep-Pak cartridge Light (WAT 023501, 130 mg; Waters).
2. Disposable syringe (2 mL) with Luer slip tip.
3. Vacuum manifold.
4. Methanol (p.a.).

2.3. Analytical-Scale NP-HPLC

1. HPLC system for binary gradient with a pressure limit of 150 bar or higher, containing a manual injector with a 1-mL sample loop.
2. Fluorescence detector (e.g. Jasco FP-2020plus or Waters 2475).
3. Control station with HPLC software, which is capable of registering the fluorescence signal (with analog/digital converter).
4. Formic acid (Fluka 94318, puriss. p.a., for mass spectrometry, 98%).
5. Ammonium hydroxide solution (Fluka 09857, 25% in water).
6. NP-HPLC column (TSKgel Amide-80; 4.6 mm × 25 cm, 5 μm; TOSOH Bioscience, Stuttgart, Germany).

2.4. Nano-RP-HPLC-ESI-MS/MS

1. RP-nano-LC-MS/MS may be performed on quite different systems which are used for the analysis of tryptic digests in proteomics settings. In the following, details of the system used by us are given and possible variations thereof are mentioned:
2. Autosampler (Dionex/LC Packings, Amsterdam, The Netherlands) equipped with a 10-μL sample loop.
3. Micropump system (Switchos; Dionex/LC Packings) with a switching valve equipped with an RP-trap column (PepMap 100, 300 μm × 5 mm, 3 μm, 100Å; Dionex/LC Packings).

4. Nano-HPLC (HPLC with nano-splitter; Ultimate or Ultimate 3000 Dionex/LC Packings) supplying a flow of approximately 150 nL/min.
5. RP-nano-column (PepMap 100, 75 μm × 150 mm, 3 μm, 100Å; Dionex/LC Packings).
6. Z-type nano-flow UV detector (Dionex/LC Packings).
7. Tandem mass spectrometer. Esquire High Capacity Trap (HCT ultra) ESI-IT-MS (Bruker Daltonics, Bremen, Germany) equipped with an online nano source.
8. Electrospray needles (360 μm OD, 20 μm ID with 10 μm opening; New Objective, Cambridge, MA, USA).
9. Sodium hydroxide solution.

3. Methods

Reducing-end glycans are labeled with 2-AB and may be analyzed by RP-nano-LC-MS/MS (*see* **Subheading 3.4**) in one of the following ways: if the total 2-AB-labeled glycan mixture is to be analyzed in a single nano-LC-MS/MS run, a cartridge purification step is included (*see* **Subheading 3.2**; *see* **Fig. 1**). Alternatively, for very complex glycan mixtures that are available in sufficient quantities (high picomole amounts), a very detailed characterization may be achieved using the two-dimensional separation approach. The cartridge work-up step can be skipped then, and the 2-AB-labeled glycans together with the labeling reagents can be fractionated in an NP-HPLC run (*see* **Subheading 3.3**; *see* **Fig.1**). All the individual fractions may then be analyzed by RP-nano-LC-MS/MS.

In RP-nano-LC-MS/MS of AB-labeled glycans, different mass spectrometers may be chosen, with quadrupole time of flight instruments and ion traps being the most widely distributed instruments. With these two types of instruments, collision-induced dissociation (CID) is the fragmentation method applied *(9–13)*. CID may be performed on deprotonated 2-AB-labeled glycans (negative mode), or on protonated or sodiated species (positive mode). These three forms of ionized 2-AB glycans differ in their fragmentation characteristics and the obtained MS/MS spectra may provide different sets of information. It is often desirable, therefore, to perform multiple LC-MS/MS runs and fragment the same 2-AB-labeled glycans in protonated, sodiated, and deprotonated forms. Fragmentation of sodiated and deprotonated species will often result in carbohydrate ring fragmentation, providing valuable information about linkage positions. Regarding the fragmentation of protonated species, it should be pointed out that rearrangements may be observed in MS/MS spectra *(19, 25–27)*, which may be misleading and suggest "false" structures of the precursor. This phenomenon is predominantly observed for fucosylated oligosaccharides (fucose migration/rearrangements), but also other monosaccharides may be involved.

3.1. 2-AB Labeling

1. Dry the samples in 1.5 mL Eppendorf vials by lyophilization.
2. Add 150 µL glacial acetic acid to 500 µL DMSO (*see* **Note 2**) in a separate Eppendorf vial and mix by pipette action (= solvent mixture).
3. Dissolve an aliquot of the labeling dye 2-AB in the solvent mixture by pipette action (100 µL solvent mixture per 4.8 mg labeling dye; 0.35 M) (*see* **Note 3**).
4. Add an aliquot of the dye solution to $NaCNBH_3$ (100 µL of dye solution per 6.3 mg of $NaCNBH_3$; = reducing agent; 1 M) (*see* **Note 2** and **4**). Mix by pipette action until dissolved, which may take several minutes (= labeling reagent).
5. Add 5 µL of the labeling reagent to each dried sample, put the lid on, and vortex briefly.
6. Spin down the samples.
7. Incubate for 3 h at 65°C. After 1 and after 2 h, briefly vortex and spin down the samples again.
8. Centrifuge the samples briefly and let them cool down. Store them at –20°C until usage.

3.2. Sample Clean-Up with RP Cartridge

1. Place the C18 RP cartridge on the vacuum manifold with a 2-mL disposable syringe on top (= solvent reservoir).
2. Wash the cartridge with 1 mL methanol under vacuum (*see* **Note 5** and **6**).
3. Wash the cartridge with 2 mL water under vacuum.
4. Switch off the vacuum. Add 100 µL water to the 2-AB-labeled sample and vortex. Apply the sample to the solvent reservoir. Rinse the Eppendorf vial with 100 µL water and also apply this solution to the solvent reservoir. Load the sample completely onto the cartridge under vacuum (*see* **Note 5**).
5. Wash the cartridge with 2 mL water.
6. Elute 2-AB-labeled glycans with 1 mL methanol/water (1:1) mixture and collect in glass tubes. Maintain the vacuum until all the solvent is retrieved from the cartridge (dry cartridge).
7. Dry the eluate by vacuum centrifugation (heating temperature of 40°C max.). (*see* **Note** 7) Dissolve in 100 µL water. Optionally perform a first screening by MALDI-TOF-MS in the positive-ion and negative-ion modes. Analyze the eluate by RP-nano-LC-MS(/MS).

3.3. Analytical-Scale NP-HPLC (Fig.2)

1. *Preparation of solvent A* (50 mM ammonium formate solution, pH 4.4): Add 2.6 mL formic acid to 950 mL water. Adjust pH to 4.4 with ammonium hydroxide (approximately 6 mL). Add water to a total volume of 1 L. Degas by sonication in a sonifier bath for 5 min.
2. *Preparation of solvent B* (20% solvent A, 80% acetonitrile). Add 800 mL acetonitrile to 200 mL solvent A. Degas by sonication in a sonifier bath for 5 min.
3. Set up the HPLC system with manual injector, NP-HPLC column, and fluorescence detector.
4. Wash the system for 10 min at 100% solvent A at a flow of 1 mL/min.
5. Equilibrate the system for 20 min at 100% solvent B at a flow of 1 mL/min.

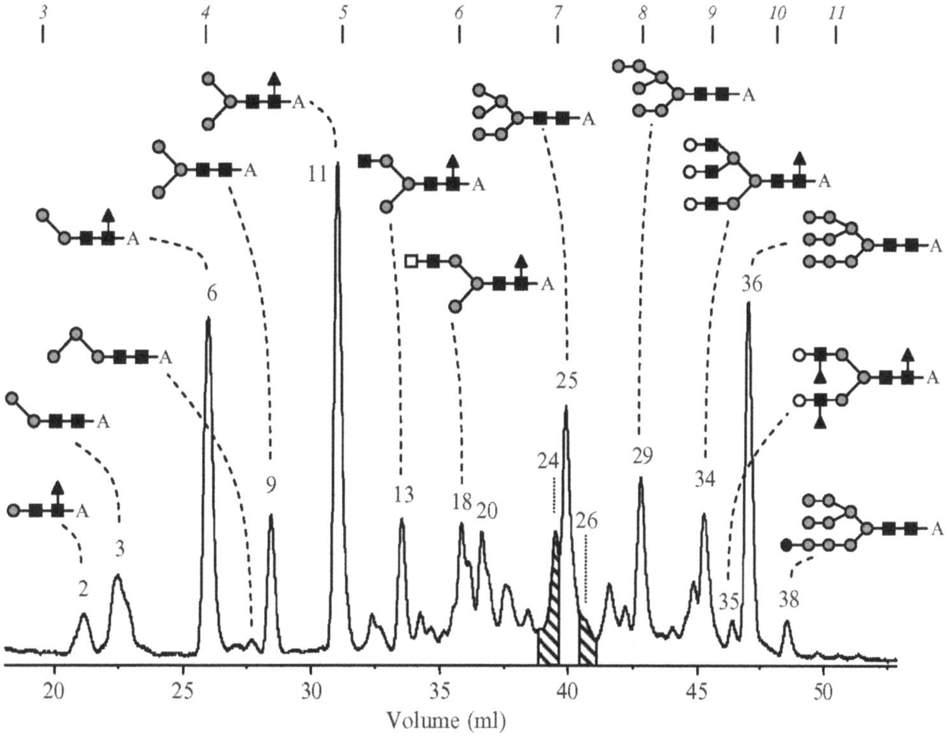

Fig. 2. First-dimension NP-HPLC separation of *N*-glycans from *S. mansoni* adult worms labeled with AB. Peaks are marked with fraction numbers and schematic representations of the AB-labeled glycans detected. Elution positions of AB-labeled dextran oligomers are indicated in italics at the top of the figure. All fractions were subjected to second dimension RP-nano-LC-MS/MS. RP-nano-LC-MS/MS data obtained for fractions 24 and 26 (striated) revealing three isobaric glycan structures are presented in **Figs. 2** and **3**. *Dark square* *N*-acetylglucosamine; *light square* *N*-acetylgalactosamine; *dark circle* mannose or glucose; *light circle* galactose; *triangle* fucose; *A* 2-aminobenzamide. Reproduced from *(19)* in modified form with permission from Blackwell Publishing, Oxford, UK.

6. Set up the following parameters in your HPLC software: flow 0.4 mL/min; injection: run system in injection position with 2 mL (two loop volumes; 5 min); gradient: *t*= 0 min, 100% solvent B; *t*= 152 min, 52.5% solvent B; *t*= 155 min, 0% solvent B; *t*= 162 min, 0% solvent B; and *t*= 163 min, 100% solvent B. The total run time is 180 min.
7. Adjust the settings of the fluorescence detector: excitation wavelength 360 nm, emission wavelength 425 nm, bandwidth 13 nm, maximum sensitivity.
8. Rinse the sample loop of the injector extensively with solvent A and subsequently with solvent B (*see* **Note 8**).
9. Run 2-AB-labeled dextran hydrolysate sample (dextran ladder): Add 1 μL of the stock solution of 2-AB-labeled dextran hydrolysate (1 mg/mL) to 1 mL of solvent B. Vortex. Use 100 μL for the HPLC run. Compare the fluorescence profile to literature data and assign the peaks. The first peak observed is the excess label. Compared to the label peak, monomeric and dimeric glucose will elute with a delay of approximately 3.5 and 9.5 mL, respectively.
10. Run aliquots of the 2-AB-labeled oligosaccharide standards for quantification purposes (*see* **Note 9**).
11. Run a blank (inject 100 μL of solvent B).
12. Perform an analytical run of the sample of interest: Bring the volume of your 2-AB-labeled glycan pool to 1 mL with solvent B. Take 1 μL, mix with 100 μL solvent B, and use this for an HPLC run. On the basis of this analytical run, calculate the amount of the major oligosaccharide species in the sample.
13. On the basis of the calculated amount, adjust the sensitivity of the fluorescence detector. If necessary, perform another analytical run (*see* **Note 9**).
14. Perform a preparative run with your sample of interest. Manually collect the peaks into Eppendorf vials (*see* **Note 10**).
15. Dry the peak fractions by vacuum centrifugation. Dissolve the peak fractions in 100 μL water. Optionally, perform a first screening of the fractions with MALDI-TOF-MS in positive- and negative-ion modes. Analyze peak fractions by RP-nano-LC-MS(/MS).

3.4. Nano-RP-HPLC-ESI-MS/MS (Figs. 3 and 4)

1. Prepare eluent A (0.4% acetonitrile, 0.1% formic acid) and eluent B (H_2O/acetonitrile 5:95, v/v, containing 0.1% formic acid) for the gradient pump. Degas the solvents.
2. Prepare the solvent for the microflow pump (0.4% acetonitrile, 0.1% formic acid). Degas the solvent.
3. Set up the nano-HPLC system with autosampler, micropump with switching valve and trap column, HPLC pump with nanosplitter and RP-PepMap column, and Z-type UV-detector.

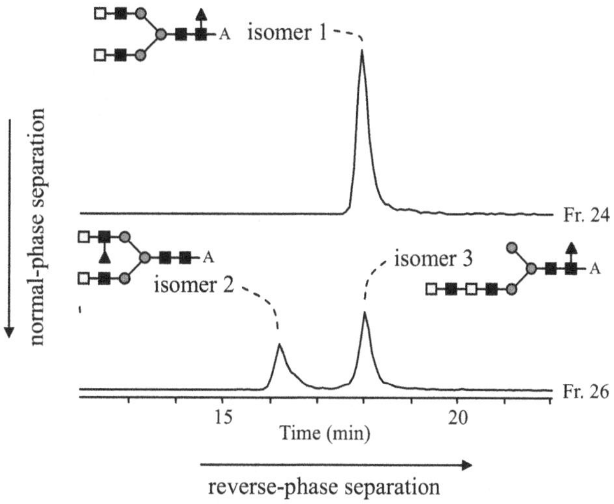

Fig. 3. Second-dimension RP-nano-LC-MS/MS of NP-HPLC fractions 24 and 26. Extracted ion chromatograms of *m/z* 1,017 displayed for fractions 24 and 26 indicated three isomers separated by the two-dimensional HPLC system. The assigned structures are based on the MS/MS spectra shown in Fig. 3. *Light square N*-acetylgalactosamine; *dark square N*-acetylglucosamine; *circle* mannose; *triangle* fucose; *A* 2-aminobenzamide. Reproduced from *(19)* in modified form with permission from Blackwell Publishing, Oxford, UK.

4. Switch on the pumps and wash the trap-column system and the nano-column.
5. Set up the following method in your HPLC software. Flow of the nano-HPLC, 150 nL/min; flow of the micro-pump, 10 μL/min; both trap-column and nano-column (eluent A) equilibrated; injection (5 μL aqueous sample) on trap column; wash using micropump (5 min with 10 μL/min); switching of the trap-column valve into the in-line position with nano-column (flow 150 nL/min). The following gradient is then applied: *t*= 0 min, 0% solvent B; *t*= 30 min, 50% solvent B; *t*= 31 min, 100% solvent B; *t*= 36 min, 100% solvent B; and *t*= 37 min, 0% solvent B. The total run time is 60 min.
6. Prepare a dilution of the 2-AB-labeled standard oligosaccharides in autosampler vials (between 50 and 500 fmol/μL).
7. Run a blank.
8. If necessary, tune the MS with direct infusion using a syringe pump (flow 18 μL/h), the online nano-source, and a solution of a 2-AB-labeled oligosaccharide standard in 50% acetonitrile, 0.1% formic acid.
9. Put the MS with the online nano-source in line. Choose positive- or negative-ion detection mode. In the case of positive-mode analysis, you may choose to add sodium hydroxide to solvent A to a final concentration of 0.6 mM in order

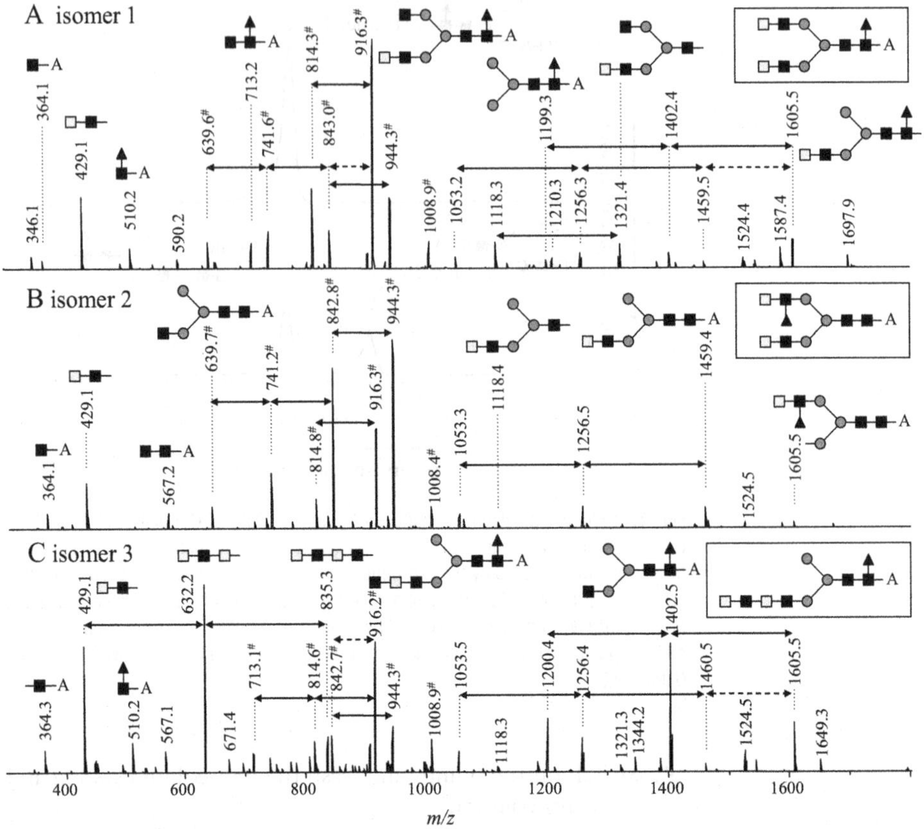

Fig. 4. Fragment ion spectra (RP-nano-LC-ESI-IT-MS/MS) of the $[M + 2Na]^{2+}$species (*m/z* 1,017) of the three isomers displayed in **Fig. 2**. The deduced structures are boxed. *Open square N*-acetylgalactosamine; *closed square N*-acetylglucosamine; *circle* mannose; *triangle* fucose; *A* 2-aminobenzamide; *double-headed arrow with continuous line N*-acetylhexosamine; *double-headed arrow with dashed line* deoxyhexose; # double-sodiated fragment. Reproduced from *(19)* in modified form with permission from Blackwell Publishing, Oxford, UK.

to generate sodium adducts instead of proton adducts (*see* **Note 11**).

10. Run 2-AB-labeled standard oligosaccharides in auto-MS/MS mode followed by a blank (*see* **Note 12**).
11. Run your samples (NP-HPLC fractions) (*see* **Note 12** and **13**).
12. Analyze the data with a processing software. Check for example for reoccurring (terminal) structural motifs using extracted ion chromatograms of the MS/MS data, which are characteristic for B ions (oxonium ions) (*see* **Note 11**). Compare the obtained MS/MS data to literature data or databases for interpretation *(9–13)*.

4. Notes

1. Use Milli-Q water throughout.
2. Prepare aliquots of 2-AB and sodium cyanoborohydride in separate Eppendorf vials. Perform the work with sodium cyanoborohydride in a hood, as it is toxic. Weigh the aliquots, seal them with parafilm, and store them at –20°C until usage (up to 6 months).
3. Store under inert gas (nitrogen or argon) in the dark.
4. Store stock in an exsiccator with phosphorous pentoxide (P_2O_5). Be careful: very poisonous for humans and environment.
5. During the whole procedure, do not let the cartridge run dry.
6. Control the vacuum. Perform the wash and elution steps with a maximum flow of 1 mL/min.
7. Alternatively, reduce sample volume under a stream of nitrogen to about one-fourth, followed by lyophilization.
8. Before sample loading, make sure that the sample loop is filled with 100% solvent B. All samples are injected with an acetonitrile content of 80%.
9. For large amounts of sample, the fluorescence detector may still be too sensitive, so that the most abundant oligosaccharide species would be out of scale. This can be prevented by choosing alternative excitation/emission wavelengths (e.g., 280/500 nm).
10. The HPLC profiles of analytical and preparative runs are usually very similar, which helps peak fractionation.
11. Sodium adducts are preferable for fragment ion analysis (MS/MS mode), as they do not show the rearrangements that are often observed in MS/MS analysis of proton adducts of glycoconjugates, mostly associated with fucose and xylose residues *(19, 25–27)*.
12. Check the performance of the nano-LC system by registering a UV trace at 254 nm. This should allow the detection of 2-AB-labeled oligosaccharides at a sensitivity of several hundred femtomoles.
13. Preferably, samples should be run automatically in a series, with controls (standard oligosaccharides or reference fractions) in between to check the performance of the instrument. You may decide to analyze your samples again under different conditions. After adjustment of the MS settings and solvents, automatic fragment ion analysis of the series of samples may be performed in negative-ion mode, positive-ion mode (proton adducts), and positive-ion mode (sodium adducts).

Acknowledgments

We thank Drs Louise Royle and Pauline M. Rudd from the Oxford Glycobiology Institute for their support with setting up various glycoanalytical methods. We thank Dr Cornelis H. Hokke and Marjolein M. L. Robijn for their involvement in optimizing the analytical procedures.

References

1. Merry, A. H. & Merry, C. L. (2005) Glycoscience finally comes of age. *EMBO Rep.*, 6, 900–903.
2. Kleene, R. & Schachner, M. (2004) Glycans and neural cell interactions. *Nat. Rev. Neurosci.*, 5, 195–208.
3. Endo, T. (2005) Glycans and glycan-binding proteins in brain: galectin-1-induced expression of neurotrophic factors in astrocytes. *Curr. Drug Targets.*, 6, 427–436.
4. Freeze, H. H. & Aebi, M. (2005) Altered glycan structures: the molecular basis of congenital disorders of glycosylation. *Curr. Opin. Struct. Biol*, 15, 490–498.
5. Kobata, A. & Amano, J. (2005) Altered glycosylation of proteins produced by malignant cells, and application for the diagnosis and immunotherapy of tumours. *Immunol. Cell Biol*, 83, 429–439.
6. Nyame, A. K., Kawar, Z. S., & Cummings, R. D. (2004) Antigenic glycans in parasitic infections: implications for vaccines and diagnostics. *Arch. Biochem. Biophys.*, 426, 182–200.
7. Hokke, C. H. & Deelder, A. M. (2001) Schistosome glycoconjugates in host-parasite interplay. *Glycoconj. J.*, 18, 573–587.
8. Hokke, C. H. & Yazdanbakhsh, M. (2005) Schistosome glycans and innate immunity. *Parasite Immunol.*, 27, 257–264.
9. Mechref, Y. & Novotny, M. V. (2002) Structural investigations of glycoconjugates at high sensitivity. *Chem. Rev.*, 102, 321–369.
10. Zaia, J. (2004) Mass spectrometry of oligosaccharides. *Mass Spectrom. Rev.*, 23, 161–227.
11. Morelle, W. & Michalski, J.-C. (2005) The mass spectrometric analysis of glycoproteins and their glycan structures. *Curr. Anal. Chem.*, 1, 29–57.
12. Harvey, D. J. (2003) Matrix-assisted laser desorption/ionization mass spectrometry of carbohydrates and glycoconjugates. *Int. J. Mass Spec.*, 226, 1–35.
13. Wuhrer, M., Deelder, A. M., & Hokke, C. H. (2005) Protein glycosylation analysis by liquid chromatography-mass spectrometry. *J. Chromatogr. B*, 825, 124–133.
14. Novotny, M. V. (1996) Glycoconjugate analysis by capillary electrophoresis. *Methods Enzymol.*, 271f, 320–347.
15. Karlsson, N. G., Wilson, N. L., Wirth, H. J., Dawes, P., Joshi, H., & Packer, N. H. (2004) Negative ion graphitised carbon nano-liquid chromatography/mass spectrometry increases sensitivity for glycoprotein oligosaccharide analysis. *Rapid Commun. Mass Spectrom.*, 18, 2282–2292.
16. Wuhrer, M., Koeleman, C., Deelder, A. M., & Hokke, C. H. (2004) Normal-phase nanoscale liquid chromatography-mass spectrometry of underivatized oligosaccharides at low-femtomole sensitivity. *Anal. Chem.*, 76, 833–838.
17. Takegawa, Y., Deguchi, K., Keira, T., Ito, H., Nakagawa, H., & Nishimura, S. I. (2006) Separation of isomeric 2-aminopyridine derivatized *N*-glycans and *N*-glycopeptides of human serum immunoglobulin G by using a zwitterionic type of hydrophilic-interaction chromatography. *J Chromatogr. A.*, 1113, 177–181
18. Bruggink, C., Wuhrer, M., Koeleman, C. A., Barreto, V., Liu, Y., Pohl, C., Ingendoh, A., Hokke, C. H., & Deelder, A. M. (2005) Oligosaccharide analysis by capillary-scale high-pH anion-exchange chromatography with on-line ion-trap mass spectrometry. *J. Chromatogr. B*, 829, 136–143.
19. Wuhrer, M., Koeleman, C. A. M., Deelder, A. M., & Hokke, C. H. (2006) Repeats of LacdiNAc and fucosylated LacdiNAc on *N*-glycans of the human parasite Schistosoma mansoni. *FEBS J.*, 273, 347–361.
20. Guile, G. R., Rudd, P. M., Wing, D. R., Prime, S. B., & Dwek, R. A. (1996) A rapid high-resolution high-performance liquid chromatographic method for separating glycan mixtures and analysing oligosaccharide profiles. *Anal. Biochem.*, 240, 210–226.

21. Royle, L., Mattu, T. S., Hart, E., Langridge, J. I., Merry, A. H., Murphy, N., Harvey, D. J., Dwek, R. A., & Rudd, P. M. (2002) An analytical and structural database provides a strategy for sequencing *O*-glycans from microgram quantities of glycoproteins. *Anal. Biochem.*, 304, 70–90.
22. Rudd, P. M., Colominas, C., Royle, L., Murphy, N., Hart, E., Merry, A. H., Hebestreit, H. F., & Dwek, R. A. (2001) A high-performance liquid chromatography based strategy for rapid, sensitive sequencing of N-linked oligosaccharide modifications to proteins in sodium dodecyl sulphate polyacrylamide electrophoresis gel bands. *Proteomics*, 1, 285–294.
23. Rudd, P. M., Guile, G. R., Küster, B., Harvey, D. J., Opdenakker, G., & Dwek, R. A. (1997) Oligosaccharide sequencing technology. *Nature*, 388, 205–207.
24. Wing, D. R., Garner, B., Hunnam, V., Reinkensmeier, G., Andersson, U., Harvey, D. J., Dwek, R. A., Platt, F. M., & Butters, T. D. (2001) High-performance liquid chromatography analysis of ganglioside carbohydrates at the picomole level after ceramide glycanase digestion and fluorescent labeling with 2-aminobenzamide. *Anal. Biochem.*, 298, 207–217.
25. Harvey, D. J., Mattu, T. S., Wormald, M. R., Royle, L., Dwek, R. A., & Rudd, P. M. (2002) "Internal residue loss": rearrangements occurring during the fragmentation of carbohydrates derivatized at the reducing terminus. *Anal. Chem.*, 74, 734–740.
26. Wuhrer, M., Koeleman, C. A. M., Hokke, C. H., & Deelder, A. M. (2006) Mass spectrometry of proton adducts of fucosylated *N*-glycans: fucose transfer between antennae gives rise to misleading fragments. *Rapid Commun. Mass Spectrom.*, 20, 1747–1754.
27. Wuhrer, M., Balog, C. I. A., Catimel, B., Jones, F. M., Schramm, G., Haas, H., Doenhoff, M. J., Dunne, D. W., Deelder, A. M., & Hokke, C. H. (2006) IPSE/alpha-1, a major secretory glycoprotein antigen from schistosome eggs, expresses the Lewis X motif on core-difucosylated *N*-glycans. *FEBS J.*, 273, 2276–2292.

Chapter 7

Capillary Lectin-Affinity Electrophoresis for Glycan Analysis

Kazuaki Kakehi and Mitsuhiro Kinoshita

Summary

Glycosylation is one of the most important post-translational events for proteins, affecting their functions in health and disease, and plays significant roles in various information trafficking for intercellular and intracellular biological events. The glycans which show such important effects are generally present as quite complex mixtures in minute amounts. The approach described here makes it possible to profile glycans for the analysis of post-translational modification of proteins with carbohydrates. The method is based on high-resolution separation of fluorescent-labeled carbohydrates by capillary electrophoresis with laser-induced fluorescent detection in the presence of carbohydrate-binding proteins at different concentrations. The technique affords simultaneous determination of glycans having similar structures even in complex mixtures.

Key words: Capillary affinity electrophoresis, Lectin, Glycoprotein, Milk, Oligosaccharides, Fluorescent detection.

1. Introduction

Carbohydrate chains in glycoconjugates play an important role in the modulation of their structures and functions. In the extracellular environment, carbohydrate chains exert effects on cellular recognition in infection, cancer and immune response, but details of the specific mechanisms still remain as unsolved targets. Therefore, analysis of carbohydrate chains is of primary importance for understanding the role of carbohydrate chains.

Capillary electrophoresis (CE) allows high-resolution analysis of carbohydrate chains. A combination of CE and laser-induced

Nicolle H. Packer and Niclas G. Karlsson (eds.), *Methods in Molecular Biology, Glycomics: Methods and Protocols, vol. 534*

DOI: 10.1007/978-1-59745-022-5_7

fluorescence (LIF) detection is a powerful technique for ultrahigh sensitive detection of carbohydrate chains labeled with fluorescent tags such as 8-aminopyrene-1,3,6-trisulfonate (APTS) *(1–4)* and 2-aminobenzoic acid (2-AA) *(5)*. We applied this technique to the analysis of carbohydrate chains derived from various glycoprotein samples *(6–8)*.

CE can be also employed to observe the interaction between carbohydrates and carbohydrate-binding proteins (i.e. lectins). We call this technique for the study of such interactions "capillary affinity electrophoresis (CAE)". CAE is an analytical approach by which the migration of the carbohydrate molecules is observed in the presence of lectins. We can evaluate the specific binding between the carbohydrate and lectin based on the change of migration. Taga et al. *(9)* reported a model study for the determination of the interactions between a mixture of simple oligosaccharides and a lectin.

We have developed this technique to glycomics studies for the profiling of complex mixtures of oligosaccharides derived from glycoproteins, glycosaminoglycans and animal milk samples *(10–14)*. It should be noted that the method can be used for the analysis of the binding reaction between each carbohydrate and a lectin even in a mixture of carbohydrates. Another important point is that the interactions are observed in the solution state, whilst most other methods are based on the interactions between the carbohydrate and immobilized lectins.

The principle of CAE for the application of glycan profiling is shown in **Fig. 1**.

In the first step, a mixture of fluorescently labeled carbohydrate chains (A, B and C) is analyzed by capillary electrophoresis in an electrolyte that does not contain a lectin (**Fig. 1a**). In the following step, the same sample is analyzed in the presence of a lectin whose specificity is well established. When the lectin recognizes carbohydrate A (peak A in **Fig. 1a**), the peak is observed later than in the first step due to the equilibrium based on the interaction with the lectin. On the contrary, carbohydrate C (peak C) does not show affinity to the lectin, and is observed at the same migration time as that observed in the absence of the lectin. Carbohydrate B (peak B) shows weak affinity to the lectin and the peak is observed slightly later than without the lectin. Thus, the migration order of the carbohydrate chains changes as shown in **Fig. 1b**.

By repeating the procedures described above using an appropriate set of lectins, one can categorize (profile) all carbohydrate chains in the mixture. As mentioned previously, the most important advantage of CAE is that the binding specificity of each carbohydrate in the complex mixture can be determined and subtle differences in binding specificities of glycans having similar structural characteristics can be compared.

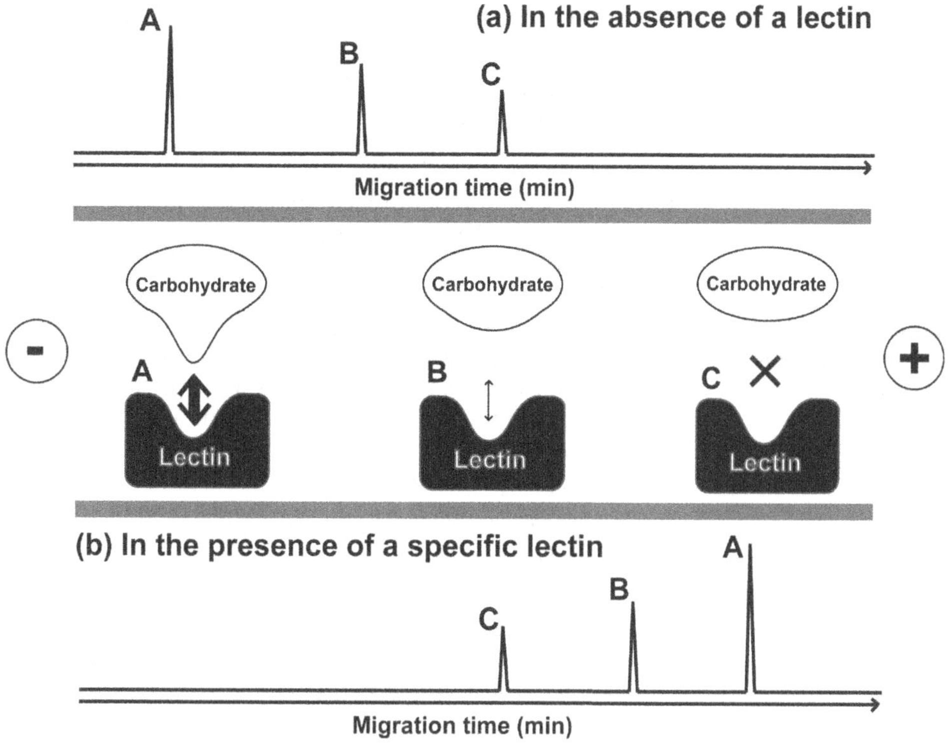

Fig. 1. Principle of capillary affinity electrophoresis.

2. Materials

2.1. Preparation of Fluorescently Labeled Oligosaccharides for CAE

1. Peptide-N^4-(acetyl-β-D-glucosaminyl) asparagine amidase (*N*-glycoamidase F) from Roche Molecular Biochemicals (Minato-ku, Tokyo, Japan).
2. *Buffer for* N-glycoamidase *digestion.* 20 mM phosphate buffer (pH 7.0).
3. Neuraminidase (*Arthrobacter ureafaciens*) from Nakalai Tesque (Nakagyo-ku, Kyoto, Japan).
4. 8-Aminopyrene-1,3,6-trisulfonate (APTS) for carbohydrate analysis by CE (Beckman-Coulter, Fullerton, CA) (*see* **Note 1**).
5. 2-Aminobenzoic acid (2-AA) and 3-aminobenzoic acid (3-AA) (Tokyo Kasei, Chuo-ku, Tokyo, Japan) (*see* **Note 2**).
6. Sodium cyanoborohydride ($NaBH_3CN$) (Sigma-Aldrich, Shinagawa-ku, Tokyo, Japan) (*see* **Note 3**). *Caution*: This compound is toxic; injurious to skin, eyes and mucosa. Avoid inhalation!

7. Sephadex G-25 (fine grade) and Sephadex LH-20 (GE Healthcare Bio-Sciences KK, Shinjuku-ku, Tokyo, Japan).
8. Mineral oil, n_D 1.4670, d 0.838 (Sigma-Aldrich).
9. Milli-Q water (18 MeΩ).
10. 100 mM Tris–acetate buffer (pH 7.4) for CAE (*see* **Note 4**).
11. Polyethylene glycol (average molecular masses 70,000) (Wako Pure Chemicals, Dosho-machi, Osaka, Japan) (*see* **Note 5**).

2.2. Lectins for Capillary Affinity Electrophoresis

1. Concanavalin A (Con A), wheat germ agglutinin (WGA), *Datsura stramoium* agglutinin (DSA), *Aleuria aurantia* lectin (AAL), *Ulex europaeus* agglutinin (UEA-1), *Ricinus communis* agglutinin (RCA_{120}), soybean agglutinin (SBA), and *Macckia amurensis* lectin (MAM) are obtained from Seikagaku Kogyo (Chuo-ku, Tokyo).
2. *Pseudomonas aeruginosa* lectin (PA-I) (Sigma Aldrich).
3. *Psathyrella velutina* lectin, PVL (Wako).
4. *Tulipa gesneriana* agglutinin (TGA), *Crocus sativus* lectin (CSL) are isolated and purified from the bulbs of tulip and crocus, respectively, and *Rhizopus stolonifer* lectin (RSL) is prepared from *R. stronifer* (IFO 30816) *(10, 13, 15)*.

2.2.1. Preparation of the Electrolyte Solutions Containing Lectins for CAE

1. *Con A.* Con A requires calcium ion in the binding reaction. Calcium chloride dihydrate (0.7 g) is dissolved in 100 mM Tris–acetate buffer (pH 7.4). The solution (250 µl) is added to 100 mM Tris–acetate buffer (pH 7.4, 24.75 ml), and used for the electrolyte solution (100 mM Tris–acetate buffer containing 0.5 mM Ca^{2+}) of Con A. Con A (1.0 mg, 104 kDa) is dissolved in the buffer containing calcium ion (96 µl). The solution (100 µM Con A) is diluted with the same buffer and is used as the electrolyte.
2. *PVL.* The electrolyte containing glycerol is used as the running buffer, because PVL aggregates in water. Glycerol (0.5 g) is dissolved in 100 mM Tris–acetate buffer (pH 7.4), and is used as the electrolyte. A portion (10 µl) of the commercially available PVL solution (1 mg/ml, 25 µM, 40 kDa) is diluted with the glycerol-containing buffer (40 µl), and is used as the electrolyte after dilution.
3. A portion (250 µl) of the commercially available AAL (2 mg/ml, 27 µM, 72 kDa) solution is diluted with 100 mM Tris–acetate buffer (pH 7.4, 28 µl) to make up 24 µM solution.
4. Other lectins are dissolved in 100 mM Tris–acetate buffer (pH 7.4) at the following concentrations (100 µM for all lectins). DSA (1.0 mg, 86 kDa) in 116 µl: RCA_{120} (1.0 mg, 120 kDa) in 83 µl: SBA (1.0 mg, 120 kDa) in 83 µl: SSA (1.0 mg, 160 kDa) in 62 µl: MAM (1.0 mg, 130 kDa) in 76 µl: PA-I (1.0 mg,

110 kDa) in 91 μl: PHA-E_4 (1.0 mg, 130 kDa) in 77 μl: WGA (1.0 mg, 43 kDa) in 232 μl: UEA-1 (1.0 mg, 26.7 kDa) in 375 μl: TGA (1.0 mg, 43.2 kDa) in 232 μl: RSL (1.0 mg, 28 kDa) in 357 μl: CSL (1.0 mg, 48 kDa) in 208 μl. Store the lectin solutions in aliquots at −20°C until use (*see* **Note 6**).

2.3. Equipment

1. A centrifugal vacuum evaporator (SpeedVac, Savant, Farmingdale, NY).
2. A CE system equipped with a laser-induced fluorescence detection system (Beckman-Coulter). An Ar-laser (488 nm) and a He-Cd laser (325 nm) are available for this system.
3. An eCAP N-CHO coated capillary (Beckman-Coulter, 50 μm i.d., 375 μm o.d.) (*see* **Note 7**).

3. Methods

3.1. Preparation of Fluorescently Labeled Carbohydrate

Most carbohydrates have no chromophores or fluorophores in their molecules, and can be labeled with chromogenic or fluorogenic reagents for high sensitivity detection. Derivatization of the carbohydrates with charged compounds is also important for their separation by CE, because the charged carbohydrates can migrate in the electric field. In CAE analysis, APTS or 2- and 3-AA are often used for labeling of carbohydrates *(4, 5, 8)*. These fluorescent labeling reagents strongly fluoresce under irradiation with an Ar-laser for APTS and with a He-Cd laser for 2- and 3-AA. The negative charges of these reagents, due to their sulfonic acid and carboxylic acid residues, are also the driving force in the electric field.

3.1.1. Preparation of APTS-Labeled Carbohydrates (see **Notes 8** and **9**: Releasing of Carbohydrates from Glycoproteins)

1. To a sample of oligosaccharide (ca. 0.1–1 nmol) or a mixture of oligosaccharides (typically obtained from 10 to 1,000 μg of glycoprotein samples), is added a solution (5 μl) of 100 mM APTS in 15% aqueous acetic acid.
2. A freshly prepared solution of 1 M $NaBH_3CN$ in tetrahydrofuran (5 μl) is added to the mixture.
3. The mixture is overlaid with mineral oil (100 μl) to prevent evaporation of the reaction solvent and to obtain good reproducibility *(16)*.
4. The mixture is kept for 90 min at 55°C.
5. Water (200 μl) is added to the mixture. The fluorescent yellowish aqueous phase is collected, and applied to a column of Sephadex G-25 (1 cm i.d., 50 cm length) previously equilibrated with water.

6. The earlier eluting fluorescent fractions are collected and evaporated to dryness. The residue is then dissolved in water (100 μl), and a portion (10–20 μl) is used for capillary affinity electrophoresis (*see* **Note 10**).

3.1.2. Preparation of 2-AA Labeled Carbohydrates

1. To a sample of oligosaccharide (ca. 0.1–1 nmol) or a mixture of oligosaccharides (typically obtained from 10 to 1,000 μg of glycoprotein samples), is added a solution (200 μl) of 2-AA and $NaBH_3CN$, prepared by freshly dissolving both reagents (30 mg each) in methanol (1 mL) containing 4% CH_3COONa and 2% boric acid.
2. The mixture is kept at 80°C for 60 min. After cooling, water (200 μl) is added to the reaction mixture.
3. The reaction mixture is applied to a small column (1 cm × 50 cm) of Sephadex LH-20 equilibrated with 50% aqueous methanol. The earlier eluting fractions which contain labeled oligosaccharides are collected and evaporated to dryness. The residue is dissolved in water (100 μl), and a portion (typically 10 μl) is used for CAE.

3.1.3. Preparation of 3-AA Labeled Carbohydrates

1. To a sample of oligosaccharide (ca. 0.1–1 nmol) or a mixture of oligosaccharides (typically obtained from 10 to 1,000 μg of glycoprotein samples), is added a solution (30 μl) of 0.7 M 3-AA in dimethylsulfoxide-acetic acid (7:3 by volume).
2. A solution (30 μl) of freshly prepared 2 M $NaBH_3CN$ in the same solvent is added to the mixture and incubated for 1 h at 50°C.
3. The reaction mixture is diluted with water (200 μl) and applied to a small column (1 cm × 50 cm) of Sephadex LH-20 equilibrated with 50% aqueous methanol. The earlier eluted fractions which contain labeled oligosaccharides are collected and evaporated to dryness. The residue is dissolved in water (100 μl), and a portion (typically 10 μl) is used for CAE.

3.2. Choice of Lectins for CAE

It is important to select an appropriate set of lectins for the observation of clear interactions with various oligosaccharides. A list of the lectins available for such a purpose is shown in **Table 1**, and the interactions of each lectin with glycan/monosaccharide residues are illustrated in **Fig. 2** using typical *N*-glycans and milk oligosaccharides as model structures.

The lectins which show specificities toward (1) mannose (Man), (2) *N*-acetyl glucosamine (GlcNAc), (3) galactose (Gal), (4) *N*-acetylgalactosamine (GalNAc), (5) fucose (Fuc) and (6) sialic acids (NeuAc) are selected. Some of these lectins can recognize bi-, tri- and tetra-antennary structures of *N*-glycans. A set of lectins is also available for the analysis of other glycans such as those derived from mucin-type glycoproteins and glycolipids.

Table 1
List of the lectins for CAE

Lectin	Specificity	Concentrations used for CAE (μM)
Con A	High-mannose, bi-antennary	0.2, 3.0
WGA	Hybrid-type, complex-type, Galβ1-4GlcNAc	3.0, 6.0, 12.0
TGA	Tri-antennary and complex-type with coreα1-6Fuc	2.0, 12.0
CSL	Man3GlcNAc in the *N*-glycan core structure	0.5, 4.0
DSA	Tri- and tetra-antennary	0.01, 0.2
PVL	GlcNAc at the non-reducing terminal	0.8, 3.0
PHA-E_4	Bisecting GlcNAc	0.8, 6.0
RCA_{120}	Galβ1-4 residues at the non-reducing terminal	0.8, 3.0
PA-I	α-Gal, α-GalNAc at the non-reducing terminal	0.8, 3.0
SBA	Gal(GalNAc)α1-3*(4)*Gal(GalNAc)	0.8, 3.0
AAL	Fucα1-residues	0.2, 0.8, 3.0, 12.0
UEA	Fucα1-2Gal (H type2-antigen)	3.0, 12.0
RSL	Complex type with coreα1-6Fuc	3.0, 12.0
SSA	NeuAcα2-6 residues	0.5, 2.0
MAM	NeuAcα2-3 residues	0.5, 2.0

All the solutions of lectins are diluted with 100 mM Tris-acetate buffer (pH 7.4), and a portion (5 μl) of the solution is mixed with the running buffer (5 μl) containing 1% polyethylene glycol. The mixed solution is used as the electrolyte for CAE.

3.3. Capillary Affinity Electrophoresis

3.3.1. Capillary Affinity Electrophoresis for APTS-Labeled Oligosaccharides

1. Analytical conditions. Capillary affinity electrophoresis is performed with a P/ACE MDQ glycoprotein system (Beckman) equipped with a laser-induced fluorescence detection system. For the analysis of APTS-labeled carbohydrates, an argon-laser (488 nm) is used for detection with an emission band path filter at 520 nm. Separation is performed using an eCAP N-CHO coated capillary (10 cm effective length (30 cm total length), 50 μm i.d.). Injections are performed automatically using the pressure method (1.0 psi., 10 s). Separation is performed using negative polarity at 25°C. Data are collected and analyzed with a standard 32 Karat software (version 7.0, Beckman-Coulter) on Windows 2000.

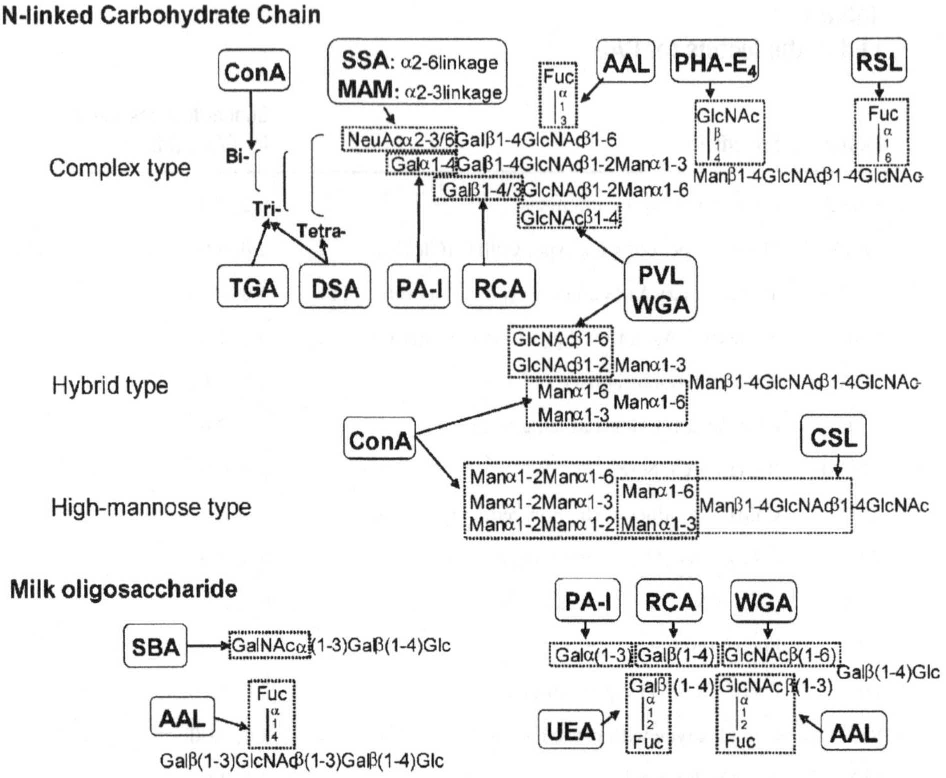

Fig. 2. Selection of lectins for capillary affinity electrophoresis. Abbreviations used for the structures: *GlcNAc* *N*-acetylglucosamine; *GalNAc* *N*-acetylgalactosamine; *Man* mammose; *Gal* galactose; *Fuc* fucose; *NeuAc* *N*-acetylneuraminic acid.

2. Prior to the analysis, the capillary is rinsed with 100 mM Tris–acetate buffer (pH 7.4) for 1 min at 20 psi.
3. The capillary is filled with the same buffer containing 0.5% polyethylene glycol for 1 min at 20 psi.
4. A sample solution containing APTS-labeled oligosaccharides is introduced for 10 s at 1.0 psi.
5. Analysis is performed under a constant voltage mode at an electric field of 600 V/cm.
6. The capillary is then rinsed with the same buffer containing 0.5% polyethylene glycol for 1 min at 20 psi.
7. The capillary is filled with a lectin solution at the specified concentration in the same buffer containing 0.5% polyethylene glycol.
8. Analysis is performed under a constant voltage mode at an electric field of 600 V/cm.

9. When necessary, **steps 7** and **8** are repeated using different concentrations of the lectin or different lectins.
10. The electropherogram observed in the presence of the lectin is compared with that observed in the absence of the lectin.

3.3.2. Capillary Affinity Electrophoresis for 2-/3-AA Labeled Oligosaccharides

Analytical conditions. A Beckman Coulter P/ACE MDQ Glycoprotein System with a He-Cd laser-induced fluorescence detection system is used. Detection is performed with a 405-nm filter for emission and a 325-nm He-Cd laser for excitation. An eCAP N-CHO capillary (Beckman. 20 cm effective length, 30 cm total length, 50 μm i.d.) is used. A DB-1 capillary of the same size is also available.

The procedures are the same as described for CAE analysis of APTS-labeled oligosaccharides (*see* **Subheading 3.3.1**).

A typical example for CAE is shown in **Fig. 3** using a mixture of asialo-*N*-glycans derived from human α1-acid glycoprotein as model *(11)*. The upper-left box shows the migrations of five asialo-*N*-glycans derived from α1-acid glycoprotein in the absence of lectin. In this analysis, the glycans are labeled with APTS. All glycans are well resolved within 8 min. It should be noticed that good resolutions between AII and AIII and also between AIV and AV are achieved. AIII (a triantennary glycan) and AV (a tetraantennary glycan) contain a fucose residue as a Lewis^X type epitope. To confirm the structural characteristics of these glycans, we can use the set of Con A, TGA and AAL lectins. Con A interaction obviously decreases the peak of biantennary glycan (AI), which disappears at a 3.0 μM concentration of Con A (*see* **Note 11**). The other tri- and tetraantennary glycans did not show changes in migration time in the presence of Con A. The presence of triantennary glycans is confirmed using the electrolyte which contains TGA. At 2.0 μM TGA, both triantennary glycans AII and AIII are observed eluting slightly later than in the absence of lectin. At 12.0 μM TGA, AII and AIII are fused into a broad peak and elute after the tetraantennary glycans (AIV and AV). Finally, the glycans having fucose residues attached to the GlcNAc residue of lactosamine (AIII and AV) are determined using the electrolyte containing AAL. At 0.8 μM AAL, peak intensities of AIII and AV were obviously decreased, and these peaks disappeared at 3.0 μM AAL (*see* **Note 12**). As shown in **Fig. 3**, the set of Con A, TGA and AAL can determine the structural characteristics of the *N*-glycans present on human α1-acid glycoprotein.

Figure 4 shows another example by the analysis of milk oligos-accharides using capillary affinity electrophoresis. In this example, milk oligosaccharides are labeled with 3-AA. The electropherogram at the bottom shows the separation of six milk oligosaccharides: 2-fucosyllactose (2-FL), Fucα1-2Galβ1-4Glc; Lewis^X trisaccharide, Galβ1-4(Fucα1-3)GlcNAc; difucosyllactose

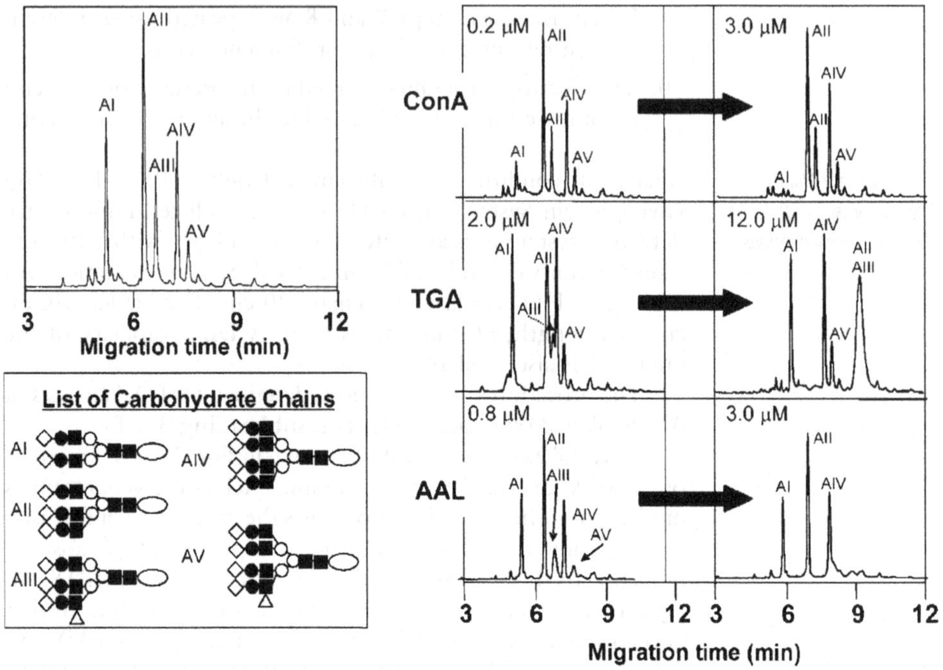

Fig. 3. Profiling of N-linked glycans derived from human α1-acid glycoprotein by capillary affinity electrophoresis. Analytical conditions: capillary, eCAP N-CHO capillary (30 cm total length, 10 cm effective length, 50 μm i.d.); running buffer, 100 mM Tris–acetate (pH 7.4) containing 0.5% PEG70000; applied voltage, 6 kV (reverse polarity); sample injection, pressure method (1.0 psi., 10 s). Fluorescence detection was performed with a 520 nm emission filter by irradiating with an Ar laser-induced 488 nm light. The symbols were: GlcNAc (*filled square*), Man (*open circle*), Gal (*filled circle*), Fuc (*open triangle*), and NeuAc (*open rhombus*).

(LDFT), Fucα1-2Galβ1-4(Fucα1-3)Glc; Lacto-*N*-neotetraose (LNnT), Galβ1-4GlcNAcβ1-3Galβ1-4Glc; Lacto-*N*-fucopentaose (LNFPIII), Galβ1-4(Fucα1-3)GlcNAcβ1-3Galβ1-4Glc; Digalactosyl lacto-*N*-neohexaose (DGLNnH),Galα1-3Galβ1-4GlcNAcβ1-6(Galα1-3Galβ1-4GlcNAcβ1-3Galβ1-4Glc (*see* **Note 13**). In the presence of RCA_{120} at 0.8 μM, the peak intensity of LnNT is dramatically decreased, and indicates the presence of LacNAc residue (*see* **Note 14**). PA-1 at 3.0 μM shows retardation of DGLNnH, and indicates the presence of an αGal residue at the non-reducing terminal of the DGLNnH molecule. UEA which recognizes the fucose residue linked through an α1-2 linkage clearly interacts with 2-FL and LDFT. In contrast, AAL which also recognizes the fucose residue interacts only with the $Lewis^X$ glycan at 0.2 μM (*see* **Note 15**). However, LDFT and LNFPIII do show interaction with AAL at higher concentrations (12.0 μM) (data not shown).

As shown in **Figs. 3** and **4**, use of an appropriate set of lectins is a powerful tool for profiling of glycans. The time required for a single CAE run is less than 15 min. Thus we can characterize

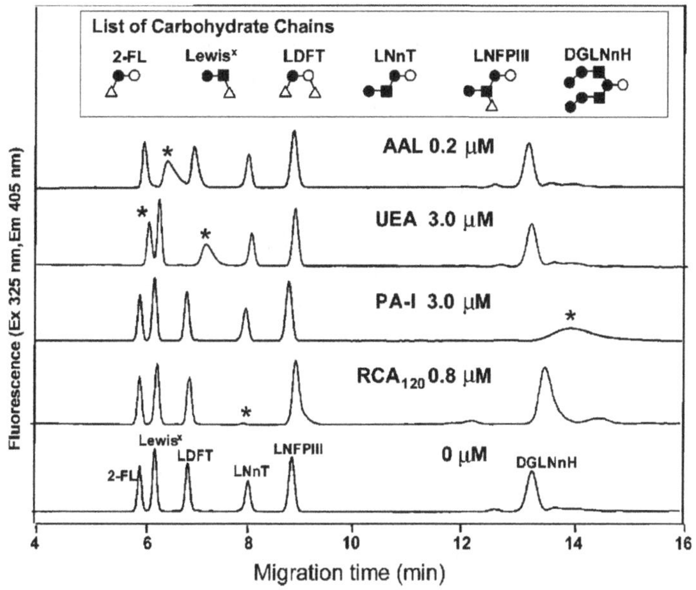

Fig. 4. Capillary affinity electrophoresis for the analysis of 3-AA labeled milk oligosaccharides. Analytical conditions: capillary, eCAP N-CHO capillary (30 cm total length, 20 cm effective length, 50 µm i.d.); running buffer, 100 mM Tris–acetate (pH 7.4) containing 0.5% PEG70000; applied voltage, 18 kV (reverse polarity); sample injection, pressure method (0.5 psi., 5 s). Fluorescence detection was performed with a 405 nm emission filter by irradiating with a He–Cd laser-induced 325 nm light. The symbols were: Glc (*open circle*), GlcNAc (*filled square*), Gal (*filled circle*), and Fuc (*open triangle*).

the total glycans derived from a biological sample in a few hours using CAE techniques.

4. Notes

1. APTS reagent is also available from other commercial sources, but the reagent from other suppliers often contains SO_3^--positional isomers which give minor peaks due to isomers in the analysis by CE.
2. Both reagents are of the highest grade commercially available, and used without further purification.
3. The reagent is stable in the desiccator at room temperature for several months, but fresh reagent should be used.
4. Running buffer for CAE is passed through an ultrafiltration membrane (0.40 µm pore size) and degassed under reduced pressure before use.

5. Polyethylene glycol (PEG70000) is added in the electrolyte to minimize the electro-osmotic flow. In addition, protein adsorption on the capillary surface is suppressed when polyethylene glycol is added in the running buffer at low concentrations.
6. Lectin solution is stable at –20°C for several months.
7. The inner surface of the capillary is chemically modified with polyvinyl alcohol and is employed for CAE, because lectins are often adsorbed onto a capillary surface. A DB-1 capillary with dimethylpolysiloxane (DB-1) of the same size (GL Science Co. Ltd., Nishi-Shinjuku, Tokyo, Japan) is also available and can be used instead of the eCAP N-CHO capillary. A DB-1 capillary shows weak electro-osmotic flow (EOF) at neutral pH, and polyelthylene glycol should be added to suppress EOF.
8. Typical release of the N-linked carbohydrates from glycoproteins was performed according to the method reported previously *(7, 17)*. A sample of glycoprotein (1 mg) is dissolved in 20 mM phosphate buffer (pH 7.0, 40 μl). *N*-Glycoamidase F (5 mU, 5 μl) is added, and the solution is incubated for 24 h at 37°C. The reaction is stopped by immersion in a boiling water bath for 5 min and then centrifuged at 10,000 × *g* for 10 min. The supernatant containing the released oligosaccharides is evaporated to dryness by a centrifugal vacuum evaporator.
9. Neuraminidase (2 munits, 4 μL) is added to the mixture of sialo-oligosaccharides in 20 mM acetate buffer (pH 5.0, 20 μL), and the mixture is incubated at 37°C for 12 h. After keeping the mixture in the boiling water bath for 5 min followed by centrifugation, the supernatant is used as asialo-oligosaccharides.
10. Removal of the reagent (APTS) from the reaction mixture is necessary to keep the carbohydrate derivatives at –20°C without decomposition.
11. High-affinity interaction with biantennary glycan is not prevented, even if outer Gal residues are substituted at C-3/6 position by sialic acids.
12. Complex type glycans with core α1–6 Fuc residues are observed later using the electrolyte containing RSL.
13. The smaller oligosaccharides migrated faster than the larger ones based on their charge/mass ratios.
14. The interaction between RCA_{120} and Gal residues is inhibited when terminal Gal residues at the nonreducing ends are substituted at C2 position by Fuc residues. Affinity of RCA_{120} toward Gal residue is decreased if Glc/GlcNAc residues on lactose/lactosamine are substituted with Fuc residues.
15. AAL recognizes Fucα1-3 residue more specifically than Fucα1-2 residue.

References

1. Chen, F.T. and Evangelista, R.A. (1995) Analysis of mono- and oligosaccharide isomers derivatized with 9-aminopyrene-1,4,6-trisulfonate by capillary electrophoresis with laser-induced fluorescence. *Anal. Biochem.* 230, 273–280
2. Guttman, A. and Pritchett, T. (1995) Capillary gel electrophoresis separation of high-mannose type oligosaccharides derivatized by 1-aminopyrene-3,6,8-trisulfonic acid. *Electrophoresis* 16, 1906–1911
3. Chen, F.T., Dobashi, T.S. and Evangelista, RA. (1998) Quantitative analysis of sugar constituents of glycoproteins by capillary electrophoresis. *Glycobiology* 8, 1045–1052
4. Chen, F. and Evangelista, R. (1998) Profiling glycoprotein n-linked oligosaccharide by capillary electrophoresis. *Electrophoresis* 19 (15), 2639–2644
5. Anumula, K.R. and Dhume, S.T. (1998) High resolution and high sensitivity methods for oligosaccharide mapping and characterization by normal phase high performance liquid chromatography following derivatization fluorescent anthranilic acid. *Glycobiology* 8 (7), 685–694
6. Kamoda, S., Nomura, C., Kinoshita, M., Nishiura, S., Ishikawa, R., Kakehi, K., Kawasaki, N. and Hayakawa, T. (2004) Profiling analysis of oligosaccharides in antibody pharmaceuticals by capillary electrophoresis. *J. Chromatogr. A* 1050 (2), 211–216
7. Kakehi, K., Kinoshita, M., Kawakami, D., Tanaka, J., Sei, K., Endo, K., Oda, Y., Iwaki, M. and Masuko, T. (2001) Capillary electrophoresis of sialic acid-containing glycoprotein. Effect of the heterogeneity of carbohydrate chains on glycoform separation using an alpha1-acid glycoprotein as a model. *Anal. Chem.* 73, 2640–2647
8. Kakehi, K., Funakubo, T., Suzuki, S., Oda, Y. and Kitada, Y. (1999) 3-Aminobenzamide and 3-aminobenzoic acid, tags for capillary electrophoresis of complex carbohydrates with laser-induced fluorescent detection. *J. Chromatogr. A* 863, 205–218
9. Uegaki, K., Taga, A., Akada, Y., Suzuki, S. and Honda, S. (2002) Simultaneous determination of the association constants of oligosaccharides to a lectin by capillary electrophoresis. *Anal. Biochem.* 309, 269–278
10. Oda, Y., Senaha, T., Matsuno, Y., Nakajima, K., Naka, R., Kinoshita, M., Honda, E., Furuta, I. and Kakehi, K. (2003) A new fungal lectin recognizing alpha(1–6)-linked fucose in the *N*-glycan. *J. Biol. Chem.* 278 (34), 32439–32447
11. Nakajima, K., Oda, Y., Kinoshita, M. and Kakehi, K. (2003) Capillary affinity electrophoresis for the screening of post-translational modification of proteins with carbohydrates. *J. Proteome Res.* 2, 81–88
12. Nakajima, K., Kinoshita, M., Matsushita, N., Urashima, T., Suzuki, M., Suzuki, A. and Kakehi, K. (2006) Capillary affinity electrophoresis using lectins for the analysis of milk oligosaccharide structure and its application to bovine colostrum oligosaccharides. *Anal. Biochem.* 348 (1), 105–114
13. Nakajima, K., Kinoshita, M., Oda, Y., Masuko, T., Kaku, H., Shibuya, N. and Kakehi, K. (2004) Screening method of carbohydrate-binding proteins in biological sources by capillary affinity electrophoresis and its application to determination of *Tulipa gesneriana* agglutinin in tulip bulbs. *Glycobiology* 14 (9), 793–804
14. Kinoshita, M. and Kakehi, K. (2005) Analysis of the interaction between hyaluronan and hyaluronan-binding proteins by capillary affinity electrophoresis: significance of hyaluronan molecular size on binding reaction. *J. Chromatogr. B Analyt. Technol. Biomed. Life Sci.* 816, 289–295
15. Oda, Y., Nakayama, K., Abdul-Rahman, B., Kinoshita, M., Hashimoto, O., Kawasaki, N., Hayakawa, T., Kakehi, K., Tomiya, N. and Lee, YC. (2000) *Crocus sativus* lectin recognizes Man3GlcNAc in the *N*-glycan core structure. *J. Biol. Chem.* 275, 26772–26779
16. Sei, K., Nakano, M., Kinoshita, M., Masuko, T. and Kakehi, K. (2002) Collection of alpha1-acid glycoprotein molecular species by capillary electrophoresis and the analysis of their molecular mass and carbohydrate chains. *J. Chromatogr. A.* 958, 273–281
17. Ma, S. and Nashabeh, W. (1999) Carbohydrate analysis of a chimeric recombinant monoclonal antibody by capillary electrophoresis with laser-induced fluorescence detection. *Anal. Chem.* 71, 5185–5192

Chapter 8

Analysis of Methylated *O*-Glycan Alditols by Reversed-Phase NanoLC Coupled CAD-ESI Mass Spectrometry

Franz-Georg Hanisch and Stefan Müller

Summary

The structural diversity of mucin-type *O*-glycans (*O*-GalNAc core-based) exceeds that of N-linked glycans by far. Structural analysis of this type of protein modification is hampered, however, by the unavailability of specific endo-acetylgalactosaminidases that would cleave the intact oligosaccharide from threonine or serine residues. Chemical cleavage is either performed by hydrazinolysis resulting in reducing glycans ready for terminal labeling reactions or by the classical reductive β-elimination, which yields the chemically inert alditols. Composition and sequence of the *O*-glycans can be analyzed by mass spectrometry using FAB-, MALDI-, or ESI-ionization technology. A significant increase in sensitivity and a facilitated fragmentation of the sugar chains can be achieved by methylation of free hydroxyl groups according to a modified procedure based on NaOH/DMSO and methyl iodide. Permethylated glycan alditols are readily separated on reversed-phase capillary columns under conditions similar to those used in peptide chromatography, and their sequences can be deduced from MS/MS spectra registered after collision-induced fragmentation of parental proton or sodium adduct ions using a Q-TOF mass spectrometer.

Key words: Glycomics, O-linked glycans, Methylation, Mass spectrometry, Reversed-phase liquid chromatography.

1. Introduction

The structural aspects of mucin-type O-linked glycans are characterized by complexity of cores and peripheral variations. Eight different di- and trisaccharide cores starting with α-linked GalNAc were described so far. These are extended by linear or branched (poly)lactosamine chains and terminate at the nonreducing end by the addition of mostly α-linked sialic acids, fucose, galactose, GalNAc, or GlcNAc resulting in tremendous heterogeneity *(1)*.

Nicolle H. Packer and Niclas G. Karlsson (eds.), *Methods in Molecular Biology, Glycomics: Methods and Protocols, vol. 534*

DOI: 10.1007/978-1-59745-022-5_8

Early strategies in the elucidation of these complex structures were based on reductive β-elimination of the glycans *(2)*, HPLC separation of the respective oligosaccharide alditols *(3)* and the structural characterization by 1H-NMR or FAB mass spectrometry of the methylated derivatives *(4–6)* combined with linkage analysis by GC-MS of partially methylated alditol acetates. FAB ionization was advantageous, because it generated in-source fragmentation of methylated glycans, which revealed sequence information by preferred cleavage of glycosidic bonds at the reducing side of HexNAc residues. On addition of sodium acetate to the sample on-target, ion intensities could be shifted to the molecular ions and allowed in this way a sequencing of glycans in moderately complex mixtures. With advent of modern soft ionization techniques in mass spectrometry and the development of two-dimensional mass spectrometry (MS) on the basis of collision-induced-dissociation (CID) experiments and various configurations of quadrupole (Q), time-of-flight (TOF) or ion trap (trap) analyzers (Q-Q-Q, Q-TOF, Q-trap) the sequencing of glycans in complex mixtures became possible at much higher sensitivity. Recent publications referred to sequencing of methylated glycans by MALDI-TOF/TOF and ESI-Q-TOF mass spectrometry *(7, 8)*. The coupling of MS/MS analysis with nano-LC separation of methylated glycans further enhances sensitivity of the analysis. The described method is based on previous work by the groups of Albersheim *(9)* and Dell *(10)*. The advantage of the described protocol for LC-MS/MS of methylated glycan alditols is given in its compatibility with standard peptide analysis by allowing automatic batch injection using the same column and solvent system.

2. Materials

2.1. Reductive β-Elimination

1. Reductive β-elimination is performed in 0.5 ml Eppendorf vials.
2. Sodium borohydride ($NaBH_4$, p.a.) (Fluka, Sigma-Aldrich, St. Louis, MO) and dissolved in 50 mM NaOH immediately before use as a 1 M solution. The solution is saturated with argon by gassing.
3. Heating facility (50°C).
4. Acetic acid (p.a.) (Fluka).
5. Dowex 50 Wx8 (200–400 mesh) from BioRad (Hercules, CA) is rinsed prior to use with 1N HCl and washed with doubly distilled water to reach a pH of approximately 5.0.
6. Desalted sugars are further processed in 0.5 or 1.0 ml conical, screw-capped glass vials (Pierce, Rockford, IL).

7. All organic solvents including methanol, dimethyl sulfoxide (DMSO) and chloroform are of highest available grade (HPLC-grade) and purchased from Sigma-Aldrich.
8. Heating block (40°C).
9. Nitrogen (for evaporation of organic solvents).
10. Screw-capped conical glass tubes.

2.2. Methylation of Glycan Alditols

1. Aqueous 50% NaOH.
2. Argon for storage of base solution.
3. Vortex mixer.
4. Screw-capped conical glass tubes.
5. Bench-top centrifuge for conical glass tubes.
6. Base solution prepared by a procedure described by Anumula and Taylor *(11)*. A 0.1 ml aliquot of aqueous 50% NaOH is mixed with 0.2 ml methanol in a screw-capped tube prior to the addition of 6 ml dry DMSO. The suspension is mixed on a Vortex mixer and sonicated for 3–5 min. to obtain a fine suspension. The NaOH precipitate is collected by centrifugation (5 min, 7,000 ×g and resuspended in 6 ml of fresh DMSO. Washing of the precipitate with DMSO is repeated three more times to obtain an essentially anhydrous suspension. The NaOH precipitate is finally suspended in 2 ml of DMSO for use in methylation reactions. The suspension is stored under argon at –20°C.
7. Methyl iodide (Fluka) and stored under argon at 4–8°C.
8. Acetic acid (p.a.) (Fluka).
9. All organic solvents including DMSOand chloroform are of highest available grade (HPLC-grade) and purchased from Sigma-Aldrich.
10. Nitrogen (for evaporation of organic solvents)

2.3. Nano-LC and CAD-ESI-Mass Spectrometry

1. Q-TOF 2 quadrupole-time of flight mass spectrometer (Waters, Milford, MA) controlled by MassLynx 4.0 software. The instrument is equipped with a Z-spray nanoflow source, which allows the use of the glass capillary assembly (Waters, Eschborn, Germany) or the PicoTip holder assembly (New Objective, Woburn, MA).
2. Static nanospray assembly comprising a stainless steel glass capillary holder (Waters, Eschborn, Germany) and 2 μm tip i.d. nanospray capillaries coated with a thin layer of gold/palladium (Proxeon, Odense, Denmark).
3. UltiMate Nano-LC system (Dionex, Sunnyvale, CA) equipped with a Famos autosampler and a Switchos column switching module.

4. 0.35 mm by 5 mm trapping column packed with 5 μm Atlantis dC18 (Waters) and a 0.075 mm by 150 mm analytical column, packed with 3 μm Atlantis dC18 (Waters).
5. Nano LC-MS interface comprising a 360 μm o.d. and 10 μm tip i.d. uncoated Picotip spray emitter (New Objective) linked to the HPLC outlet capillary using a 7 nl dead volume stainless steel union mounted onto a PicoTip holder assembly (New Objective).

3. Methods

The methylation procedure by Ciucanu and Kerek *(12)*, which uses finely suspended NaOH in DMSO as base, is much less affected by moisture and oxygen compared to the classical protocol based on the sodium dimsyl base. Despite its convenience and robustness a few critical points need to be mentioned. Some degree of oxidative degradation starting at the reducing end was noticed quite early and assigned to the formation of a dimethyl methoxy sulphonium salt from reaction of DMSO with methyl iodide. This oxidative degradation is particularly observed on methylation of reducing glycans and less pronounced with corresponding alditols. To overcome the problem two modifications of the protocol were published later on *(13, 14)*. The protocol by Needs et al. avoids formation of the oxidizing agent by prolongation of base treatment in the absence of methyl iodide *(13)*. The modification described more recently by Ciucanu *(14)* uses a defined small proportion of water in the DMSO solution of the carbohydrates, while base treatment and methylation with methyl iodide are still performed simultaneously and during short reaction times. A further modification was introduced by neutralization of the reaction mixture with dilute acetic acid prior to chloroform-water extraction of the methylated glycans. We have applied the various protocols for methylation of glycan alditols and revealed finally a combination of the modified procedures as optimal.

3.1. Reductive β-Elimination and Methylation of Glycan Alditols

Reductive β-elimination follows a micromethod described by Schulz et al. *(15)*.

1. The *O*-glycoprotein (10–30 μg) is dried in a 0.5 ml Eppendorf vial.
2. Freshly prepared $NaBH_4$ (1 M in 50 mM NaOH) is added (20 μl) and the reaction mixture is covered with argon.
3. After overnight incubation at 50°C in an oven, the reaction is stopped by placing the vial on ice and adding 1 μl of acetic acid to destroy excessive reducing reagent.
4. A 50 μl aliquot of Dowex 50Wx8(H^+) is added and the suspension is rotated for 5 min at RT.

5. After a brief centrifugation step the supernatant is transferred into a conical glass vial (0.5 or 1.0 ml).
6. The beads are washed with 50 μl water, centrifuged, and the supernatant is combined with the first aliquot.
7. The desalted glycan alditol solution is dried by vacuum centrifugation.
8. To remove boric acid 50 μl aliquots of 1% acetic acid in methanol are added (5×) and evaporated under nitrogen at 40°C.

3.2. Methylation of Glycan Alditols

1. The methylation procedure (*see* **Note 1**) starts after extensive drying of the sample in a dessicator over P_2O_5/KOH for several hours or overnight. All handlings are performed under argon atmosphere.
2. To the dry sample 50 μl dry DMSO is added.
3. The sample is briefly sonicated for 1–2 min.
4. The same volume of base (NaOH/DMSO) is added and the reaction mixture is incubated for 30 min at RT with occasional shaking.
5. An aliquot of 25 μl methyl iodide is pipetted into the frozen reaction mixture followed by incubation for further 30 min at RT.
6. Reaction mixture is neutralized with 40 μl of 1 M acetic acid.
7. Extraction of methylated glycans is performed by stepwise addition of 0.3 ml chloroform and 0.2 ml water. After vigorous mixing and phase separation, the water layer is removed and repeatedly replaced by at least two further aliquots of 0.2 ml.
8. The chloroform phase is dried under nitrogen and the glycans are solubilized in solvent mixtures suitable for further analysis by nano-LC or static nanospray-ESI mass spectrometry.

3.3. CAD-ESI Mass Spectrometry of Methylated Glycan Alditols (Fig. 1)

3.3.1. Analysis by Static Nanospray

1. Electrospray ionization **Fig. 1** (ESI) mass spectrometry data are acquired on a Q-TOF 2 quadrupole-time-of-flight mass spectrometer equipped with a Z-spray source. ESI(Q-TOF) mass spectrometry of methylated glycans is performed in the positive ion mode.
2. The permethylated glycans are dissolved in 80% methanol containing 1% acetic acid before loading 3 μl into a nanospray capillary.
3. A potential of 800 V is applied to a nanoflow tip. Nitrogen is used as drying gas and argon as the collision gas, with the collision gas pressure maintained at 0.5 bar. The cone voltage is set at 50 V. Collision energies vary in accordance with the type of molecular ion (M + Na: 50–75 V; M + H: 15–30 V).

3.3.2. Analysis by Nano-LC-MS

1. *LC-parameters*. 10 μl permethylated glycans dissolved in 20% methanol are injected onto the trap column and are desalted for 3 min with 5% acetonitrile in 0.1% TFA at a flow rate of

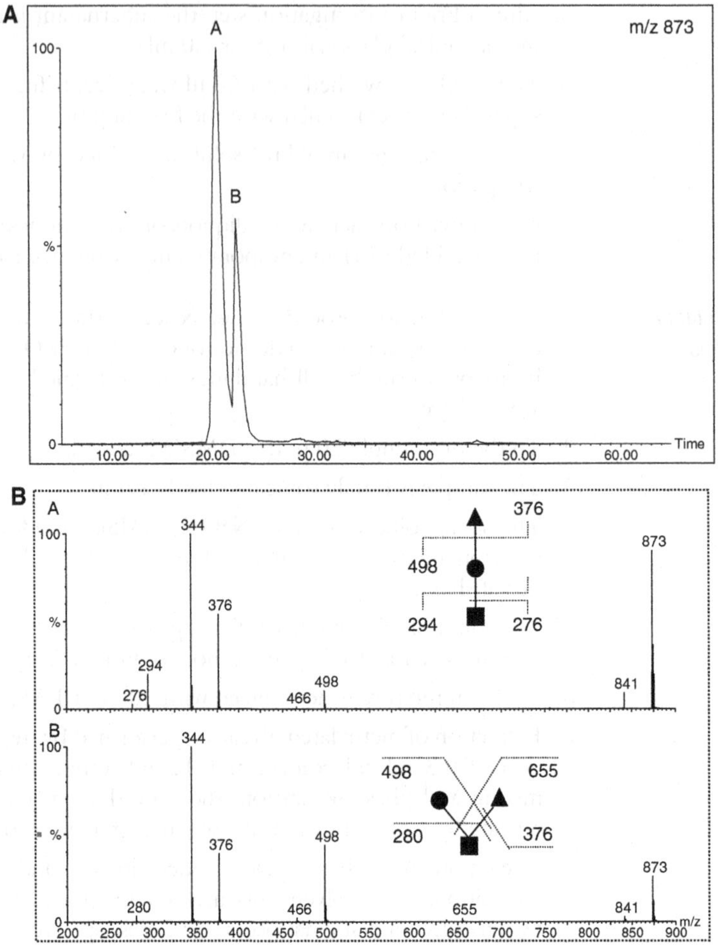

Fig. 1. NanoLC-ESI-MS/MS of permethylated glycan alditols from fusion protein MUC1-S expressed in EBNA-293 cells (**A**) A comprehensive study of O-linked glycans on MUC1-S was published previously *(16)*. The ion trace at *m/z* 873 corresponding to the relative mass of protonated trisaccharide alditol $NeuAc_1Hex_1HexNAc\text{-}ol_1$ as its permethylated derivative demonstrates effective separation of two isomeric glycan alditols. Peaks A and B contain the linear NeuAc-Hex-HexNAc-ol or the branched Hex-(NeuAc-) HexNAc-ol, respectively. (**B**) MS/MS spectra represent two isomeric trisaccharides in peaks A and B (see upper panel) of the ion specific chromatogramm. The two isomeric forms can be discriminated on the basis of specific Y_1 and Z_1 ions at *m/z* 294 and 276 indicating a monosubstituted HexNAc-ol and the Y_1 ion at *m/z* 655 indicating a disubstituted HexNAc-ol (refer to the schematic structural assignments: NeuAc; Hex; HexNAc-ol).

10 μl/min before the trap column is switched into the analytical flowpath. The analytical column flow rate is approximately 200 nl/min, resulting from a 1 : 1,000 split of the 200 μl/min flow delivered by the system pump. The samples are eluted onto the analytical column by using a gradient of 27.5% acetonitrile (ACN) in 0.1% formic acid (FA) to 81% ACN in 0.1% FA over 30 min. The column is washed with 81% ACN for

10 min before the concentration is decreased to 27.5% over 2.5 min and the system is reequilibrated to start conditions for 17.5 min. The trap column is switched back to the loading pump flowpath and the system is ready for a new injection.

2. *MS-parameters*. Stable nanospray is established by the application of +1.7 kV to +1.9 kV to the stainless steel union in the LC-MS interface. The cone voltage is set to 40 V and the source temperature is 80°C. No drying gas is used in nano LC-MS experiments. A collision energy setting of 4 V is used for the acquisition of survey scans. The argon pressure in the collision cell is 0.5–0.6 bar for all experiments. Automatic MS/MS scans are acquired with collision energy settings ranging from 10 –30 V depending on the mass and charge state of the selected precursor ion (*see* **Note 2**). The data-dependent acquisition of MS and tandem MS (MS/MS) spectra is controlled by Masslynx 4.0 software. Survey scans of 1 s cover the range from *m/z* 500 to 1,800. Single and double charged ions rising above a given threshold are selected for MS/MS experiments. In MS/MS mode the mass range from *m/z* 40 to 1,400 is scanned in 1 s, and 5 scans are added up for each experiment.

3.3.3. Evaluation of Mass Spectrometric Data

Under electrospray ionization conditions methylated glycans form preferentially sodiated molecular ions in addition to less intense protonated species. The relative abundance of the two molecular species change upon desalting of samples by using the above described LC-MS protocol resulting in a preponderance of protonated molecular ions. In collision induced dissociation (CID) experiments these molecular species yield simple and predictable fragmentation, which is described by the nomenclature of Domon and Costello *(17)*. According to this nomenclature, Bi^+, Yi^+, Ci^+, and Zi^+ ions correspond to fragments of the nonreducing (Bi^+, Ci^+) or reducing ends (Yi^+, Zi^+) generated by fission of the glycosidic bonds at C_1–O (Bi^+, Yi^+) or *O*–C_n (Ci^+, Zi^+). The protonated molecular species yield largely Bi^+ ion series fragments with a preference of cleavage at the reducing side of HexNAc residues similar to the fragmentation in FAB mass spectrometry. Sodiated molecular species need higher collision energies (twofold to threefold) to induce significant fragmentation. Their fragmentation pattern is more complex and covers formation of Ci^+, Zi^+, Yi^+ and less intense Bi^+ ions. Accordingly, the fragmentation patterns of protonated and sodiated molecular ions yield complementary data for sequencing of the oligosaccharides. Intense fragment ions of the sodiated molecular species are found at m/z 259 (C_1: nonreducing terminal Hex), 398 (B_1: non-reducing terminal NeuAc), 211 (B_1: non-reducing terminal dHex), 298 (Z_1: reducing terminal HexNAc-ol; core1 structure), 520 ($Y_{1\beta}$: Hex-HexNAc-ol; core2 structure, if the ion at *m/z* 298 is absent), 747 ($Z_{1\alpha}$: reducing terminal Hex-HexNAc-HexNAc-ol; core2 structure), 486 (B_2:

nonreducing terminal Hex-HexNAc), 660 (B_3: non-reducing terminal dHex Hex HexNAc), 847 (B_3: NeuAc-Hex-HexNAc).

4. Notes

1. The described methylation protocol avoids the formation of oxidizing by-products, which can partially degrade the glycans from their reducing ends. However, sometimes the protocol may yield under-methylated glycans in significant quantities. In case of glycan alditols alternative protocols can be applied, where the base and methyl iodide are mixed 2 : 1 (vol/vol), and the reaction time can be increased to over 1 h.
2. The use of collision energy settings, which have been optimized for peptide work will result in low quality fragment spectra when used for automatic MS/MS measurements of permethylated glycans. Proton adducts of permethylated glycans are readily fragmented at very low collision energy settings in the range of 10–30 V, whereas the fragmentation of sodium adducts requires energy settings in the range of 60–80 V. A list with optimal energy settings for the fragmentation of single- and double-charged ions in the mass range from *m/z* 600 to 1,800 should be deduced from MS/MS measurements in the static nanospray mode and can later be used with the respective MS-methods for data-dependent acquisition of MS/MS spectra under Masslynx.

References

1. Hanisch, F.-G. (2001) O-Glycosylation of the mucin-type. *Biol. Chem.* 382, 143–149
2. Iyer, R.N., Carlson, D.M. (1971) Alkaline borohydride degradation of blood group H substance. *Arch. Biochem. Biophys.* 142, 101–105
3. Lamblin, G., Boersma, A., Lhermitte, M., Roussel, P., Mutsaers, J.H., van Halbeek, H., Vliegenthardt, J.F. (1984) Further characterization, by combined high-performance liquid chromatography/1H-NMR approach, of the heterogeneity displayed by the neutral carbohydrate chains of human bronchial mucins. *Eur. J. Biochem.* 143, 227–236
4. Klein, A., Lamblin, G., Lhermitte, M., Roussel, P., Breg, J., van Halbeek, H., Vliegenthardt, J.F.G. (1988) Primary structure of neutral oligosaccharides derived from respiratory-mucus glycoproteins of a patient suffering from bronchiectasis, determined by combination of 500-MHz 1H-NMR spectroscopy and quantitative sugar analysis. *Eur. J. Biochem.* 171, 631–642
5. Hounsell, E.F., Lawson, A.M., Feeney, J., Gooi, H.C., Pickering, N.J., Stoll, M.S., Lui, S.C., Feizi, T. (1985) Structural analysis of the O-glycosidically linked coreregion oligosaccharides of human meconium glycoproteins which express oncofetal antigens. *Eur. J. Biochem.* 148, 367–377
6. Hanisch, F.-G., Uhlenbruck, G., Peter-Katalinic, J., Egge, H., Dabrowski, J., Dabrowski, U. (1989)

Structures of neutral O-linked polylactosaminoglycans on human skim milk mucins. *J. Biol. Chem.* 264, 872–883

7. Morelle, W., Faid, V., Michalski, J.-C. (2004) Structural analysis of permethylated oligosaccharides using electrospray ionization quadrupole time-of-flight tandem mass spectrometry and deutero-reduction. *Rapid Commun. Mass Spectrom.* 18, 2451–2464
8. Morelle, W., Slomianny, M.-C., Diemer, H., Schaeffer, C., Dorsselaer, A., Michalski, J.-C. (2004) Fragmentation characteristics of permethylated oligosaccharides using a matrix-assisted laser desorption/ionization two-stage time-of-flight (TOF/TOF) tandem mass spectrometer. *Rapid Commun. Mass Spectrom.* 18, 2637–2649
9. McNeil, M., Darvill, A.G., Aman, P., Franzen, L.-E., Albersheim, P. (1982) Structural analysis of complex carbohydrates using high-performance liquid chromatography, gas chromatography, and mass speectrometry. *Meth. Enzymol.* 83, 3–45
10. Dell, A. (1990) Preparation and desorption mass spectrometry of permethyl and peracetyl derivatives of oligosaccharides. *Meth. Enzymol.* 193, 647–660
11. Anumula, K.R., Taylor, P.B. (1992) A comprehensive procedure for preparation of partially methylated alditol acetates from glycoprotein carbohydrates. *Anal. Biochem.* 203, 101 108
12. Ciucanu, I., Kerek, F. (1984) A simple and rapid method for the permethylation of carbohydrates. *Carbohydr. Res.* 131, 209–217
13. Needs, P.W., Selvendran, R.R. (1993) Avoiding oxidative degradation during sodium hydroxide/methyl iodide-mediated carbohydrate methylation in dimethyl sulfoxide. *Carbohydr. Res.* 245, 1–10
14. Ciucanu, I., Costello, C.E. (2003) Elimination of oxidative degradation during the per-O-methylation of carbohydrazes. *J. Am. Chem. Soc.* 125, 16213–16219
15. Schulz, B.L., Packer, N.H., Karlsson, N.G. (2002) Small-scale analysis of O-linked oligosaccharides from glycoproteins and mucins separated by gel electrophoresis. *Anal. Chem.* 74, 6088–6097
16. Engelmann, K., Kinlough, C.L., Müller, S., Razawi, H., Baldus, S.E., Hughey, R.P., Hanisch, F.-G. (2005) Transmembrane and secreted MUC1 probes show trafficking-dependent changes in O-glycan core profiles. *Glycobiology* 15, 1111–1124
17. Domon, B., Costello, C.E. (1988) A systematic nomenclature for carbohydrate fragmentations in FAB-MS/MS spectra of glycoconjugates. *Glycoconj. J.* 5, 397–409

Chapter 9

High-Throughput and High-Sensitivity Nano-LC/MS and MS/MS for *O*-Glycan Profiling

Hasse Karlsson, Jessica M. Holmén Larsson, Kristina A. Thomsson, Iris Härd, Malin Bäckström, and Gunnar C. Hansson

Summary

Sensitive and fast methods for the profiling of biologically important molecules are highly demanded. Mucins are densely *O*-glycosylated glycoproteins found at mucosal surfaces and are of great medical interest. Here we describe sensitive methods for the analysis of *O*-glycans from mucins using gel electrophoresis, and chromatography by nanoLC on graphite columns and structural analysis by electrospray mass spectrometry on a linear trap mass spectrometer.

Key words: Mucin, Glycoprotein, *O*-Glycan, Mass spectrometry, nanoLC, HPLC.

1. Introduction

Glycosylation by *O*-glycans on the amino acids serine and threonine is found in many proteins that pass the secretory apparatus of the cell. This takes place in the Golgi apparatus where each monosaccharide is sequentially attached after the initial addition of a GalNAc. The added monosaccharides and sulfate groups typically generate very complex mixtures of glycan structures. *O-glycosylations* typical for mucins, a group of very large glycoproteins found at mucosal surfaces where they protect the epithelial cells *(1)*. The *O*-glycans of mucins typically make up more than 80% of the mucin mass and these are functionally important for the mucins. However, as *O*-glycans are also found on many other extracellular proteins their importance spans outside of the mucin

Nicolle H. Packer and Niclas G. Karlsson (eds.), *Methods in Molecular Biology, Glycomics: Methods and Protocols, vol. 534*

DOI: 10.1007/978-1-59745-022-5_9

glycoproteins. Due to the heterogeneity of *O*-glycans, they are the least explored in glycobiology.

We have a long-term interest in profiling *O*-glycans using the combination of chromatography and mass spectrometry. We described high-temperature gas chromatography-MS for such profiling almost 20 years ago *(2)*. The development of *electrospray* mass spectrometry has given new possibilities. The analysis of *O-glycosylation* on mucins separated by gel electrophoresis was pioneered at Proteome Systems, Sydney, by Drs. Niclas Karlsson, Nicki Packer, and coworkers *(3)*. Their approach was based on composite agarose-polyacrylamide gel electrophoresis and miniaturized release of the *O*-glycans after transfer to PVDF membranes. The glycans were then separated on a graphite column and analyzed by electrospray MS and MS^n. We have essentially used their approach and optimized the analytical step, something that has allowed us to reach a very high sensitivity. The adaptation of columns and electrospray interface to allow for columns down to 50–75 μm i.d. and flow rates in the nanoliter/min range has been especially important. Our nano-LC/MS interface is modified from one previously described *(4)*.

Here we describe how to prepare the guanidinium chloride soluble and insoluble fractions of mucins. This is especially relevant for the MUC2 mucin that exists in both forms in the small and large intestine *(5–7)*. We have recently successfully analyzed the glycosylation of the MUC2 mucin in 50 routine biopsies from patients undergoing coloscopy (unpublished). We have also used an approach where the glycoprotein of interest is purified by immunoprecipitation from a crude lysate. Using this approach, we have analyzed the *O*-glycans of less than 10^6 MUC1 transfected dendritic cells (unpublished).

2. Materials

2.1. Mucin Preparation

1. Eppendorf tube (1.7 ml, Maxymum recovery, Axygen).
2. Magnets, 3 mm diameter.
3. Mini pestle.
4. *Extr-GuHCl.* 6 M Guanidinium chloride, 5 mM EDTA and 0.01 M NaH_2PO_4, pH 6.5.
5. *Protease inhibitor.* 100 mM Phenylmethylsulfonyl fluoride (PMSF) in 2-propanol, stored at +4°C.
6. Magnetic stirrer.
7. Centrifuge for Eppendorf tubes, minimum 16,000 × g.
8. *Reducing agent.* 1.0 M Dithiotreitol (DTT), stored at −20°C.
9. *Red-GuHCl.* 6 M Guanidinium chloride, 5 mM EDTA, 0.1 M Tris-HCl, pH 8.0.

10. 4-vinylpyridine (97% solution, Sigma).
11. Dialysis cups (10 kDa MWCO, Slide-A-Lyzer MINI dialysis cups, Pierce, Rockford, IL).
12. Speed vac (DNA120, Thermo-Savant or equivalent).
13. *2× Sample buffer.* 0.75 M Tris–HCl pH 8.1, 2% SDS, 0.01% bromophenol blue, 60% glycerol).

2.2. Composite Agarose-Polyacrylamide Gel Electrophoresis and Mucin Transfer to PVDF Membranes

1. Agarose (Type 1-B: LowEEO, Sigma).
2. *Acrylamide.* 40% solution, Acrylamide/Bis 19:1, (BioRad).
3. 40% Ammoniumpersulfate solution (APS), stored at −20°C.
4. TEMED (BioRad).
5. *5× Tris–HCl buffer.* 1.875 M, pH 8.1.
6. 50% Glycerol solution.
7. Casting equipment (BioRad Miniprotean or equivalent).
8. Peristaltic pump, for example Pharmacia Biotech P-1 pump.
9. Large and small glass plate, two 1.5 mm spacers, two 0.75 mm combs.
10. *Lower gel solution.* Agarose (0.08 *g*) is mixed with 5× Tris–HCl buffer pH 8.1 (1.6 ml), 50% glycerol (1.6 ml) and H_2O (1.6 ml).
11. *Upper gel solution.* Agarose (0.04 *g*) is mixed with 5× Tris–HCl buffer pH 8.1 (1.6 ml) and H_2O (6.4 ml).
12. Gradient mixer, pump (Pharmacia Biotech or equivalent).
13. Oven, 60°C.
14. Gel electrophoresis equipment (BioRad or equivalent).
15. *Running buffer* 192 mM boric acid, 1 mM EDTA, 0.1% SDS, pH 7.6, adjusted with Tris base.
16. Wet blot equipment (Mini Trans-Blot electrophoretic transfer cell, Bio-Rad or equivalent).
17. Immobilon membrane, poly vinylidene fluoride (PVDF) P^{SQ}, (Millipore).
18. Filter papers, 2 mm thickness (Schleicher & Schuell).
19. *Wet blot buffer.* 25 mM Tris, 192 mM glycine, 0.04% SDS, 20% methanol.
20. *Alcian blue solution.* 0.625% Alcian blue in 25% Ethanol/10% Acetic Acid.

2.3. Release of O-Glycans from Blot Membranes

1. Glass plate, scalpel blade.
2. 96-well microtiter plate (C12, Maxisorp, Nunc).
3. Methanol (Merck).
4. *Release solution.* 50 mM Potassium hydroxide and 0.5 M sodium borohydride (*see* **Note 1**).

5. Glacial acetic acid (Merck).
6. Centrifuge for Eppendorf tubes.
7. Zip-Tips C_{18} (Millipore).
8. AG50Wx8 cation exchange resin (BioRad).
9. HCl (1 M solution).
10. Eppendorf tube (0.6 ml, Maxymum recovery, Axygen).
11. Desalting solution: 1% glacial acetic acid in methanol.
12. Speedvac (DNA120, Thermo-Savant or equivalent).

2.4. Nano-LC Column Preparation

1. A fused silica capillary 75 μm i.d., 375 μm o.d. (Polymicro Technologies, Phoenix, AZ).
2. 1/16 in. union (ZU1C, 0.25 mm bore, Vici AG, CH-6214 Schenkon, Switzerland) with a steel screen (1μm pores, Vici AG).
3. PEEK tubing, 25 mm, 380 μm i.d., Upchurch Scientific, Oak Harbor, WA.
4. Hypercarb, Thermo Hypersil-Keystone, CO.
5. Tetrahydrofuran.
6. 2-propanol.
7. Ultrasonic bath.
8. Small magnet (7 × 2 mm).
9. Compressed air driven packing pump (Maximator, type M72, Zorge/Harz, Germany). Binary HPLC pump (Agilent 1100 binary pump, Agilent Technologies, Palo Alto, CA).
10. Valco-T (Vici AG).
11. Acetonitrile.
12. Ammonium hydroxide.

2.5. Nano-LC/MS Interface

1. Through bore 1/16 in. union (ZU1T, Vici AG).
2. Steel screen (1 μm pores, Valco).
3. Fused silica capillary 20 μm i.d. and 150 μm o.d. (Polymicro Technologies, Phoenix, AZ).
4. PEEK tubing, 20 mm (150 μm i.d., Upchurch Scientific, Oak Harbor, WA).

2.6. Nano-LC/MS

(a) Agilent 1100 binary pump (Agilent Technologies, Palo Alto, CA).

(b) Mass spectrometer, LTQ (Thermo Electron Corp., San Jose, CA) equipped with a nanospray module.

1. For manual injection.
   (a) Valco-T (Vici AG).

(b) A manual microinjector M-485 (Upchurch Scientific, Oak Harbor, WA, USA) equipped with a 1 µl sample loop prepared from 13 cm × 100 µm i.d. fused silica capillary.

(c) 1/16 in. through bore union (ZU1T, Vici AG).

(d) Solvent A: 0.04% NH_3

(e) Solvent B: acetonitrile (HPLC grade):water, 80:20.

2. For injection with an autosampler.

   (a) HTC-PAL (CTC Analytics AG, Zwingen, Switzerland).

   (b) Microvolume-T (0.15 mm bore, Vici AG).

   (c) Injector (Cheminert valve, 0.25 mm ports, C2V-1006D-CTC) and equipped with a 1 µl sample loop made from 13 cm × 100 µm i.d. fused silica capillary.

   (d) Transfer line of 20 cm × 50 µm i.d. fused silica capillary.

   (e) Through bore union (ZU1T, Vici AG).

3. For injection with an autosampler using a precolumn configuration

   (a) Two microvolume-T's (0.15 mm bore, Vici AG).

   (b) Fused silica capillary 60 cm × 50 µm i.d. and 25 cm × 100 µm i.d.

   (c) Rheodyne valve controlled by the spectrometer software (Xcalibur for the LTQ).

   (d) Packed fused silica capillary, 5 cm × 100 µm i.d.

3. Methods

3.1. Mucin Preparation

1. The biopsy/tissue/scraped mucosa (~10 mg) is homogenized in an Eppendorf tube (1.7 ml, Maxymum recovery) in Extr-GuHCl (40 µl) with PMSF (4 µl, 2 mM final concentration) using a plastic mini pestle. Add 160 µl Extr-GuHCl (total volume 200 µl). Add magnet.
2. The homogenized material is stirred on a magnetic stirrer gently over night in +4°C.
3. The solution is centrifuged for 30 min at minimum 16,000×g.
4. The GuHCl soluble (supernatant) and the insoluble (pellet) fractions are separated by collecting the supernatant. The soluble fraction is stored at −20°C.
5. To the insoluble fraction (the pellet), Extr-GuHCl (100 µl) is added followed by stirring for ~4 h or over night at +4°C. Centrifugation according to **step 3**. **Step 4** is repeated and the two soluble fractions are pooled.

6. Red-GuHCl (120 μl) is added to the insoluble fraction. Reducing agent DTT (1.0 M stock solution) is added to the insoluble and soluble fraction to a final concentration of 50 mM. The solutions are stirred for 1–2 h at 37°C.
7. Alkylating agent 4-vinylpyridine is added to a final concentration of 125 mM. Stirring for 1 h in RT in the dark.
8. Sample is transferred to a dialysis cup (10,000 MWCO) and dialyzed against water (>4 l) at 4°C over night (*see* **Note 2**).
9. Sample is transferred to an Eppendorf tube (1.7 ml, Maxymum recovery) and dried almost to dryness using a Speedvac.
10. Sample is dissolved in 2× sample buffer in small aliquots with heating to 100°C in between. Water is added to a final concentration of 1× sample buffer (~500 μl final volume) plus additional 1.0 M DTT (25 μl) for complete reduction (*see* **Note 3**).

3.2. Immunoprecipitation of Glycoproteins (Optional for Mucins)

Mucins are large and usually do not require any further purification by immunoprecipitation. However, this is an option for especially smaller glycoproteins not readily separated from other contaminants (*see* **Note 4**).

3.3. Composite Agarose-Polyacrylamide Gel Electrophoresis and Mucin Transfer to PVDF Membranes

1. The gel casting equipment (glass plates and spacers) is mounted and placed in a 60°C oven, together with a gradient mixer and a pump. The pump is connected to the gradient mixer and a tube connected to the pump is placed in between the glass plates of the casting equipment.
2. A gel containing agarose (0.5–1% gradient), acrylamide (0–6%) and glycerol (0–10%) is prepared. The lower and upper gel solutions are prepared as described in **Subheading 2.3, items 9** and **10**. Both solutions are boiled in a microwave oven until the agarose has melted and then immediately placed in the 60°C oven.
3. When the lower gel solution has cooled down to ~60°C, acrylamide (40%, 1.2 ml) is added.
4. The upper and lower gel solutions (5 ml of each) are added to each chamber of the gradient mixer, respectively. APS (2 μl) and Temed (2 μl) are added to the two chambers, respectively, and the solutions are mixed. The lower solution is stirred with a magnet and a magnetic stirrer.
5. The pump is started on maximum rate (10 on a Pharmacia Biotech P-1 pump) and the two valves of the gradient mixer are opened. The gel must be cast within two minutes.
6. Two combs (2 × 0.75 mm) are put into the gel which is polymerized in room temperature for ~1 h. The gels can be stored (with combs) at +4°C wrapped in plastic with wet papers for at least a week.

7. Prior to electrophoresis, remove the combs carefully, place the gel in the electrophoresis equipment and add running buffer. The wells are carefully washed with running buffer prior to use. Fill the outer and inner container with running buffer to keep the gel cool during electrophoresis.
8. The electrophoresis container is placed on ice in a cold room (+4°C). After samples are loaded, the gel is run at 12–15 mA per gel over night (16 h).
9. The gel is wet blotted in a Mini Trans-Blot electrophoretic transfer cell. Immobilon (PVDF P^{SQ}) membrane and filter papers (2 × 2 mm thickness) are cut out in the size of 6 × 9 cm. The wells of the gel are cut off before mounting the wet blot sandwich (mounting according to BioRad manual).
10. The sandwich is placed in the transfer cell (with the gel facing the cathode), which is filled with transfer buffer and put on stirring on ice in the cold room (+4°C).
11. The blotting is performed at 40 W (gives ~350–450 mA over time) for 2.5 h (human MUC5B mucin) up to 5 h (human MUC2 mucin) (*see* **Note 5**).
12. After blotting, the bands are visualized with carbohydrate or protein stains like Alcian blue, Coomassie blue or immunostaining (*see* **Note 6**). A band that stains well with Alcian blue usually has enough amount of glycans to be analyzed with LC-MS^n (*see* **Note 7**).
13. The membrane is dried and stored safely until glycan release.

3.4. Release of O-Glycans from Blotted Membranes

1. Cut out glycoprotein bands from a stained blot membrane and further divide into smaller pieces (~2 × 2 mm). Place the membrane pieces in one well of a 96-well microtiter plate.
2. Wet the membranes with methanol (20 µl), remove the methanol and directly add freshly made alkaline borohydride solution (20 µl, 50 mM KOH with 0.5 M $NaBH_4$).
3. Seal the microtiter plate tightly with adhesive plastic and place the plate in a box with a lid. Add about 1 cm of water to the bottom of the box (*see* **Note 8**). Seal the box and incubate at 50°C over night (16 h).
4. Add glacial acetic acid (1 µl) to each well to neutralize the solution. The solution should then fizz due to the production of H_2 gas (*see* **Note 9**).
5. Prepare cation exchange resin AG50Wx8 according to the BioRad instructions (15 repeated washes).
6. Prepare a column by adding AG50Wx8 cation exchange resin suspension (25–50 µl) to a Zip-Tip C_{18}. Place the column

in a 1.5 ml Eppendorf tube and micro centrifuge (2000×g) for 10 s.

7. Wash the column with methanol (2 × 60 μl) and HCl (1.0 M, 2 × 60 μl) with 10 s centrifugation in between. Discard the flow-through and wash the column twice with water (2 × 60 μl) with centrifugation in between (2 × 10 s).
8. Place the column in an Eppendorf tube (0.6 ml, Maxymum recovery) and add the neutralized sample from **step 4**. Centrifuge for 10 s.
9. Wash the well with water (2 × 25 μl) and load onto the column with centrifugation in between (2 × 10 s).
10. Elute residual glycans from the column with ddH_2O (30 μl). Centrifuge for 10 s.
11. Dry the eluted glycans in a Speedvac to dryness (~45 min).
12. Remove the borate complexes (BH_4, white salt) by the addition of 1% glacial acetic acid in methanol, 50 μl. Dry the samples in a Speedvac to dryness. Repeat the washing step an additional four times with drying in speedvac in between. The amount of white salt should be substantially reduced by this treatment (*see* **Note 10**).
13. Dissolve the glycans in ddH_2O (6–15 μl, *see* **Note 11**). Store the samples at −20°C.

3.5. Nano-LC Column Preparation

1. A fused silica capillary of 20 or 25 cm length (75 μm i.d., 375 μm o.d.) is connected to a 1/16 in. union (ZU1C, 0.25 mm bore) with a steel screen (1 μm pores) using 25 mm of PEEK tubing (380 μm i.d.) as a sleeve.
2. A slurry container is made of a piece of normal 55 × 4.6 mm stainless steel column (volume around 1 ml) with an external 1/16 in. Swagelock connection in one end. A Vespel ferrule is used to seal the fused silica capillary in the slurry container.
3. A slurry of around 5 mg of 5 μm graphite particles (Hypercarb) is prepared in 1.1 ml tetrahydrofuran:2-propanol (9:1) and sonicated in an ultrasonic bath for 4 min before transfer to the slurry container. The slurry also contains a small magnet (7 × 2 mm) for stirring during packing.
4. The slurry container is connected to a compressed air driven packing pump (Maximator) and the pressure is set to 300 bar at the start and later increased to 560 bar. Methanol is used as pusher solvent. The packing takes about 20 min and the packed bed can be followed as the black graphite particles are clearly visible in the fused silica column.
5. After the packing is complete, the column is placed in an ultrasonic bath where it is conditioned for 10 min, still maintained under a pressure of 560 bar.

6. The column is depressurized, connected to a standard binary HPLC pump via a Valco-T (Vici AG) with a 50 cm × 50 μm fused silica capillary as splitter and washed with 80% acetonitrile before connected to mobile phase A (0.04% NH_3). A flow rate of 200 μl/min from the binary pump gives around 300 nl/min through the packed fused silica column.

3.6. Nano-LC Interface

1. A through bore 1/16 in. union (ZU1T, Vici AG) is connected to the end of the packed fused silica capillary column instead of the one used for packing, a steel screen (1 μm pores, Valco) is placed directly on top of the sleeved capillary column in the union and above that a tapered nanoESI-emitter tip is connected.
2. The nanoESI-emitter tips (around 30 mm length) are made from 20 μm i.d. and 150 μm o.d. fused silica capillary (Polymicro Technologies) and held in position by 20 mm PEEK tubing (150 μm i.d., Upchurch) as a sleeve. This connection produces a negligible dead volume for nano-column flow rates. *See* **Fig. 1A** and **B** for pictures of the nano-LC/MS interface. This nano-LC/MS interface is similar to the one previously described *(4)*.

3.7. Nano-LC/MS

It is possible to setup the nano-LC in different ways; either using manual injection (A) or using an autosampler (B, C). In the B variant, the sample is loaded directly on the column, whereas in C a precolumn is used.

1. The pump is set to deliver a flow of 200–250 μl/min and the flow split in a T using a 50 cm × 50 μm fused silica capillary as restrictor. The flow rate through the 25 cm × 75 μm i.d. column packed with 5 μm graphite particles is then about 300 nl/min.
2. Injections are made at 0% B.
3. The elution gradient in 5 min is 0% B followed by a gradient to 50% B in 50 min

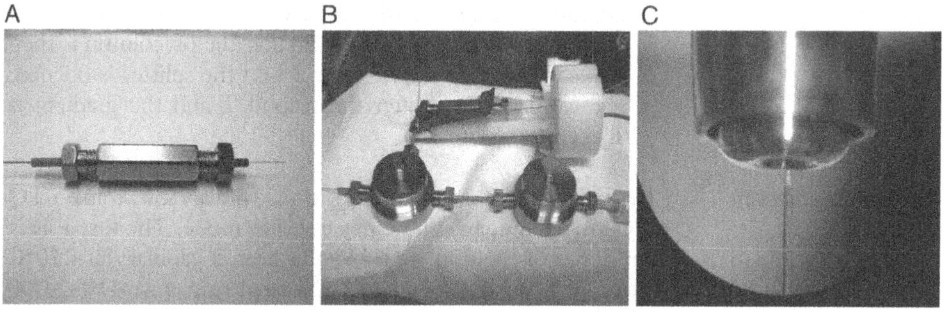

Fig. 1. **(A)** The nano-LC/MS interface. **(B)** Precolumn and home made nano-LC/MS interface probe for the LTQ mass spectrometer. **(C)** The fused silica emitter in front of the heated capillary of the LTQ mass spectrometer.

4. The end of the column is sealed against a steel screen (1 μm pores) and connected to the ESI-emitter tip in a 1/16 in. through bore union which is kept at –1.6 kV in front of the heated capillary ion inlet (<1 mm) of the spectrometer, *see* **Fig. 1C**.

Nano-LC could be performed with different injectors:

A. For manual injection.

1. A manual microinjector is equipped with a 1 μl sample loop prepared from 13 cm × 100 μm i.d. fused silica capillary. The packed fused silica capillary column is connected to the injector via a 4 cm × 50 μm i.d. fused silica capillary and a 1/16 in. through bore union (*see* **Note 12**).

B. For injection with an autosampler

1. The injection is made with an autosampler, HTC-PAL. Split the flow from the pump in a microvolume-T (0.15 mm bore) prior to the injector equipped with a 1 μl sample loop made from 13 cm × 100 μm i.d. fused silica capillary. A transfer line of 20 cm × 50 μm i.d. fused silica capillary is connected to the outlet port of the injector to a 1/16 in. through bore union where the packed fused silica column is connected.

C. For injection with an autosampler using a precolumn configuration.

1. The injection is made with an autosampler, HTC-PAL in a precolumn configuration with a larger sample loop. A packed 5 cm × 100 μm i.d. fused silica capillary is used as precolumn to trap the oligosaccharides in a setup between two microvolume-T's (0.15 mm bore), *see* **Fig. 1B**. From the first T a 60 cm × 50 μm i.d. fused silica capillary is connected to a Rheodyne valve controlled by the spectrometer software (Xcalibur for the LTQ) and this acts as a splitter. From the second T after the precolumn a 25 cm × 100 μm i.d. fused silica capillary is connected to the same Rheodyne valve. The analytical 25 cm × 75 μm i.d. packed capillary column is connected after the second T. In the loading step the valve to the splitter is closed and the precolumn is then loaded with the sample. After 3 min the splitter is opened, the valve is closed after the second T and the gradient is started (*see* **Note 13**).

3.8. Mass Spectrometry

i. Nano-LC/MS^n is performed on a linear ion trap, LTQ (Thermo Finnigan) in the negative ion mode. The fused silica emitter tip is held at –1.6 kV, the heated capillary at 220°C and capillary voltage at –44 V. Normal scan *m/z* 410–2,000 (2 microscans, 100 ms), data dependent scans of the most abundant ion in each scan, MS^2, MS^3 and MS^4 also with 2 microscans (100 ms), normalized collision energy of 35%, isolation window of 4.0 u and an activation time of 30 ms. All data are acquired as profile data (*see* **Note 14**).

4. Notes

1. Prepare fresh KOH/$NaBH_4$ solution on the day of use. Use dry $NaBH_4$ powder (store in exsiccator) and be aware if the powder starts to become moist. The top layer of powder in the $NaBH_4$ can easily become destroyed by moisture from the surrounding air and the powder then forms lumps. If this occurs, use the powder under the damaged layer or use fresh $NaBH_4$.
2. After dialysis, proteins and the excess amount of 4-vinylpyridine might have precipitated. Both the protein solution and the precipitates are transferred to an eppendorf tube (Maxymum recovery) and solubilized in 2× sample buffer.
3. The precipitates after dialysis are not always possible to dissolve completely. Try to make small additions of sample buffer and boil the sample in between. To keep the SDS concentration as high as possible during dissolving procedure, add the water to give a final concentration of 1× sample buffer as a last step. Add 1.0 M DTT (25 μl) and boil the solution a final time for complete reduction. If precipitates are still present after reduction, it is probably due to the excess amount of 4-vinylpyridine. Centrifuge the solution at 2,000×g and transfer the solubilized protein to a new Eppendorf tube (maximum recovery).
4. As the sensitivity for *O*-glycan analysis of glycoprotein from mixtures is often sufficient, one can include an immunoprecipitation step as purification before, for example, polyacrylamide gel electrophoresis and transfer to PVDF membranes. The immunoprecipitation can be performed with any protocol including sufficient precautions to limit the amount of contaminants. The precipitated glycoproteins can be visualized either by Coomassie (Imperial protein stain, Pierce) or Alcian Blue (8GX, Sigma). If no specific bands (as compared to the controls) show up, immunostain a small strip cut out from each sample lane on the membrane. The glycoprotein of interest is localized on this strip and the corresponding band cut out and subjected to *O*-glycan analysis. As the sensitivity is high, it is important to include negative controls to make sure that the released *O*-glycans detected emanate from the glycoprotein of interest. One such important control is an immunoprecipitation with only the precipitating antibody and no

cell-lysate as *O*-glycans from an antibody can give considerable background of *O*-glycans also outside of the immunoglobulin bands. The sensitivity of this approach is high and we have shown the *O*-glycans of a specific mucin, MUC1, from less than a million cells.

5. When wet blotting MUC2 for 5 h, approximately half of the protein amount is blotted to the membrane according to the Alcian blue stained gel after blotting as compared to the stained membrane. It has not been possible to improve this result by blotting for a longer time (e.g. over night on lower voltage). Blotting over night can make the membrane slightly dry and the gel starts to melt due to the produced heat in the blotting cell.
6. Mucins or other glycoproteins are poorly stained by Coomassie. We therefore use Imperial protein stain (Pierce) which in our hands has a higher sensitivity for glycoproteins. Alcian blue will stain negatively charged glycoproteins, i.e. with a high content of sialylated or sulfated oligosaccharides.
7. Staining with Alcian blue: The membrane is washed in methanol for 1 min and Alcian blue solution is added. The membrane is stained for 10–20 min. The membrane is destained with methanol repeatedly (2 × 5 min) until the bands are visible.
8. The water in the box keeps the air surrounding the plate hydrated.
9. When working with small sample volumes in the miniaturized release, it can be difficult to visualize the fizz due to H_2 gas production. In some cases it is possible to see small bubbles in the wells due to H_2 gas, but there is no fizzing. Even when it is not possible to spot any bubbles, the glycan release might have worked, but consider the quality of the $NaBH_4$ (*see* **Note 1**).
10. There is sometimes a white shadow of BH_4 still present in the tube. This does not, normally, cause any problems. If the $NaBH_4$ powder has become moist (*see* **Note 1**) it can be difficult to destroy the borate complexes and a white salt precipitate is still present in the tube after treatment with desalting solution. The remaining BH_4 is usually dissolved in water, but can give rise to problems during analysis, since BH_4 clogs the tubing or column of the LC-MS system.
11. No particles should be present in the samples due to the risk of clogging the LC-MS system. Therefore, carefully inspect the samples for particles before transferring the solutions to

proper tubes for analysis (e.g. with a magnifier). If particles are present, centrifuge the solutions at 10,000 rpm for 1 min prior to transfer.

12. This is used to electrically isolate the injector from the high voltage at the union with the emitter tip (−1.6 kV). The graphite particles in the nano-column have very good electrical conductivity.
13. Unfortunately, due to the relatively low retention of smaller oligosaccharides on graphite particles these have a tendency to be discriminated on a precolumn so this setup cannot be recommended. The precolumn configuration is common for analysis of tryptic peptides on a reverse phase C_{18} column since the peptides are better retained on this stationary phase.
14. The result from the analysis of the mucin MUC5B from human cervix is illustrated in **Figs. 2** and **3**. The base peak chromatogram of the *O*-glycans is shown in **Fig. 2** and the MS^2, MS^3, and MS^4 of the component with a retention time of 34 min and *m/z* 1,121.5 in **Fig. 3**.

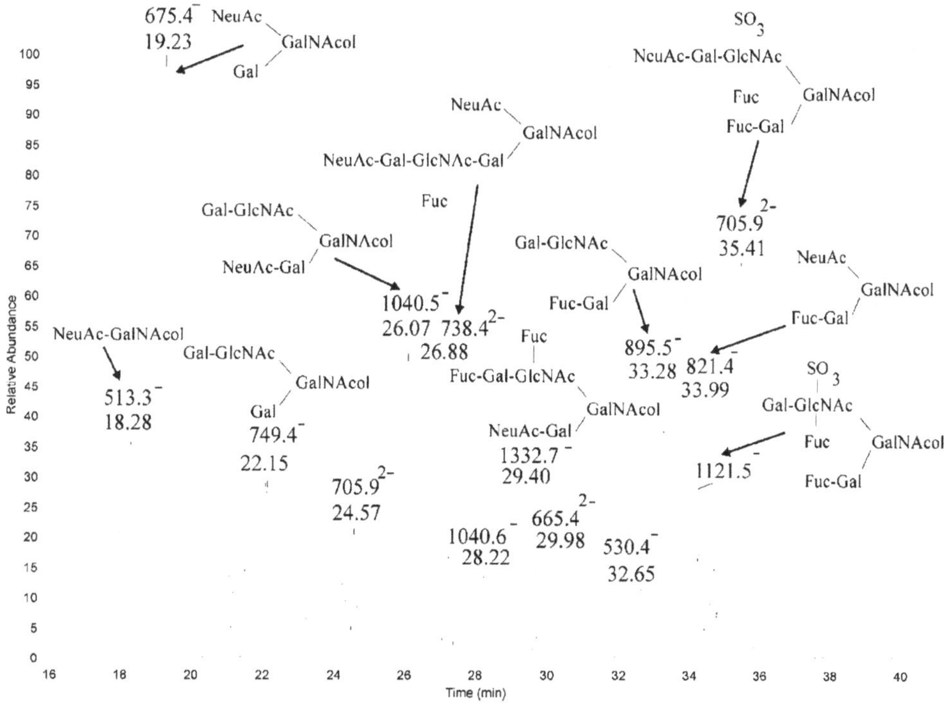

Fig. 2. Base peak chromatogram of *O*-glycans released from the MUC5B mucin from endocervix.

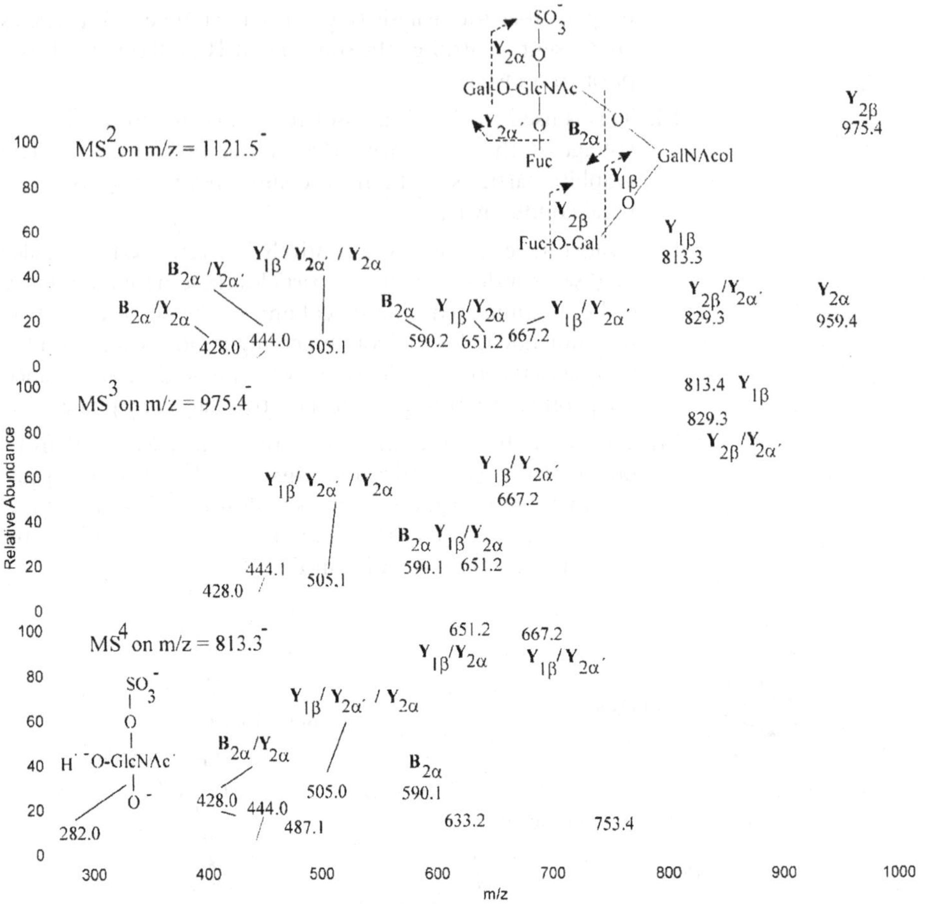

Fig. 3. MS², MS³ and MS⁴ of the components with a retention time of 34 min and parent ion at *m/z* 1,121.5 from the MUC5B mucin analysis shown in Fig. 1. The nature of the fragment ions is indicated.

Acknowledgements

The work was supported by the Swedish Research Council (No. 7491 and an expensive equipment grant) and IngaBritt and Arne Lundberg Foundation.

References

1. Hollingsworth, M. A. and Swanson, B. J. (2004) Mucin in cancer: protection and control of the cell surface. *Nat. Rev. Cancer* 4:45–60.
2. Karlsson, H., Carlstedt, I. and Hansson, G. C. (1989) The use of gas chromatography and gas chromatography-mass spectrometry for the characterization of permethylated oligosaccharides with molecular mass up to 2300. *Anal. Biochem.* 182:438–446.
3. Schulz, B. J., Oxley, D., Packer, N. H. and Karlsson, N. G. (2002) Identification of two

highly sialylated human tear-fluid DMBT1 isoforms: the major high-molecular-mass glycoproteins in human tears. *Biochem. J.* 366:511–520.

4. Shen, Y., Zhao, R., Berger, S. J., Anderson, G. A., Rodriguez, N. and Smith, R. D. (2002) High-Efficiency Nanoscale Liquid Chromatography Coupled On-Line with Mass Spectrometry Using Nanoelectrospray Ionization for Proteomics. *Anal. Chem.* 74:4235–4249.
5. Carlstedt, I., Herrmann, A., Karlsson, H., Sheehan, J. K., Fransson, L. and Hansson, G. C. (1993) Characterization of two different glycosylated domains from the insoluble mucin complex of rat small intestine. *J. Biol. Chem.* 268:18771–18781.
6. Axelsson, M. A. B., Asker, N. and Hansson, G. C. (1998) O-glycosylated MUC2 monomer and dimer from LS 174T cells are water-soluble, whereas larger MUC2 species formed early during biosynthesis are insoluble and contain nonreducible intermolecular bonds. *J. Biol. Chem.* 273:18864–18870.
7. Herrmann, A., Davies, J. R., Lindell, G., Martensson, S., Packer, N. H., Swallow, D. M. and Carlstedt, I. (1999) Studies on the "Insoluble" glycoprotein complex from human colon. *J. Biol. Chem.* 274:15828–15836.

Chapter 10

Collision-Induced Dissociation Tandem Mass Spectrometry for Structural Elucidation of Glycans

Bensheng Li, Hyun Joo An, Jerry L. Hedrick, and Carlito B. Lebrilla

Summary

The complexity of glycans poses a major challenge for structure elucidation. Tandem mass spectrometry is currently an efficient and powerful technique for the structural characterization of glycans. Collision-induced dissociation (CID) is most commonly used, and involves first isolating the glycan ions of interest, translationally exciting them, and then striking them with inert target gas to fragment the precursor ions. The structural information of the glycan can be obtained from the fragment ions of the tandem MS spectra.

In this chapter, sustained off-resonance irradiation-collision-induced dissociation (SORI-CID) implemented with matrix-assisted laser desorption/ionization Fourier transform ion cyclotron resonance mass spectrometry (MALDI FT ICR MS) is demonstrated to be a useful analysis tool for structural elucidation of mucin-type *O*-glycans released from mucin glycoproteins. The mechanisms by which the glycans undergo fragmentations in the tandem mass analysis are also discussed.

Key words: *O*-Linked glycans, Glycoproteins, MALDI, FTICR MS, SORI-CID, Tandem mass spectrometry.

1. Introduction

As one of the most widespread posttranslational modifications of proteins in eukaryotes, glycosylation of proteins is recognized for biological versatility and indispensability. Glycans attached to a glycoprotein play important roles in many biological processes including the proper folding of proteins, molecular recognition involved in specific inter- and intracellular interactions, and cellular adhesion *(1–4)*. To better understand the roles of oligosaccharides in glycoproteins, it is important to elucidate their structures

Nicolle H. Packer and Niclas G. Karlsson (eds.), *Methods in Molecular Biology, Glycomics: Methods and Protocols, vol. 534*

DOI: 10.1007/978-1-59745-022-5_10

and distributions in glycoproteins. Unlike proteins and nucleic acids where the macromolecular chains are linear, the oligosaccharides can be connected in many ways and branched with as many as four carbohydrate residues linked to a central monosaccharide.

The *O*-glycosylation of proteins can result in the formation of mucin-type macromolecules. Mucins are *O*-glycosylated proteins mostly found on the cell surface or in the secretions of cells *(5)*. The *O*-glycans have specific functions, such as protecting underlying proteins as well as epithelial cell surfaces from pathogen attack, involving in sperm–egg recognition during fertilization, participating in the immune system, and in clotting blood *(6, 7)*. Recently the roles played by *O*-glycans in biology, physiology and immunology of cancer have become more and more realized, thus the investigation on *O*-glycans has been one of the most attractive areas in oncology and clinical study of cancers. *(6–10)*.

The structural complexity and variety of *O*-linked glycans have posed a challenge for structural elucidation. Traditional structure analysis of glycans by NMR has some limitations in that it requires a relatively large amount of sample and it is time consuming. Mass spectrometric analysis of *O*-glycans has proved to be a very powerful and efficient tool not only in profiling the structural distribution of glycans *(11–13)* but for specific structure elucidations as well *(14–20)*, due to its high sensitivity and potentially high throughput.

Collision-induced dissociation (CID, also called collision-activated dissociation, CAD) tandem mass spectrometry has been the most widely employed technique for structural elucidation and continues to play a prominent role in the analyses of oligosaccharides and other molecules *(21–26)*.

The advantage of FTICR (and ion traps in general) is that CID is temporally rather than spatially resolved. Furthermore, it is easily implemented with either MALDI- or ESI-produced ions. CID in FTICR is performed by isolating the desired ion. This procedure involves the resonance excitation of all other ions to the point where they are ejected from the cell or collide with the cell walls. In the past, ejection of unwanted ions was performed with selective resonance ejection of individual ions. More sophisticated methods have been developed using arbitrary waveform generators that can be programmed for the retention of desired masses *(27)*.

There are various CID techniques that vary not only in collision energy but also in the amount of internal energy deposited in precursor ions upon collision, the collision number prior to fragmentation, and the time scale between collision activation and detection. They can be broadly grouped into two categories based on the translational energy (collision energy) possessed by the precursor ions just prior to collision with the target inert gas molecules: low (1–300 eV) and high (1–25 keV) collision energy CIDs.

With FTICR MS, CID generally involves low collision energy. In the ICR cell the ions of interest can be excited on-resonance, i.e., at a frequency equal to the ions' cyclotron frequency. The event can increase the translation energy to about 100 eV. In this method, the ions are translationally excited to a larger cyclotron orbit. The precursor ions can also be periodically excited by the application of a sustained (typically 600 ms) off-resonance irradiation (SORI) of alternating electric field pulse with a frequency slightly offset from the ions' natural cyclotron resonance frequency. A constant RF level is applied to the excite electrodes throughout the CID event. As a consequence, the ions undergo acceleration–deceleration cycles and thus a sequential activation of ions by multiple collisions of low translational energy (<10 eV) with the target gas throughout the duration of the electric field pulse. Spatially, the ions experience the corresponding cycles of being excited to a small radius away from the center of ICR cell, then relaxing back to the center. The precursor and fragment ions hover near the center of the cell allowing additional stages of CID and stronger intensities during detection. SORI-CID can facilitate the lowest energy pathway of fragmentation of precursor ions as only small increments of internal energy are deposited on the ions throughout the duration of the event *(28, 29)*. In this way, the SORI-CID is an analog to infrared multiphoton dissociation (IRMPD), which is discussed in the Chapter 2 *(30–43)*.

2. Materials

2.1. Release and Purification of Oligosaccharides

1. Release solution: 1.0 M $NaBH_4$ in 0.1 M NaOH, prepared freshly.
2. Lyopholized glycoprotein (e. g. egg jelly coat of *Xenopus tropicalis* (*see* **Note 1**).
3. Heating facility (42°C).
4. Neutralizing solution: 1.0 M HCl.
5. Solid phase extraction (SPE) graphitized carbon cartridges (150 mg, 4 mL) from Alltech Associates, Inc. (Deerfield, IL).
6. Activation solution: nano pure water and 80% acetonitrile (AcN) in 0.1% trifluoroacetic acid (TFA) (v/v).
7. Washing solution: nano pure water.
8. Elution solutions: 10% and 20% AcN in H_2O and 40% AcN with 0.05% TFA in H_2O.

9. Purification with HPLC as described in the Chapter 2 "Infrared Multiphoton Dissociation Mass Spectrometry for Structural Elucidation of Oligosaccharides."

2.2. MALDI-FTICR Analyses of Oligosaccharides

1. MALDI-FT mass spectrometer (IonSpec, Irvine, CA).
2. Released and/or HPLC fractionated oligosaccharides (1–6 μL) (*see* the Chapter 2 "Infrared Multiphoton Dissociation Mass Spectrometry for Structural Elucidation of Oligosaccharides").
3. MALDI matrix solution: 0.4 M matrix, 2,5-dihydroxy benzoic acid (DHB), in 50:50 H_2O/AcN.
4. Positive mode dopant: 0.1 M NaCl in 50:50 H_2O/AcN.

3. Methods

3.1. Release and Purification of Oligosaccharides (See* Note 2*)

1. Freshly prepared release solution is added into a 15 mL plastic tube containing the lyophilized glycoprotein to a volume of 267 μL release solution/mg of glycoprotein *(44)*. The solution is gently vortexed to mix well.
2. The tube is wrapped with parafilm and placed at 42°C for 16–20 h. The release of *O*-linked glycans from glycoproteins is a β-elimination reaction under strongly basic conditions (*see* **Note 3**).
3. After incubation, the excess $NaBH_4$ is slowly neutralized with 1.0 M HCl acid on ice until the solution is acidic (pH 2–5).
4. The salts and deglycosylated proteins are removed from the solution by solid phase extraction (SPE) with graphitized carbon cartridge. The cartridges are washed sequentially with nano pure water, 80% acetonitrile (AcN) in 0.1% trifluoroacetic acid (TFA) (v/v) and then pure water.
5. The flow rate during the SPE extraction should be kept at 200 μL/min after the sample loading to maximize the interaction of oligosaccharides with the stationary phase of the cartridge.
6. The subsequent wash with nano pure water is made 5–6 times (one full column each time) to remove salts and deglycosylated proteins.
7. After the wash, the glycans are eluted with 10%, 20% AcN in H_2O and 40% AcN with 0.05% TFA in H_2O, respectively. The collected glycan solution is dried and reconstituted in 10–30 μL of nano pure water. Oligosaccharides were further fractionated using HPLC, as described in this book (the Chapter 2

"Infrared Multiphoton Dissociation Mass Spectrometry for Structural Elucidation of Oligosaccharides").

3.2. MALDI-FTICR Analyses of Oligo-saccharides

A commercial MALDI-FT mass spectrometer (IonSpec, Irvine, CA) with an external ion source was used to perform the analysis. The instrument is equipped with a 7.0-T shielded, superconducting magnet and an Nd:YAG laser at 355 nm. MALDI sample was prepared by loading 1–6 µL of analyte and 1 µL of MALDI matrix solution on a stainless steel target plate. For the positive mode analyses, 1 µL of 0.1 M NaCl in 50:50 H_2O/AcN was applied to the spot to enrich the Na^+ concentration and thus produce primarily sodiated species. The plate was placed in ambient air to dry the sample spots before insertion into the ion source.

3.3. Sustained Off-Resonance Irradiation-Collision-Induced Dissociation Tandem Mass Spectrometry for Structure Elucidation of Glycans

A typical pulse sequence for SORI-CID in an FTCIR mass spectrometer is shown in **Fig. 1**. A desired ion is readily selected in the analyzer with the use of an arbitrary waveform generator and a frequency synthesizer. All CID experiments were performed at +1,000 Hz off-resonance from the cyclotron frequency of the precursor ion. The CID excitation time was 1,000 milliseconds (ms). Two pulses of argon were introduced into analyzer chamber at 0 and 500 ms for collisions. The excitation voltages ranged from 2.5 to 8.0 V (base-to-peak) depending on the desired level of fragmentation and the size of oligosaccharides.

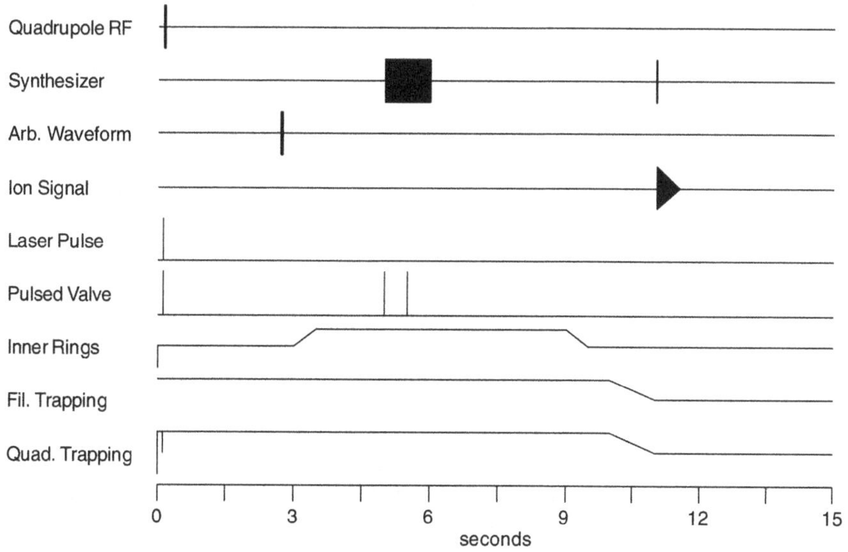

Fig. 1. A typical pulse sequence employed for sustained off-resonance irradiation collision-induced dissociation (SORI-CID) in an FTICR mass spectrometer.

3.4. Interpretation of SORI-CID Spectra of O-Linked Oligosaccharides

Manual interpretation of SORI-CID spectra involves knowledge of fragmentation pathways as well as understanding of the biosynthetic pathway of oligosaccharides. In addition, tables of masses of common monosaccharide building blocks are essential in order to work out the monosaccharide sequence of unknown structures. (for tables, *see* the Chapter 1 “Analysis of *N*- and *O*-Linked Glycans from Glycoproteins Using Maldi-TOF Mass Spectrometry”). Here, two examples are given of the interpretation and the logical reasoning for the assessment of an unknown structure. The first example of an *O*-linked oligosaccharide being probed by SORI-CID is shown in **Fig. 2**. The MS/MS spectrum of an *O*-linked glycan, *m/z* 1,268.463 ($[M + Na]^+$) is shown in **Fig. 2A**. The sodium cation allows a fine balance between abundant quasimolecular ions and CID fragment ions. The sequence of the glycan can be elucidated based on the CID spectra. The glycan consists of 2 hexoses (Hex), 2 deoxyhexoses (dHex), and 3 *N*-acetyl hexosamines (HexNAc) based on the accurate mass (experimental mass 1,268.463 Da, theoretical mass 1,268.475 Da, $\Delta m = 0.012$ or 9 ppm). **Figure 2B** corresponds to the MS^3 spectrum (1,268 to 1,122), while **Fig. 2C** corresponds to MS^4 spectrum (1,268 to 1,122 to 976). **Figure 2A, B** indicate that two dHex residues are readily lost from quasimolecular ion (*m/z* 1,268) to give the base peaks at *m/z* 1,122 and 976, respectively. The ion at *m/z* 1,268 loses a dHex (*m/z* 1,122) as the major loss, a Hex (*m/z* 1,106) as a minor, and a HexNAc (*m/z* 1,065) also minor, suggesting that the three residues are non-reducing termini (**Fig. 2A**). The ion (*m/z* 1,122) loses a dHex (*m/z* 976), a Hex (*m/z* 960), and a HexNAc (*m/z* 919). The latter two ions subsequently lose a HexNAc and a Hex, respectively, to yield *m/z* 757. The resulting ion further loses the reducing end GalNAc-ol to yield *m/z* 534 $[1dHex + 1Hex + 1HexNAc + Na]^+$. The fragmentation pattern suggests that the reducing terminus, GalNAc-ol, is linked to both HexNAc-Hex and Hex-HexNAc. The presence of the ion *m/z* 388 (**Fig. 2C**) corresponds to a $[HexNAc\text{-}Hex + Na]^+$ species while the absence of *m/z* 347 and *m/z* 429 which would correspond to a $[Hex\text{-}Hex + Na]^+$ and a $[HexNAc\text{-}HexNAc + Na]^+$, respectively, confirm the structural assignment. The position of the dHex is likely to be on an internal residue as the fragment ions indicate the presence of terminal Hex and HexNAc. In addition, dHex is almost never found on the core residue. Based on these considerations, the structure for the *m/z* 1,268 is proposed as shown in the inset of **Fig. 2C.**

As shown in **Fig. 2**, *O*-glycan alditols fragment to yield only B ions and Y ions under the SORI-CID conditions (*see* **Note 4**). This observation has been confirmed by extensive CID studies of many *O*-linked oligosaccharides *(45)*. The final residue is often the alditol core (*m/z* 246), which suggests that the reduced core or the alditol binds the most strongly to the sodium ion.

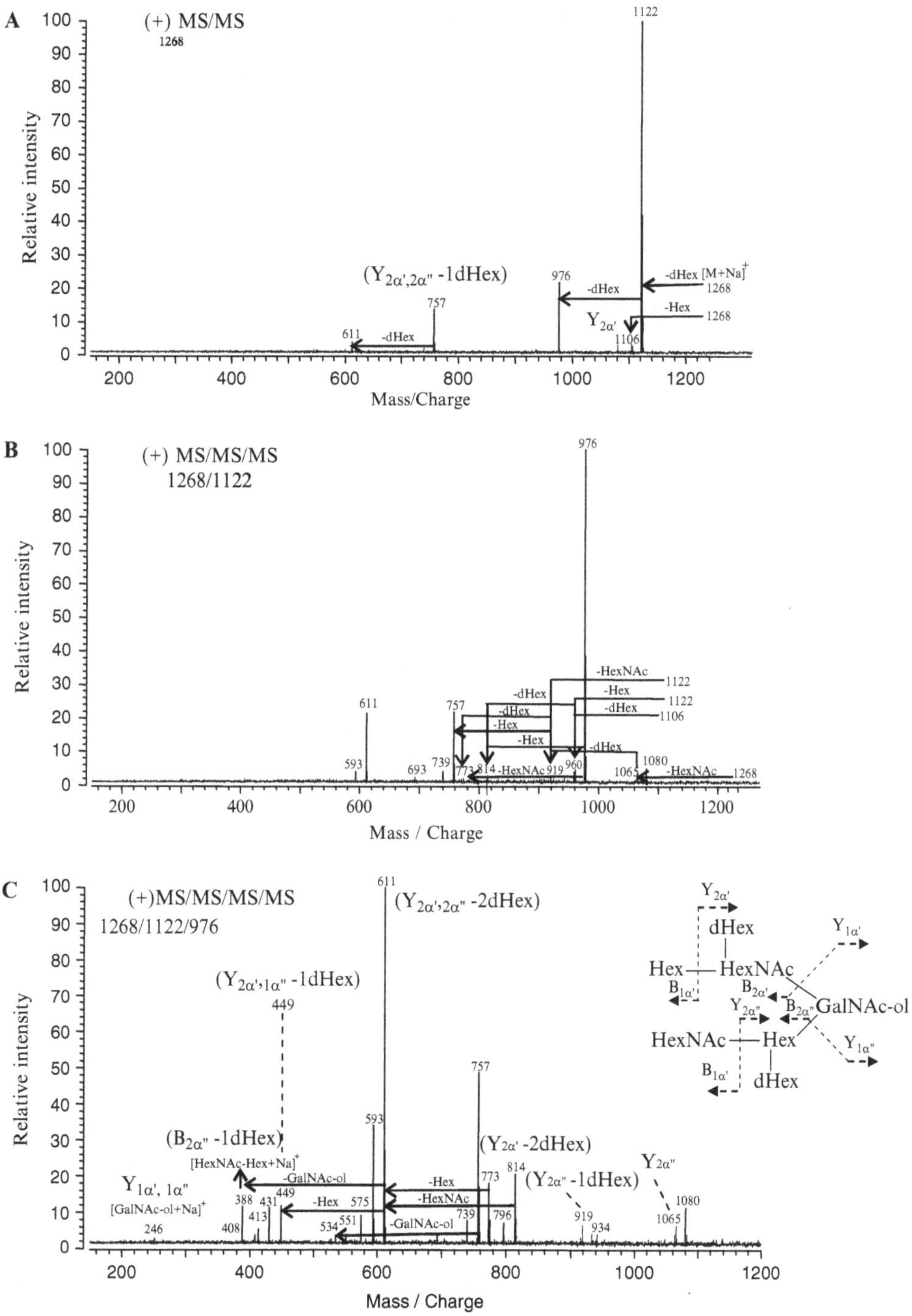

Fig. 2. MALDI SORI-CID tandem mass spectra of an *O*-linked glycan, *m/z* 1,268 ([M + Na]$^+$), from egg jelly glycoprotein of *Xenopus tropicalis* in positive mode. (**A**) MS^2 spectrum of the precursor ion, *m/z* 1,268. (**B**) MS^3 spectrum of a product ion (*m/z* 1,122) from MS^2. (**C**) MS^4 spectrum of a product ion (*m/z* 976) from MS^3. Based on the fragmentation patterns of the tandem mass spectra, the glycan structure of the quasimolecular ion, *m/z* 1,268, has been elucidated and shown in inset of the figure (systematic nomenclatures of some fragments are also included).

The second example of a CID analysis of an *O*-linked oligosaccharide is shown in **Fig. 3**. An oligosaccharide was found with *m/z* 1,065.405 ([M + Na]+) consisting of 2 dHex, 2 Hex, and 2 HexNAc. The MS^2 spectrum is shown in **Fig. 3A**. The quasimolecular ion loses one dHex (*m/z* 919, the base peak) and one Hex (*m/z* 903) indicating that a dHex and a Hex are located at the non-reducing termini. The ion at *m/z* 919 further loses one dHex (*m/z* 773), one Hex (*m/z* 757) or one HexNAc (*m/z* 716). The loss of the HexNAc from *m/z* 919 indicates that the location of the first dHex is on the HexNAc. The simultaneous losses of Hex, HexNAc and dHex further indicate that the Hex and the dHex are bound to either the remaining Hex or the GalNAc-ol. However, it is seldom that a dHex is connected to the

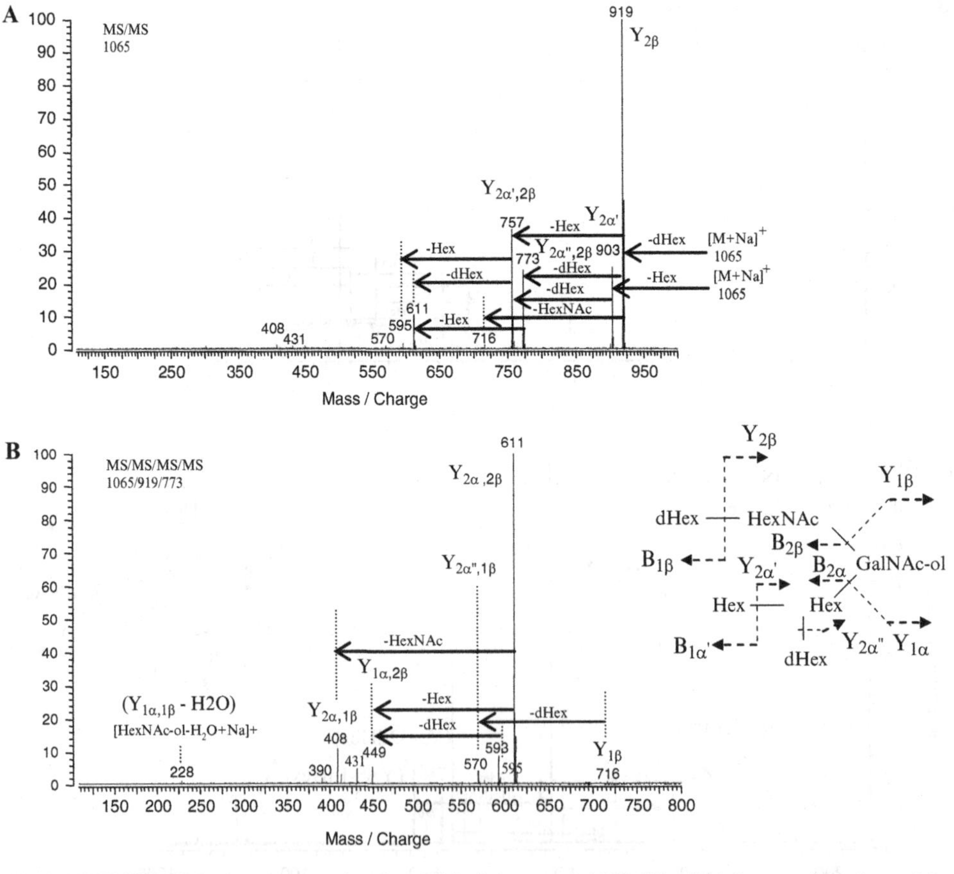

Fig. 3. MALDI SORI-CID tandem mass spectra of an *O*-linked glycan, *m/z* 1,065 ([M + Na]+), from egg jelly glycoprotein of *Xenopus laevis* in positive mode. (**A**) MS^2 spectrum of the precursor ion, *m/z* 1,065. (**B**) MS^4 spectrum of a product ion (*m/z* 773). Based on the fragmentation patterns of the tandem mass spectra, the glycan structure of the quasimolecular ion, *m/z* 1,065, has been elucidated and shown in inset of the figure (systematic nomenclatures of some fragments are also included).

core GalNAc-ol, therefore it follows that Hex and dHex are connected to the same Hex. The ion at *m/z* 611 corresponds to one HexNAc, one Hex and one GalNAc-ol, and can further fragment one Hex (*m/z* 449) to yield [HexNAc-GalNAc-ol + Na]$^+$ (**Fig. 3B**). It can also lose one HexNAc (*m/z* 408) to yield the fragment ion [Hex-GalNAc-ol + Na]$^+$ suggesting that the reducing end is branched. The structure assigned for this ion is shown in the inset of **Fig. 3B**, while the other peaks in the tandem MS spectra (**Fig. 3A, B**) confirm the assignment.

In both examples, glycosidic bond cleavages are abundant, while cross-ring cleavages are absent in the conditions used in the study.

4. Notes

1. Oligosaccharides were obtained from the lyophilized glycoproteins by the procedure described in earlier publications *(14–18)*.
2. To obtain clean and informative tandem mass spectra with high signal/noise, the glycan release reaction and sample preparations are crucial.

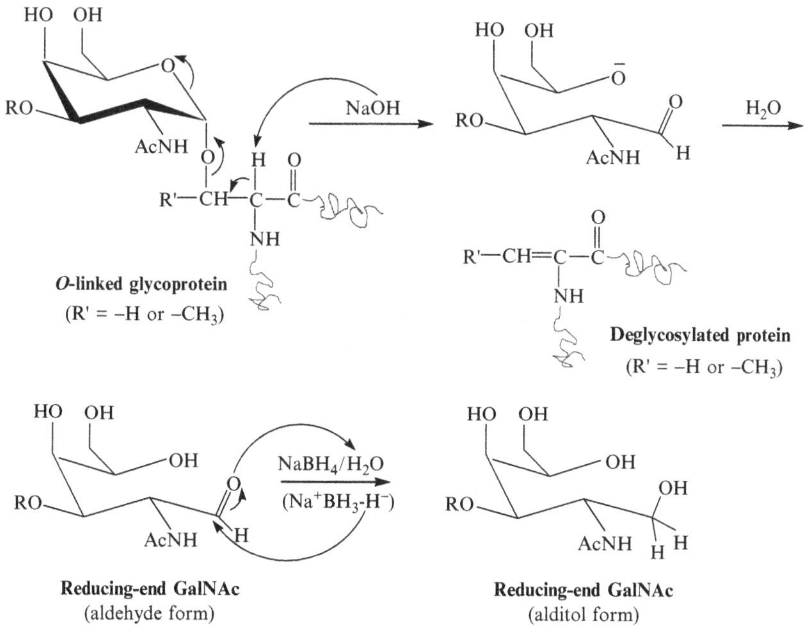

Scheme 1. The mechanism for reductive β-elimination under alkaline condition to release *O*-linked glycans from glycoprotein.  represents protein chain.

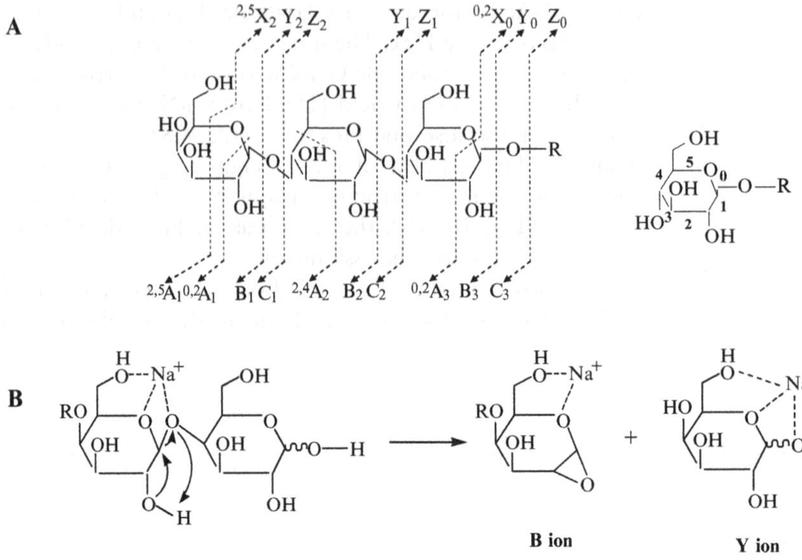

Scheme 2 **(A)**. A systematic nomenclature for carbohydrate fragmentations. **(B)** A general mechanism for B/Y ion formation in positive mode.

3. The mechanism for reductive β-elimination is shown in **Scheme 1**. First the acidic proton at α carbon of a serine or threonine residue is abstracted by existing base (OH^-) followed by the cleavage of C_β-O bond linking the glycan to the serine or threonine residue, thus removing the glycan portion from the β carbon of the specific residue in protein. The nascent sugar aldehyde is reduced by BH_4^- ion to its corresponding alditol. The alditol prevents the peeling reaction that otherwise further degrades the oligosaccharide chain. From the standpoint of the MS, the alditol has the further advantage of designating the reducing end.
4. The ion fragments of glycans follow a systematic nomenclature coined by Domon and Costello *(46)*. The mechanism for B/Y ion formation in the positive mode is shown in **Scheme 2** *(47, 48)*.

Acknowledgments

We gratefully acknowledge the financial support by the National Institutes of Health (R01 GM049077). We also wish to thank Dr. Jinhua Zhang for her kind help and useful suggestions.

References

1. Varki, A. (1993) Biological roles of oligosaccharides: all of the theories are correct. *Glycobiology* 3, 97–130.
2. Dwek, R. A. (1996) Glycobiology: toward understanding the function of sugars. *Chem. Rev.* 96, 683–720.
3. Kobata, A. (2000) A journey to the world of glycobiology. *Glycoconjugate J.* 17, 443–464.
4. Durand, G. and Seta, N. (2000) Protein glycosylation and diseases: blood and urinary oligosaccharides as markers for diagnosis and therapeutic monitoring. *Clin. Chem.* 46, 795–805.
5. Kim, Y. S., Gum, J. and Brockhausen, I. (1996) Mucin glycoproteins in neoplasia. *Glycoconjugate J.* 13, 693–707.
6. Brockhausen, I. and Kuhns, W. (1997) Glycoproteins and Human Disease, Medical Intelligence Unit, CRC Press and Mosby Year Book, Chapman and Hall, New York.
7. Kotera, Y., Fontenot, J., Pecher, G., Metzgar, R. S. and Finn, O. J. (1994) Humoral immunity against a tandem repeat epitope of human mucin MUC1 in sera from breast, pancreatic and colon cancer patients. *Cancer Res.* 54, 2856–2860.
8. Brockhausen, I. (1993) Clinical aspects of glycoprotein biosynthesis. *Crit. Rev. Clin. Lab. Sci.* 30, 65–151.
9. Brockhausen, I., Schutzbach, I. and Kuhns, W. (1998) Glycoproteins and their relationship to human disease. *Acta Anat.* 161, 36–78.
10. Kim, Y. and Varki, A. (1997) Perspectives on the significance of altered glycosylation of glycoproteins in cancer. *Glycoconjugate J.* 14, 569–576.
11. Alving, K., Paulsen, H. and Peter-Katalinic, J. (1999) Characterization of O-glycosylation sites in MUC2 glycopeptides by nanoelectrospray QTOF mass spectrometry. *J. Mass Spec.* 34, 395–407.
12. An, H. J., Ninonuevo, M., Aguilan, J., Liu, H., Lebrilla, C. B., Alvarenga, L. S. and Mannis, M. J. (2005) Glycomics analyses of tear fluid for the diagnostic detection of ocular Rosacea. *J. Proteome Res.* 4(6), 1981–1987.
13. An, H. J., Miyamoto, S., Lancaster, K. S., Kirmiz, C., Li, B., Lam, K. S., Leiserowitz, G. S. and Lebrilla, C. B. (2006) Global profiling of glycans in serum for the diagnosis of potential biomarkers for ovarian cancer. *J. Proteome Res.* (Web release at 05-27-2006).
14. Xie, Y., Tseng, K. and Lebrilla, C. B. (2001) Targeted use of exoglycosidase digestion for the structural elucidation of neutral *O*-linked oligosaccharides. *J. Am. Soc. Mass Spectrom.* 12, 877–884.
15. Xie, Y., Liu, J., Zhang, J., Hedrick, J. L. and Lebrilla, C. B. (2004) Method for the comparative glycomic analyses of o-linked, mucin-type oligosaccharides. *Anal. Chem.* 76, 5186–5197.
16. Zhang, J., Xie, Y., Hedrick, J. L. and Lebrilla, C. B. (2004) Profiling the morphological distribution of O-linked oligosaccharides. *Anal. Biochem.* 334, 20–35.
17. Zhang, J., Lindsay, L. L., Hedrick, J. L. and Lebrilla, C. B. (2004) Strategy for profiling and structure elucidation of mucin-type oligosaccharides by mass spectrometry. *Anal. Chem.* 76, 5990–6001.
18. Zhang, J., Schubothe, K., Li, B., Russell, S. and Lebrilla, C. B. (2005) Infrared multiphoton dissociation of *O*-linked mucin-type oligosaccharides. *Anal. Chem.* 77, 208–214.
19. Schulz, B. L., Packer, N. H. and Karlsson, N. G. (2002) Small-scale analysis of *O*-linked oligosaccharides from glycoproteins and mucins separated by gel electrophoresis. *Anal. Chem.* 74, 6088–6097.
20. Schulz, B. L., Oxley, D., Packer, N. H. and Karlsson, N. G. (2002) Identification of two highly sialylated human tear-fluid DMBT1 isoforms: the major high-molecular-mass glycoproteins in human tears. *Biochem. J.* 366, 511–520.
21. Orlando, R., Bush, C. A. and Fenselau, C. (1990) Structural analysis of oligosaccharides by tandem mass spectrometry: collisional activation of sodium adduct ions *Biomed. Environ. Mass Spectrom.* 19, 747–754.
22. Tseng, K., Lindsay, L. L., Penn, S., Hedrick, J. L. and Lebrilla, C. B. (1997) Characterization of neutral oligosaccharide-alditols from *Xenopus laevis* egg jelly coats by matrix-assisted laser desorption Fourier transform mass spectrometry. *Anal. Biochem.* 250, 18–28.
23. Weiskopf, A. S., Vouros, P. and Harvey, D. J. (1997) Characterization of oligosaccharide composition and structure by quadrupole ion trap mass spectrometry. *Rapid. Commun. Mass Spectrom.* 11, 1493–1504.
24. Konig, S. and Leary, J. A. (1998) Evidence for linkage position determination in cobalt coordinated pentasaccharides using ion trap mass spectrometry. *J. Am. Soc. Mass Spectrom.* 9, 1125–1134.
25. Viseux, N., de Hoffmann, E. and Domon, B. (1998) Structural assignment of permethylated oligosaccharide subunits using sequential

tandem mass spectrometry. *Anal. Chem.* 70, 4951–4959.
26. Tseng, K., Hedrick, J. L. and Lebrilla, C. B. (1999) Catalog-library approach for the rapid and sensitive structural elucidation of oligosaccharides. *Anal. Chem.* 71, 3747–3754.
27. Guan, S. and Marshall, A. G. (1996) Stored waveform inverse Fourier transform (SWIFT) ion excitation in trapped-ion mass spectrometry: Theory and applications. *Int J Mass Spectrom Ion Proc.* 157/158, 5–37.
28. Gauthier, J. W., Trautman, T. R. and Jacobson, D. B. (1991) Sustained off-resonance irradiation for collision-activated dissociation involving Fourier transform mass spectrometry. Collision-activated dissociation technique that emulates infrared multiphoton dissociation. *Anal. Chim. Acta* 246, 211–225.
29. Heck, A. J. R., Koning, L. J., Pinkse, F. A. and Nibbering, N. M. (1991) Mass-specific selection of ions in Fourier-transform ion cyclotron resonance mass spectrometry: Unintentional off-resonance cyclotron excitation of selected ions. *Rapid. Commun. Mass Spectrom.* 5, 406–414.
30. Schwartz, B. L., Bruce, J. E., Anderson, G. A., Hofstadler, S. A., Rockwood, A. L., Smith, R. D., Chilkoti, A. and Stayton, P. S. (1995) Dissociation of tetrameric ions of noncovalent streptavidin complexes formed by electrospray ionization. *J. Am. Soc. Mass Spectrom.* 6, 459–465.
31. Wu, Q. Y., Van Orden, S., Cheng, X. H., Bakhtiar, R. and Smith, R. D. (1995) Characterization of cytochrome c variants with high-resolution FTICR mass spectrometry: correlation of fragmentation and structure. *Anal. Chem.* 67, 2498–2509.
32. Little, D. P., Aaserud, D. J., Valaskovic, G. A. and McLafferty, F. W. (1996) Sequence information from 42–108-mer DNAs (complete for a 50-mer) by tandem mass spectrometry *J. Am. Chem. Soc.* 118, 9352–9359.
33. Solouki, T., Pasa-Tolic, L., Jackson, G. S., Guan, S. and Marshall, A. G. (1996) High-resolution multistage MS, MS2, and MS3 matrix-assisted laser desorption/ionization FT-ICR mass spectra of peptides from a single laser shot. *Anal. Chem.* 68, 3718–3725.
34. Kelleher, N. L., Nicewonger, R. B., Begley, T. P. and McLafferty, F. W. (1997) Identification of modification sites in large biomolecules by stable isotope labeling and tandem high resolution mass spectrometry. The active site nucleophile of thiaminase I. *J. Biol. Chem.* 272, 32215–32220.
35. Cancilla, M. T., Penn, S. G. and Lebrilla, C. B. (1998) Alkaline degradation of oligosaccharides coupled with matrix-assisted laser desorption/ionization Fourier transform mass spectrometry: a method for sequencing oligosaccharides. *Anal. Chem.* 70, 663–672.
36. Solouki, T., Reinhold, B. B., Costello, C. E., O'Malley, M., Guan, S. and Marshall, A. G. (1998) Electrospray ionization and matrix-assisted laser desorption/ionization Fourier transform ion cyclotron resonance mass spectrometry of permethylated oligosaccharides. *Anal. Chem.* 70, 857–864.
37. Kelleher, N. L., Taylor, S. V., Grannis, D., Kinsland, C., Chiu, H. J., Begley, T. P. and McLafferty, F. W. (1998) Efficient sequence analysis of the six gene products (7–74 kDa) from the *Escherichia coli* thiamin biosynthetic operon by tandem high-resolution mass spectrometry. *Protein Sci.* 7, 1796–1801.
38. Kelleher, N. L., Lin, H. Y., Valaskovic, G. A., Aaserud, D. J., Fridriksson, E. K. and McLafferty, F. W. (1999) Top down versus bottom up protein characterization by tandem high-resolution mass spectrometry. *J. Am. Chem. Soc.* 121, 806–812.
39. Maier, C. S., Yan, X., Harder, M. E., Schimerlik, M. I., Deinzer, M. L., Pasa-Tolic, L. and Smith, R. D. (2000) Electrospray ionization Fourier transform ion cyclotron resonance mass spectrometric analysis of the recombinant human macrophage colony stimulating factor β and derivatives *J. Am. Soc. Mass Spectrom.* 11, 237–243.
40. Cancilla, M. T., Wong, A. W., Voss, L. R. and Lebrilla, C. B. (1999) Fragmentation reactions in the mass spectrometry analysis of neutral oligosaccharides. *Anal. Chem.* 71, 3206–3218.
41. Gaucher, S. P., Cancilla, M. T., Phillips, N. J., Gibson, B. W. and Leary, J. A. (2000) Mass spectral characterization of lipooligosaccharides from Haemophilus influenzae 2019. *Biochemistry* 39, 12406–12414.
42. Penn, S. G., Cancilla, M. T. and Lebrilla, C. B. (2000) Fragmentation behavior of multiple-metal-coordinated acidic oligosaccharides studied by matrix-assisted laser desorption ionization Fourier transform mass spectrometry. *Int. J. Mass Spectrom.* 196, 259–269.
43. Flora, J. W., Hannis, J. C. and Muddiman, D. C. (2001) High-mass accuracy of product ions produced by SORI-CID using a dual electrospray ionization source coupled with FTICR mass spectrometry. *Anal. Chem.* 73, 1247–1251.
44. Strecker, G., Wieruszeski, J. M., Plancke, Y. and Boilly, B. (1995) Primary structure of 12 neutral oligosaccharide-alditols released from the jelly

coats of the anuran *Xenopus laevis* by reductive β-elimination. *Glycobiology* 5 (1), 137–146.

45. Lancaster, K. S., An, H. J., Li, B. and Lebrilla, B. C. (2006) Interrogation of *N*-linked oligosaccharides using infrared multi-photon dissociation in FT-ICR mass spectrometry. *Anal. Chem.* 78 (14), 4990–4997.
46. Domon, B. and Costello, C. E. (1988) A systematic nomenclature for carbohydrate fragmentations in FAB-MS/MS spectra of glycoconjugates. *Glycoconjugate* 5, 397–409.
47. Hofmeister, G. E., Zhou, Z. and Leary, J. A. (1991) Linkage position determination in lithium-cationized disaccharides: tandem mass spectrometry and semiempirical calculations. *J. Am. Chem. Soc.* 113, 5964–5970.
48. Spengler, B., Dolce, J. W. and Cotter, R. J. (1990) Infrared laser desorption mass spectrometry of oligosaccharides: fragmentation mechanisms and isomer analysis. *Anal. Chem.* 62, 1731–1737.

Chapter 11

The Structural Elucidation of Glycosaminoglycans

Vikas Prabhakar, Ishan Capila, and Ram Sasisekharan

Summary

There is accumulating evidence of the importance of linear polysaccharides in modulating biological phenomena in both the normal and the diseased states. This layer of regulation results from interactions between polysaccharides and other biomolecules, such as proteins, at the cell–extracellular matrix interface. The specific sequence of chemical modifications within the polymer backbone imparts a potential for interaction with other molecular species, and thus there exists important information within the various sulfation, acetylation, and epimerization states of such complex carbohydrates. A variety of factors have made the deciphering of this chemical code elusive. To this end, this report describes several techniques to elucidate the structural information inherent in glycosaminoglycan species. First, the use of depolymerizing enzymes that cleave polysaccharides at specific sites is described. Then, capillary electrophoretic (CE) techniques are employed to characterize the disaccharide species present in an enzymatically-cleaved polysaccharide sample. Mass spectrometry (MS) procedures can further be used to establish the length of an oligosaccharide chain and the presence of specific functional groups.

Key words: Glycosaminoglycans, Capillary electrophoresis, Mass spectrometry, Heparinase, Chondroitinase, Heparin, Heparan sulfate, Chondroitin, Dermatan.

1. Introduction

Glycosaminoglycans (GAGs) are linear acidic polysaccharides that are classified according to their structure into four distinct families: the heparin/heparan sulfate GAGs (HSGAGs), the chondroitin sulfate and dermatan sulfate galactosaminoglycans (GalAGs), keratan sulfate, and hyaluronic acid *(1)*. HSGAGs are biosynthesized as a linear chain of 20–100 disaccharide units of *N*-acetylglucosamine (GlcNAc) $\alpha(1\rightarrow4)$ linked to glucuronic acid (GlcA). Modifications to this basic disaccharide unit include

Nicolle H. Packer and Niclas G. Karlsson (eds.), *Methods in Molecular Biology, Glycomics: Methods and Protocols, vol. 534*

DOI: 10.1007/978-1-59745-022-5_11

O-sulfation at the 3-*O* and 6-*O* positions of the glucosamine and at the 2-*O* position of the uronic acid. The glucosamine can further be modified through *N*-deacetylation followed by *N*-sulfation. Epimerization at the *C*-5 position, resulting in a change from GlcA to iduronic acid (IdoA), can also occur *(2)*. All of the possible modifications relay 48 potential unique disaccharide units that can be found in an HSGAG chain *(3)*. GalAGs are similar to HSGAGs except that the hexosamine moiety is instead *N*-acetylgalactosamine (GalNAc). GalAG chains are biosynthesized as a disaccharide repeat unit of GlcA β(1→3) linked to GalNAc. These basic repeat units are in turn β(1→4) linked to form polymeric GalAG. As with HSGAGs, GalAGs can be variably modified through epimerization of the uronic acid moiety and/or sulfation. The three major classes of GalAG biomolecules are chondroitin sulfate A (chondroitin-4-sulfate), chondroitin sulfate B (dermatan sulfate), and chondroitin sulfate C (chondroitin-6-sulfate) *(4)*. Keratan sulfate has the basic disaccharide repeat unit of galactose β(1→4) linked to GlcNAc, and can be modified at the 6-*O* position of either moiety. Hyaluronic acid is the simplest of all GAG classes, comprised of GlcA β(1→3) linked to GlcNAc.

The study of GAGs is challenging for a variety of reasons *(5–8)*. Their integral isolation from biological sources is tedious. The low quantities that can be consolidated are not sufficient for many analytical platforms. The non-template-based biosynthesis of GAGs precludes the possibility of amplification, as is possible with DNA or proteins. The inherent complexity, heterogeneity, and highly anionic character of GAG chains serve as additional barriers to their examination. In order to overcome these obstacles, investigators have begun to develop and integrate various technologies to unravel the structural (and thus informational) milieu of GAGs pertinent to biology *(4)*. These include chemical and enzymatic tools with which to fragment specific GAG species; separation strategies for the analysis of these GAG fragments (capillary electrophoresis); and the characterization of the chemical groups inherent within a given oligosaccharide (mass spectrometry).

Capillary electrophoresis separates components under the influence of an electric field as they migrate through a capillary. The high resolution and sensitivity of this technique has made it a powerful analytical tool for the analysis of glycosaminoglycans *(9, 10)*. A current is applied across a fused silica capillary filled with electrolyte. The charge on the inner wall of the capillary (invariably negatively charged due to the presence of silanol groups) causes the flow of electrolyte from the anode to the cathode. This flow is called the electro-osmotic flow (EOF) and it is primarily responsible for migration of analytes through the capillary to the detector. Electrophoresis also plays a role in this separation as the electrophoretic force enhances the EOF for cations and opposes it for anions. The pH of the electrolyte controls the balance between the EOF and the force of electrophoresis. The electrolyte used for the analysis of glycosaminoglycans typically has a pH around 2.5. At this pH the capillary wall is uncharged – therefore, the primary force governing the separation is electrophoresis. Typically for the resolution of GAGs by CE the sample is applied at the cathode and detected at the anode and this mode of separation is called the reverse-polarity mode. The resolution achieved by CE is dependent mainly on the charge, mass, and molecular mobility of the analytes present. Detection is typically by ultraviolet (UV) absorbance or fluorescence emission. The procedure described in section 3.2 outlines CE analysis of a mixture of glycosaminoglycan building blocks (disaccharides and tetrasaccharides) generated by enzymatic digestion of the GAG.

Matrix-assisted laser desorption ionization mass spectrometry (MALDI-MS) can be used to identify saccharide intermediates generated upon cleavage of oligosaccharides by GAG-degrading enzymes. These analytical methodologies can be used to probe the underlying biology of GAG-protein interactions through the elucidation of GAG structural elements that are important to these phenomena *(11, 12)*.

2. Materials

2.1. Enzymatic Degradation of Glycosaminoglycans

1. A variety of GAG-depolymerizing lyases (heparinases I, II, and III and chondroitinases AC and B from *Pedobacter heparinus* and chondroitinase ABC I from *Proteus vulgaris*) are commercially available and some have been expressed in recombinant fashion *(13–18)*. Storage buffer and storage conditions for each enzyme are as directed by the supplier. For illustration, we will use chondroitinase ABC I (Seikagaku-America, East Falmouth, MA).
2. Chondroitin-6-sulfate (Seikagaku-America, East Falmouth, MA).
3. 37°C water bath.

2.2. Capillary Electrophoresis: Compositional Analysis

1. Fused silica capillaries from Hewlett-Packard (i.d. 75 μm, o.d. 363 μm, $l_{det.}$ 72.1 cm, and $l_{tot.}$ 80.5 cm).
2. Agilent ^{3D}CE (capillary electrophoresis) unit coupled with an extended path detection cell.
3. *Electrolyte solution.* 10 μM dextran sulfate and 50 mM Tris/phosphoric acid, pH 2.5.
4. Biovials (100 μl) composed of low protein adsorption CZ resin (Scientific Resources, Eatontown, NJ).

2.3. Mass Spectrometry

1. A saturated solution of caffeic acid (12 mg/ml in 30% acetonitrile/water).
2. $(Arg\text{-}Gly)_{15}Arg$ ($[RG_{15}]R$) and sucrose octasulfate for calibration.
3. Mass spectra sample plate with stainless steel flat targets (Perseptive Biosystems).
4. Glass microscope slide (Gold Seal Products, VWR, West Chester, PA).
5. Methanol.
6. Ultrasonic bath.
7. Applied Biosystems Voyager time-of-flight instrument fitted with a 337-nm laser.

3. Methods

The methods described here provide the pertinent steps needed to analyze a GAG sample once isolated from a biological source. Such a GAG sample can first be systematically fragmented through chemical/enzymatic means (**Table 1**) *(2, 4, 19–21).*

Table 1
GAG-depolymerizing enzymes

Enzyme class	Enzyme	Specificity
Heparinase	I	HNx6x(α1,4)-IdoA2S
	II	HNy6x(α1,4)-U2x
	III	HNy6x(α1,4)-GlcA & HNAc(α1,4)-IdoA
Chondroitinase	AC	GalNAc4x6x(β1,4)-GlcA2x
	B	-GalNAc4x6x(β1,4)-IdoA2x-
	ABC I	-GalNAc4x6x(β1,4)-U2x-
	ABC II	GalNAc4x6x(β1,4)-U2x-
	C	-GalNAc6x(β1,4)-GlcA-
Keratanase	I	HNAc6x-Gal(β1,4)-HNAc6x-
	II	HNAc6S(β1,3)-Gal6x-HNAc6S
Hyaluronate Lyase		-HNAc(β1,4)-GlcA-

x Sulfated or unsubstituted; *S* Sulfated; *y* Acetylated or sulfated; *Ac* Acetylated; *Gal* Galactose; *H* Hexosamine; *U* Uronic acid of either the GlcA or IdoA epimer state. Heparinases I, II, and III are from *Pedobacter heparinus.* Chondroitinases AC, B, and C are also from *Pedobacter heparinus.* Hyaluronate lyases are available from a variety of sources including *Peptostreptococcus, Staphylococcus, Streptococcus,* and *Apis mellifera*, among others. Keratanases mediate the depolymerization of keratan sulfate and are not lyases, but rather hydrolases

The resulting species can then be examined on several analytical platforms. Capillary electrophoresis can be used to determine which disaccharide or tetrasaccharide species are present in a fragmented GAG population *(22, 23)*. Mass spectrometry can be employed to determine the number and type of sulfate groups in a given GAG oligosaccharide and the length of the oligosaccharide chain.

3.1. Enzymatic Degradation of Glycosaminoglycans

As an example of GAG depolymerization via enzymatic digestion, chondroitinase ABC I is described. Other enzymes for depolymerizing GAGs are also available (as previously discussed in this chapter and in **Table** 1, Chapter "Small-Scale Enzymatic Digestion of Glycoproteins and Proteoglycans for Analysis of Oligosaccharides by LC-MS and FACE Gel Electrophoresis").

1. Chondroitinase ABC I will be placed in reaction with chondroitin-6-sulfate. A solution of 100 μl of 100 μg/ml chondroitin-6-sulfate in 50 mM Tris–HCl, 50 mM sodium acetate, pH 8.0 is mixed with 1 μg chondroitinase ABC I enzyme.
2. After gentle mixing, the reaction mixture is placed in a 37°C water bath (optimal temperature for chondroitinase ABC I-mediated catalysis) for 18 h.
3. The reaction vial can then be stored at 4°C until use.

3.2. Capillary Electrophoresis

These instructions assume use of an Agilent ^{3D}CE unit with uncoated fused silica capillaries (i.d. 75 μm, o.d. 363 μm, $l_{det.}$ 72.1 cm, and $l_{tot.}$ 80.5 cm) coupled with an extended path detection cell.

1. For CE analysis, 15 μl of the reaction mixture is placed in a 100 μl biovial from Scientific Resources (Eatontown, NJ) made of low protein adsorption CZ resin.
2. Analytes are monitored by using UV detection at 232 nm using an electrolyte solution of 10 μM dextran sulfate and 50 mM Tris/phosphoric acid, pH 2.5 (*see* **Note 1**).
3. Electrophoretic separation is performed using reverse polarity at a voltage of −30 kV (*see* **Note 2**).
   (a) Newly installed fused silica capillaries should be prepared by repeated rinsing with 0.1 N sodium hydroxide, water, and 0.1N hydrochloric acid (10 min each, 920 mbar pressure) prior to use. Wash once more for 10 min with distilled water. Finally, wash for 10 min with running buffer. In addition, capillaries are flushed with buffer for 3 min (940 mbar pressure) prior to each separation.
4. Pressure injection at 30 mbar for 20 s is used to inject samples (*see* **Note 3**). When enzymatic digests are being monitored by CE, injection of deionized water for 20 s at 10 mbar should precede sample injection. *See* **Notes 4** and **5** for additional information about capillary electrophoretic experimentation.

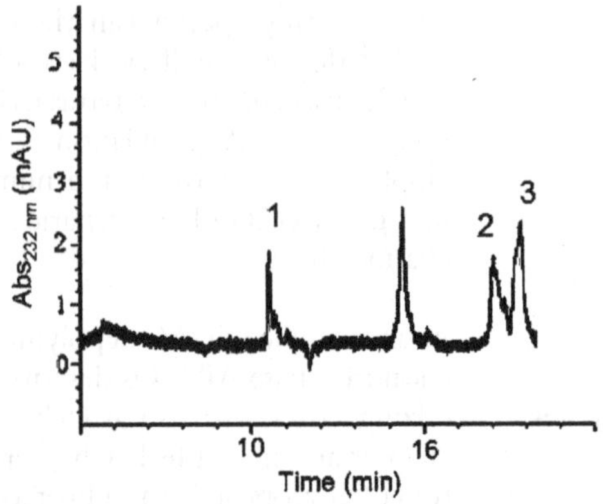

Fig. 1. *Capillary electrophoresis.* A representative capillary electrophoretogram showing the product profile of a fragmented oligosaccharide. (1) $\Delta UH_{NS, 6S}$; (2)ΔUH_{NS}; (3)$\Delta UH_{NAc, 6S}$.

Figure 1 is a representative capillary electrophoretogram depicting the disaccharides present in an enzymatic digest with heparinases I and III.

3.3. MALDI-MS

For an overview of the MALDI-MS procedure, *see* **Note 9** and **Fig. 2**.

1. The Voyager MALDI-TOF reflectron mass spectrometer from Applied Biosystems (Framingham, MA) is used. The instrument can be modified to allow for the acquisition of mass spectra by delayed extraction *(24)* (see **Note 6**).
2. Sample spots are irradiated using the N_2 laser at a repetition rate of 4 Hz. In static mode (continuous extraction), the resulting ions were immediately accelerated to 25 kV kinetic energy. Eighty-five percent of the voltage is applied between the sample stage and the grid. The guide wire is set at 0.1% of the acceleration voltage. During delayed extraction, ions are accelerated to 22 kV kinetic energy with a delay time of 150 ns and the guide wire is set at 0.15%. Ninety-three percent of the voltage is applied between the sample stage and the grid of the acceleration voltage. Low mass ions (matrix ions) are deflected by setting the low mass gate at *m/z* 1,000. Each spectrum is the average of 128 laser shots (range established by experience). The sample plate is continuously moved during acquisition for samples that have been prepared as microcrystalline layers.
3. *Matrix preparation.* Caffeic acid is used as matrix to prepare GAG oligosaccharide samples. A saturated solution of caffeic acid (12 mg/ml) in 30% acetonitrile/70% water is used (*see* **Note 7**).

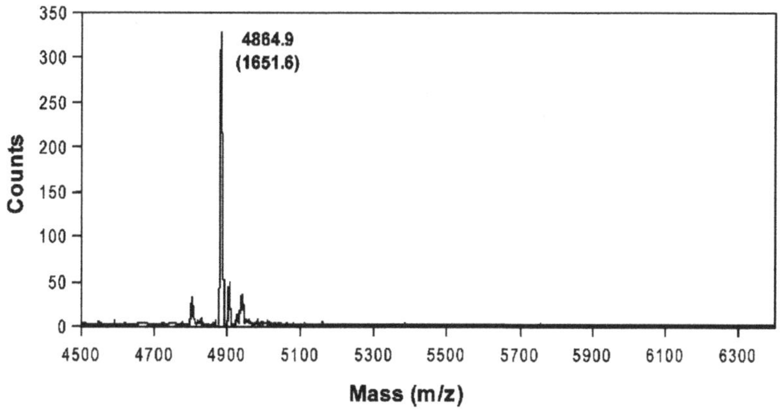

Fig. 2. *Mass spectrometry.* MALDI-MS of an oligosaccharide (shown here as a complex with RG peptide). The actual saccharide mass is shown in parentheses.

4. *Sample plate preparation.* Flat stainless steel targets (Applied Biosystems) are used for depositing samples. The targets are washed with methanol to remove any traces of the previous sample material. The plates are rinsed with water and then ultrasonicated while submerged in an acetonitrile bath. Following ultrasonication, the plates are placed vertically for efficient drying (*see* **Note 8**). Traces of methanol or water on the steel surface will inhibit the seeding of polycrystalline layers.
5. A saturated solution of caffeic acid is then deposited on the cleaned surface, with 1 μl per spot.
6. After the solvent has completely evaporated, a glass microscope slide (Gold Seal Products) is pressed against the steel surface and moved in a circular motion, creating a thin brownish film on the surface. Any loose material should be gently brushed off.
7. *Sample preparation.* Solutions of basic peptide (20–40 μM) are prepared in water. Counterions are removed by the addition of a few beads of anion exchanger AG 1-X2 (Bio-Rad Laboratories). Supernatant is diluted 1:9 with saturated caffeic acid. A volume of 9 μl peptide/matrix mixture is added to 1 μl of sample (10–40 μM oligosaccharide). A 1 μl aliquot of the latter preparation is placed on the seeded sample stage. Polycrystalline layers will form instantaneously. Excess liquid is removed after 1 min by rinsing the sample stage with deionized water. Unused parts of the sample stage must freshly be seeded with caffeic acid after every washing step in order to ensure proper formation of the microcrystalline layers. Samples are therefore deposited in batches of 2–40 prior to rinsing with water. Samples with high salt or glycerol content are

applied without prior cleanup. A twofold excess of peptide is required to obtain high sensitivity.

8. *Calibration.* Spectra are calibrated externally using sucrose octasulfate or internally using $(RG)_{15}R$.
9. MALDI-MS spectra are acquired in the linear mode on an instrument fitted with a 337-nm laser.

4. Notes

1. Dextran sulfate is added to the buffer to suppress non-specific interactions with the silica wall. This buffer is prepared by dissolving 10 mg dextran sulfate (M_{av} = 10,000) and 605 mg Tris in 80 mL deionized water. The solution is then titrated to pH 2.5 with ca. 13 mL 0.5 N phosphoric acid and diluted to 100 mL.
2. Separations are carried out at 30 kV with the anode at the detector side (reverse polarity).
3. It is crucial to carefully place the analyte at the bottom of the vial with no air bubbles. A minimum volume of 15 μl is required for the procedure.
4. The experiment should run at a constant voltage of –30 kV and a constant current between 55 and 65 μA at all times. Abort the procedure if the trial runs outside of these parameters.
5. Sample levels of the inlet vial and outlet vial should be monitored. The inlet should have a filled volume of about 2/3 and the outlet about 1/3.
6. *Overview of MALDI-MS procedures.* A saturated solution of caffeic acid (~12mg/ml in 30% acetonitrile/water) is used as the matrix solution. Seeded surfaces are prepared in an analogous fashion according to previous reports *(25, 26)*. A twofold molar excess of $(Arg\text{-}Gly)_{15}Arg$ is pre-mixed with the matrix prior to addition of the oligosaccharide sample solution. A 1 μl aliquot of the sample/matrix mixture is deposited on the seeded stainless steel surface. Following formation of the polycrystalline layer, excess liquid is removed by rinsing with deionized water. MALDI-MS spectra are acquired in the linear mode using an Applied Biosystems Voyager reflectron time-of-flight instrument fitted with a 337-nm laser. Delayed extraction is used to increase resolution (22 kV, grid at 93%, guide wire at 0.15%, pulse delay 150 ns, low mass gate at 1,000, 128 shots averaged). Mass spectra are calibrated externally by using signals for protonated $(Arg\text{-}Gly)_{15}Arg$ and its complex with oligosaccharide (**Fig. 2**).

7. The protonated complexes of peptides and GAG oligosaccharides are outside the mass range that can be isotopically resolved. Therefore only the isotope averages are measured and conditions are optimized to maximize sensitivity.
8. The properties of caffeic acid differ depending on the commercial supplier. Our studies suggest that caffeic acid from Sigma provides the greatest sensitivity, superior to material from both Aldrich and Fluka. Some batches from Sigma showed abundant +98 peaks, owing to contamination with sodium sulfate. In this case, the material can be recrystallized in two steps. First, the caffeic acid is dissolved in pure methanol at room temperature and precipitated overnight at 4°C. The precipitate is filtered, washed with water, and then dried. The large yellow needles are then redissolved in an aqueous ammonium solution and precipitated with acetic acid. The initial brown precipitate, pH 5, was discarded and only the subsequent faint yellow precipitate, pH 4, is used. A yellow powder is obtained after drying the filtrate.
9. Departures from these cleaning procedures have been observed to reduce signal intensities.

Acknowledgements

Funding was provided by the National Institutes of Health Grant GM57073. Vikas Prabhakar was the recipient of the National Institutes of Health Biotechnology Training Fellowship (5-T32-GM08334).

References

1. Sasisekharan, R., Shriver, Z., Venkataraman, G. and Narayanasami, U. (2002) Roles of heparan-sulphate glycosaminoglycans in cancer. *Nat Rev Cancer* 2, 521–528.
2. Ernst, S., Langer, R., Cooney, C. L. and Sasisekharan, R. (1995) Enzymatic degradation of glycosaminoglycans. *Crit Rev Biochem Mol Biol* 30, 387–444.
3. Sasisekharan, R. and Venkataraman, G. (2000) Heparin and heparan sulfate: biosynthesis, structure and function. *Curr Opin Chem Biol* 4, 626–631.
4. Prabhakar, V. and Sasisekharan, V. (2006) Advances in Pharmacology, vol. 53, pp. 69–115, Elsevier, Amsterdam.
5. Sasisekharan, R., Raman, R. and Prabhakar, V. (2006) Glycomics approach to structure-function relationships of glycosaminoglycans. *Annu Rev Biomed Eng* 8, 181–231.
6. Venkataraman, G., Shriver, Z., Raman, R. and Sasisekharan, R. (1999) Sequencing complex polysaccharides. *Science* 286, 537–542.
7. Turnbull, J. E. (2001) Integral glycan sequencing of heparan sulfate and heparin saccharides. *Methods Mol Biol* 171, 129–139.
8. Turnbull, J. E., Hopwood, J. J. and Gallagher, J. T. (1999) A strategy for rapid sequencng of heparan sulfate and heparin saccharides. *Proc Natl Acad Sci U S A* 96, 2698–2703.

9. Mao, W., Thanawiroon, C. and Linhardt, R. J. (2002) Capillary electrophoresis for the analysis of glycosaminoglycans and glycosaminoglycan-derived oligosaccharides. *Biomed Chromatogr* 16, 77–94.
10. Desai, U. R., Wang, H., Ampofo, S. A. and Linhardt, R. J. (1993) Oligosaccharide composition of heparin and low-molecular-weight heparins by capillary electrophoresis. *Anal Biochem* 213, 120–127.
11. Saad, O. M. and Leary, J. A. (2003) Compositional analysis and quantification of heparin and heparan sulfate by electrospray ionization ion trap mass spectrometry. *Anal Chem* 75, 2985–2995.
12. Rhomberg, A. J., Ernst, S., Sasisekharan, R. and Biemann, K. (1998) Mass spectrometric and capillary electrophoretic investigation of the enzymatic degradation of heparin-like glycosaminoglycans. *Proc Natl Acad Sci U S A* 95, 4176–4181.
13. Prabhakar, V., Capila, I., Bosques, C. J., Pojasek, K. and Sasisekharan, R. (2005) Chondroitinase ABC I from *Proteus vulgaris*: cloning, recombinant expression and active site identification. *Biochem J* 386, 103–112.
14. Prabhakar, V., Raman, R., Capila, I., Bosques, C. J., Pojasek, K. and Sasisekharan, R. (2005) Biochemical characterization of the chondroitinase ABC I active site. *Biochem J* 390, 395–405.
15. Pojasek, K., Shriver, Z., Kiley, P., Venkataraman, G. and Sasisekharan, R. (2001) Recombinant expression, purification, and kinetic characteriation of chondroitinase AC and chondroitinase B from *Flavobacterium heparinum. Biochem Biophys Res Commun* 286, 343–351.
16. Godavarti, R., Davis, M., Venkataraman, G., Cooney, C., Langer, R. and Sasisekharan, R. (1996) Heparinase III from *Flavobacterium heparinum*: cloning and recombinant expression in *Escherichia coli. Biochem Biophys Res Commun* 225, 751–758.
17. Myette, J. R., Shriver, Z., Claycamp, C., McLean, M. W., Venkataraman, G. and Sasisekharan, R. (2003) The heparin/heparan sulfate 2-*O*-sulfatase from *Flavobacterium heparinum*. Molecular cloning, recombinant expression, and biochemical characterization. *J Biol Chem* 278, 12157–12166.
18. Myette, J. R., Shriver, Z., Kiziltepe, T., McLean, M. W., Venkataraman, G. and Sasisekharan, R. (2002) Molecular cloning of the heparin/heparan sulfate delta 4,5 unsaturated glycuronidase from *Flavobacterium heparinum*, its recombinant expression in *Escherichia coli*, and biochemical determination of its unique substrate specificity. *Biochemistry* 41, 7424–7434.
19. Hernaiz, M. J. and Linhardt, R. J. (2001) Degradation of chondroitin sulfate and dermatan sulfate with chondroitin lyases. *Methods Mol Biol* 171, 363–371.
20. LeBrun, L. A. and Linhardt, R. J. (2001) Degradation of heparan sulfate with heparin lyases. *Methods Mol Biol* 171, 353–361.
21. Jandik, K. A., Gu, K. and Linhardt, R. J. (1994) Action pattern of polysaccharide lyases on glycosaminoglycans. *Glycobiology* 4, 289–296.
22. Linhardt, R. J. and Toida, T. (2002) Tech. Sight. Capillary electrophoresis. Ultra-high resolution separation comes of age. *Science* 298, 1441–1442.
23. Calabro, A., Midura, R., Wang, A., West, L., Plaas, A. and Hascall, V. C. (2001) Fluorophore-assisted carbohydrate electrophoresis (FACE) of glycosaminoglycans. *Osteoarthr Cartil* 9 Suppl A, S16–22.
24. Juhasz, P., Roskey, M. T., Smirnov, I. P., Haff, L. A., Vestal, M. L. and Martin, S. A. (1996) Applications of delayed extraction matrix-assisted laser desorption ionization time-of-flight mass spectrometry to oligonucleotide analysis. *Anal Chem* 68, 941–946.
25. Xiang, P. and Lin, P. (2006) Comment on 'Solid sampling technique for direct detection of condensed tannins in bark by matrix-assisted laser desorption/ionization mass spectrometry', Rapid Commun. *Mass Spectrom.* 2005; 19: 706. *Rapid Commun Mass Spectrom* 20, 521; author reply 522.
26. Beavis, R. C. and Chait, B. T. (1996) Matrix-assisted laser desorption ionization mass-spectrometry of proteins. *Methods Enzymol* 270, 519–551.

Chapter 12

Labelling Heparan Sulphate Saccharides with Chromophore, Fluorescence and Mass Tags for HPLC and MS Separations

Mark Skidmore, Abdel Atrih, Ed Yates, and Jeremy E. Turnbull

Summary

The analysis of heparin/HS saccharides derived from small amounts of tissue or cells is a considerable technical challenge and the development of methods to characterise these carbohydrates has progressed comparatively slowly. A number of procedures have been devised to tag glycans selectively at the reducing end with a group that will enhance the sensitivity of detection and facilitate chromatographic separations. Outlined in this chapter are two useful strategies designed specifically for the analysis of heparin/HS saccharides. The first involves a fluorophore label, Bodipy-FL-hydrazide, which permits highly sensitive (fmol level) detection of saccharides utilising high performance strong anion exchange chromatography. The second facilitates oligosaccharide separation by gel-permeation chromatography and reverse phase high performance ion-pairing chromatography (RP-HPIPC) through the use of a phenylsemicarbazide tag. The latter also serves as an effective mass tag for electrospray mass spectrometry, permitting enhanced analysis of HS saccharides. These methods provide new opportunities for the development of glycomics approaches to study the structure and function of the heparan sulfate family of glycans.

Key words: Heparin, Heparan sulfate, Fluorescence, Disaccharide analysis, High performance strong anion exchange chromatography, Reverse phase high performance liquid chromatography, Mass spectrometry, Electro-spray, Mass tag.

1. Introduction

Heparin and heparan sulfate (HS) are both members of the glycosaminoglycan family of polysaccharides which are known to mediate numerous biological processes and protein interactions. Both polysaccharides are involved in a variety of clinical conditions along with pathogen host cell invasion by many micro-organisms *(1)*. Furthermore, heparin is a widely used pharmaceutical anticoagulant *(2)*. Heparin and HS are

Nicolle H. Packer and Niclas G. Karlsson (eds.), *Methods in Molecular Biology, Glycomics: Methods and Protocols, vol. 534*

DOI: 10.1007/978-1-59745-022-5_12

both linear, acidic polysaccharides composed of the same repeating disaccharide backbone. The backbone constituents are a uronic acid (either glucuronic acid or its C_5 epimer, iduronic acid) residue and a glucosamine residue (present as the N-sulfated, N-acetylated or free amine form). O sulfation can occur at C_6 and more rarely C_3 of the glucosamine and/or C_2 of the iduronic acid residue. The presence of sulfate and/or acetyl groups in addition to the uronic acid epimer gives a theoretical possibility of 32 disaccharide building blocks, although the occurrence of all structures in nature is debatable, if not unlikely *(3, 4)*. The presence of this array of disaccharide constituents leads to high heterogeneity of these glycosaminoglycans, especially in the case of HS. Improved understanding of the structure-function relationships of the interacting oligosaccharides will help to fully appreciate their biological roles and realise their pharmaceutical and/or therapeutic potential.

The analysis of heparin/HS fragments is by no means trivial. It is hampered by its sensitivity to routine chemical procedures (being highly sensitive to extremes of pH *(5–8)* (*see* **Note 1**), solubility only in polar solvents (unless ion-paired) *(9)*, and the low abundance of tissue derived HS. The development of methods to characterise these carbohydrates has progressed comparatively slowly compared to that for proteins and nucleic acids. The hydrophilic nature of these glycans, coupled with the fact that they do not possess a suitable chromophore, has also meant that they are not readily amenable to separation techniques such as HPLC *(10, 11)*. A number of procedures have been devised to tag glycans selectively at the reducing end with a group that will enhance the sensitivity of detection, many of which also facilitate chromatographic separations *(10)*. The methodology is however, better adapted for neutral glycans. Indeed, these proceduces require low pH and heating, which can result in the loss of sulphates in glycosaminoglycans *(9, 10)*. A few methods have been developed for derivatization of glycosaminoglycans disaccharides producing fluoresecence and UV derivatives detectable at nanogram levels *(12, 13)*, but not specifically for those produced by nitrous acid digestion. Outlined in this chapter are two useful strategies for the analysis of heparin/HS saccharides. The first involves a fluorophore label (Bodipy-FL-hydrazide) which permits highly sensitive detection of disaccharides utilising high performance strong anion exchange chromato-graphy *(14)*. The second facilitates oligosaccharide separation by gel-permeation chromatography and reverse phase high performance ion-pairing chromatography (RP-HPIPC) through the use of a phenylsemicarbazide tag. The latter also serves as an effective mass tag for electrospray mass spectrometry, permitting enhanced analysis of HS saccharides. These methods provide new opportunities for the development of glycomics approaches to study the structure and function of the heparan sulfate family of glycans.

2. Materials

2.1. Fluorescent Tag Labelling with Bodipy-FL-Hydrazide

2.1.1. Enzymatic Digestion of Heparin/Heparan Sulfate Polysaccharides

1. *Lyase buffer (5×).* 1.25 M Sodium acetate, 12.5 mM Calcium acetate, pH 7.0. Store at −20°C (*see* **Note 2**).
2. Heparitinase I, II and III: Ibex (Canada). Store at −80°C
3. PD-10 Column (Amersham Biosciences)

2.1.2. Fluorescent Labelling of Heparin/Heparan Sulfate Disaccharides

1. 4,4-difluoro-5, 7-dimethyl-4-bora-3a, 4a-diaza-*s*-indacene-3-propionic acid hydrazide (Bodipy-FL-hydrazide, Molecular Probes): Dissolved in methanol at a concentration of 1.25 mg/ml. Store at −20°C.
2. Dimethyl sulfoxide (DMSO).
3. Glacial acetic acid.
4. *Labelling solution.* Dimethyl sulfoxide: glacial acetic acid (17:3, v/v). Store at 4°C.
5. *Reducing agent.* 1 M $NaBH_4$ (*see* **Note 3**).
6. *Reconstitution solution.* DMSO: water (1:1, v/v). Store at 4°C.

2.1.3. High Performance Strong Anion Exchange Chromatography

1. Gradient HPLC system (capable of mixing at least two solutions) (*see* **Note 4**).
2. In-line fluorescence detector (Excitation λ = 488 nm, emission λ = 520 nm).
3. Propac PA1 column (Dionex): Part No. 039658, 4 mm I/D × 250 mm length (*see* **Note 5**).
4. 1 M Hydrochloric acid: Filter through a 0.2 μm filter and degas using helium.
5. HPLC grade water, filter through a 0.2 μm filter and degas using helium.
6. *Sodium hydroxide (1 M).* Volumetric standard (Sigma), filter through a 0.2 μm filter and degas using helium (*see* **Note 6**).
7. *Sample loading solution.* 150 mM NaOH. Add 150 ml of 1 M NaOH to 850 ml of HPLC water. Filter through a 0.2 μm filter and degas using helium.
8. *Sodium chloride elution solution.* 2 M NaCl, 150 mM NaOH. Dissolve 116.8 g of HPLC grade sodium chloride in 150 ml of 1 M NaOH and make up to 1 l with HPLC grade water. Filter through a 0.2 μm filter and degas using helium.

9. *Column wash solution.* 1.2 M NaCl, 300 mM NaOH. Dissolve 70.08 g of HPLC grade sodium chloride in 300 ml of 1 M NaOH and make up to 1 l with HPLC grade water. Filter through a 0.2 μm filter and degas using helium (BOC).
10. Unsaturated disaccharide reference standards are available from Seikagaku Kogyo (Tokyo).

2.2. Chromophore and Mass Tag Labelling with Phenylsemicarbazide

2.2.1. Nitrous Acid Digestion of Heparin/Heparan Sulfate Polysaccharides

1. Porcine mucosal heparin (Celsus laboratories).
2. Sodium nitrite (Fluka).
3. Hydrochloric acid (BDH, Analar).
4. Sodium bicarbonate (Sigma).

2.2.2. Phenylsemicarbazide Labelling of Heparin/ Heparan Sulfate Disaccharides

1. *4-phenylsemicarbazide hydrochloride (Acros Organics).* Dissolve 93 mg of 4-phenylsemicarbazide in 500 μl of 50% ethanol, add 300 μl ammonium acetate from a stock solution of 100 mM, adjust pH to 4.5–5 with concentrated ammonia; make up to 1 ml with milliQ water and keep at room temperature before use.

2.2.3. Gel-Permeation Chromatography

1. FPLC system (Akta purifier, GE healthcare).
2. Superdex 30 prep grade column (100 × 2 cm × 30 mm).
3. *Elution solution.* Dissolve 39.5 g of ammonium bicarbonatein in 1,000 ml of milliQ water. Filter through a 0.2 μm cellulose acetate membrane (Fisher Scientific) and degas using helium.

2.2.4. High Performance Ion Pairing Reverse Phase Chromatography

1. Gradient HPLC system (capable of at least binary mixing).
2. UV detector (detection in UV at 242 nm).
3. C18 column (Supelco, 250 mm × 4 mm, 5 μm).
4. *Starting solution* 30% methanol, 10 mM dibutyl amine, 50 mM ammonium acetate. Dissolve 3.85 g of ammonium acetate in 700 ml milliQ water add 800 μl of dibutyl amine from a stock solution at 5.7 M, adjust the pH to 7 with acetic acid and make up to 1l with methanol *(5)*. Filter through a 0.2 μm nylon membrane and degas using helium.
5. *Elution solution.* 90% methanol, 10 mM dibutyl amine, 50 mM ammonium acetate. Dissolve 3.85 g of ammonim acetate in 100 ml of milliQ water, add 800 μl dibutyl amine, adjust the pH to 7 with acetic acid and *(6)* make up to 1,000 ml with methanol *(5)*. Filter through a 0.2 μm Nylon membrane and degas using helium.

3. Methods

3.1. Fluorescent Tag Labelling with Bodipy-FL-Hydrazide and Application to Disaccharide Analysis

The disaccharide analysis allows for the facile determination of the constituent disaccharides subunits present within an oligosaccharide along with their relative abundances *(15)* (*see* **Note 7**).

3.1.1. Preparation of Unsaturated Heparin/ Heparan Sulfate Disaccharides

This method relies on the total cleavage of the oligosaccharide chains into disaccharides. This is achieved by elimination using a cocktail of three bacterial derived lyase enzymes. Each enzyme has a particular cleavage site *(16)* as outlined in **Table1**. One drawback of enzymatic cleavage is the loss of information caused by the introduction of a C=C bond between C_4 and C_5 of the iduronic or glucuronic acid *(17)*. The resulting disaccharide will now contain a Δ-uronic acid monosaccharide (*see* **Note 8**). A list of the 8 common, naturally occurring Δ-disaccharides can be found in **Fig.1**. Heparin/heparin sulphated polysaccharides are commercially available or can be isolated from tissues *(18)*

1. Add 2.5 mU (10 μl aliquots) of heparitinase I, II and III per mg of sample to the polysaccharides.
2. Add 40 μl of 5× lyase buffer to the reaction mixture and make up to 1× with the addition of HPLC water to a total volume of 200 μl.
3. Incubate at 36°C for 12 h.
4. Terminate digestion by heating at 100°C for 5 min.
5. Equilibrate an Amersham PD-10 column with 30 ml of HPLC water.
6. Load sample on to the column.
7. Add 1.8 ml of HPLC water and allow breakthrough to pass to waste.
8. Add 3.5 ml of HPLC water and collect the breakthrough.
9. Lyophilize breakthrough prior to fluorophore conjugation/ analysis.

Table 1
Enzymes for depolymerising heparin and heparan sulfate

Enzyme (lyase)	Substrate specificity[a]
Heparitinase I (Heparinase III)	GlcNR(± 6S)α1-4GlcA
Heparitinase II (Heparinase II)	GlcNR(± 6S)α1-4GlcA/IdoA
Heparitinase III (Heparinase I)	GlcNS(± 6S)α1-4IdoA(2S)

[a]The specificities are shown as the glycosidic linkage specificity *(16)*

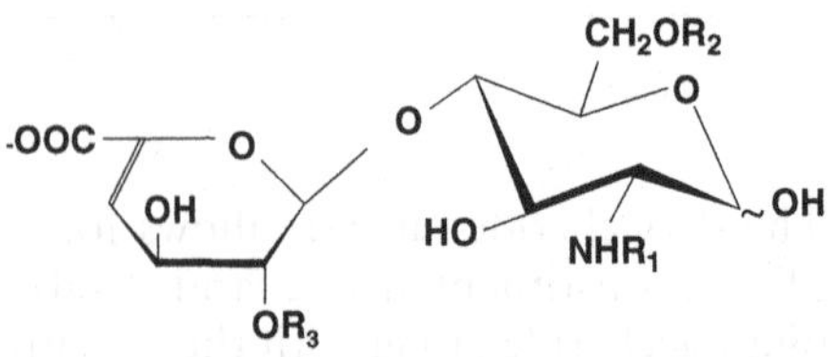

Sigma Ref.	Unit Formula	R1	R2	R3
IV-A	Δ - UA-GlcNA	Ac	H	H
II-A	Δ - UA-GlcNAc(6S)	Ac	Sulf	H
III-A	Δ - UA(2S)-GlcNAc	Ac	H	Sulf
I-A	Δ - UA(2S)-GlcNAc(6S)	Ac	Sulf	Sulf
IV-S	Δ - UA-GlcNS	Sulf	H	H
II-S	Δ - UA-GlcNS(6S)	Sulf	Sulf	H
III-S	Δ - UA(2S)-GlcNS	Sulf	H	Sulf
I-S	Δ - UA(2S)-GlcNS(6S)	Sulf	Sulf	Sulf

Fig. 1. Structure of the eight common naturally occurring Δ-disaccharides derived from digestion with a cocktail of bacterial lyase enymes. *Ac* Acetyl; *ΔUA* Unsaturated uronic acid residue; *GlcNAc N*-acetyl glucosamine; *GlcNS N*-sulfo glucosamine; *H* Proton; *2S and 6S* 2-*O*-sulfate and 6-*O*-sulfate respectively; *Sulf* Sulfate.

3.1.2. Derivatization of Unsaturated Heparin/Heparan Sulfate Disaccharides with the Fluorophore (Bodipy-FL-Hydrazide)

The coupling of the Δ-disaccharides (obtained from lyase digestion) with a fluorescent dye allows the sensitive detection of all eight common disaccharides. Derivatization proceeds via reductive amination, involving the reducing end carbonyl of the disaccharide and the hydrazide amine group present within the fluorophore. The resulting conjugate is stabilized by the reduction of the imine group (*see* **Note 9**).

1. Add 10 μl (per 10 nmol of disaccharide) of Bodipy-FL-hydrazide in methanol to the lyophilised sample.
2. Evaporate methanol by vacuum centrifugation at 36°C for 30 min.
3. Add 5 μl of the labelling solution to the disaccharide/fluorophore mixture.
4. Incubate the reaction mixture at room temperature for 4 h.
5. Add 5 μl of reducing agent to the reaction mixture.
6. Incubate the reduced reaction mixture at room temperature for 30 min.
7. Flash freeze the sample with liquid nitrogen.
8. Lyophilise the sample until dry.
9. Re-suspend the sample in 8 ul of reconstitution solution prior to analysis.

3.1.3. Separation of Bodipy Labelled Unsaturated Heparin/Heparan Sulfate Disaccharides

Prompt separation and detection of all eight fluorescently labelled Δ-disaccharides is possible using a linear sodium chloride gradient imposed over an isocratic sodium hydroxide solution mobile phase (*see* **Fig. 2**). Comparison of sample digests to that of derivatized commercially available known disaccharide standards provides a reference chromatogram for elution times (*see* **Note 10**).

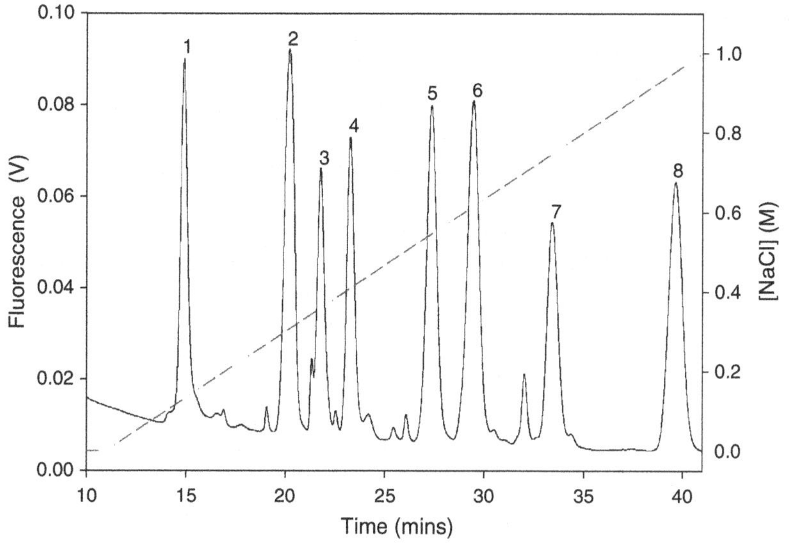

Fig. 2. Separation of heparin/heparan sulfate disaccharides on a Propac PA1 column. The profile shows separation of the eight major disaccharide subunit components of the parental polysaccharide post bacterial lyase enzyme digestion. The linear sodium chloride gradient is represented on the chromatogram by the *hashed line*. The structures of the standards resolved were (1) ΔUA-GlcNAc, (2) ΔUA-GlcNS, (3) ΔUA-GlcNAc(6S), (4) ΔUA(2S)-GlcNAc, (5) ΔUA-GlcNS(6S), (6) ΔUA(2S)-GlcNS, (7) ΔUA(2S)-GlcNAc(6S), (8) ΔUA(2S)-GlcNS(6S). *ΔUA* Unsaturated uronic acid residue produced at the non-reducing end post eliminative lyase cleavage; *GlcNAc N*-acetyl glucosamine; *GlcNS N*-sulfo glucosamine; *2S and 6S* 2-*O*-sulfate and 6-*O*-sulfate respectively.

1. Re-generate the Propac PA1 column by successive washes of:
   (a) 1 M hydrochloric acid for 60 min at a flow rate of 1 ml/min.
   (b) Water for 20 min at a flow rate of 1 ml/min.
   (c) 1 M sodium hydroxide for 60 min at a flow rate of 1 ml/min.
   (d) Water for 30 min at a flow rate of 1 ml/min.
2. Equilibrate the column in sample loading solution for 20 min at a flow rate of 2 ml/min.
3. Inject the sample and hold isocratic in sample loading solution for 11 min at a flow rate of 2 ml/min to allow the free tag to elute.
4. Elute the fluorescently labelled disaccharides with a linear gradient of sodium chloride elution solution (0–1 M over 30 min) at a flow rate of 2 ml/min.
5. Monitor eluted fluorescent disaccharides using inline fluorescent detection at λ_{exc} = 488 nm, λ_{emm} = 520 nm.
6. Wash the column in 100% (2 M) sodium chloride elution solution (2 min at a flow rate of 2 ml/min) and column wash solution (3 min at a flow rate of 2 ml/min) before re-equilibration.

3.2. Chromophore and Mass Tag Labelling with Phenylsemicarbazide

The chemical digestion of heparin/heparan sulphate with nitrous acid generates fragments which lack a chromophore, reducing the sensitivity of detection during separation. Labelling with phenylsemicarbazide provides a facile solution to this problem. In addition, it provides an effective mass tag for improved mass spectrometry of labelled saccharides.

3.2.1. Preparation of Nitrous Acid Heparin/Heparan Sulfate Fragments

Nitrous acid cleaves heparin chains at glucosamine residues where glucosamine is either N-unsubstituted or N-sulfated, and produces fragments with aldehyde containing reducing ends which can be labelled effectively with nucleophilic chemistries (*see* **Note 11**).

1. *Heparin (porcine mucosal heparin, Celsus Laboratories).* Dissolve 5 g in 5 ml milliQ water (Elga).
2. Prepare 4 ml of sodium nitrate (500 mM) and add 4 ml of 500 mM hydrochloric acid.
3. Add 5 ml of the mixture ($NaNO_2$, HCl) to 5 ml of heparin and incubate on ice for 20 min, 30 min and 40 min.
4. Take 2.5 ml aliquot and adjust the pH to 5 by adding 200 μl of 1 M sodium bicarbonate.
5. Store at –20°C prior to phenyl semicarbazide conjugation/analyis.

3.2.2. Derivatisation of Nitrous Acid Generated Fragments with Phenyl Semicarbazide

The coupling of the heparin/HS fragments with phenyl semicarbazide chromophore allows for the sensitive detection of all nitrous acid generated fragments, as well as those derived by heparin lyase cleavage (*see* **Note 12**).

1. Mix 200 μl of heparin digest with 1 ml of 4-phenylsemicarbazide.
2. Incubate the mixture at 37°C overnight.
3. Adjust the pH to 6 with 2 M ammonia and store at –20°C prior to separation.

3.2.3. Separation of Phenylsemicarbazide-Labelled Heparin Fragments by Gel Filtration

Labelled heparin fragments are fractionated and detected on a long Superdex column (200 cm × 30 mm). The separation on the gel filtration column allows the separation based on hydrodynamic volume using ammonium bicarbonate as eluting buffer.

1. Equilibrate the column in ammonium bicarbonate for 4 h at a flow rate of 0.5 ml/min.
2. Inject 1 ml (100 mg) sample and monitor eluted phenylsemicarbazide tagged fragments using UV detection at 242 nm.
3. Collect 1 ml fraction starting 2 h after injection of the sample and combine fractions of the resolved peaks.
4. Freeze-dry the different fractions and resuspend in 600 μl milliQ water prior to further purification on RP-HPIPC.

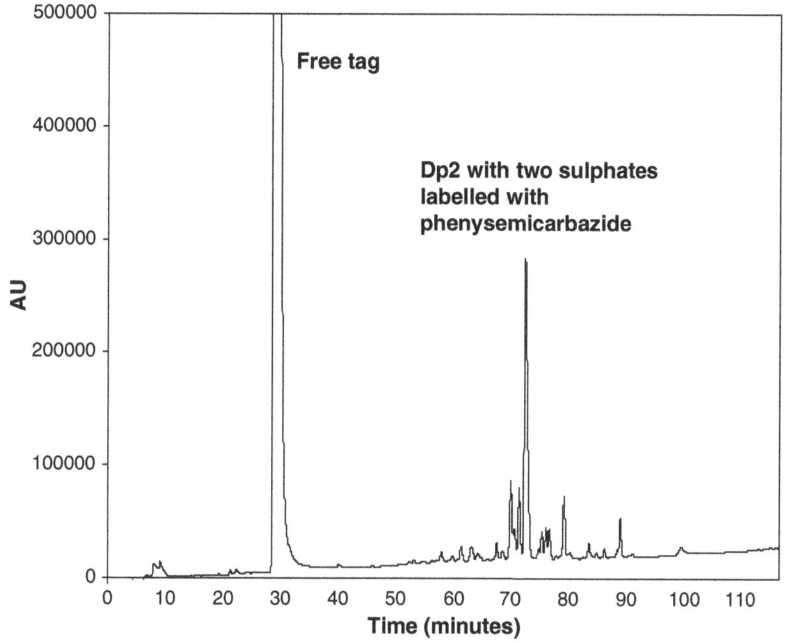

Fig. 3. Labelling and detection of DP2 (disaccharide with two sulphates generated with nitrous acid digestion) at 242 nm. The free tag is separated from the phenylsemicarbazide derivative by RP-IPHPLC as described in Subheading 3.

3.2.4. Further Separation of Phenyl Semicarbazide Heparin Fragments on RP-HPIPC

The heparin derived oligosaccharides fractionated on gel permeation column can be further separated on reverse phase high performance ion-pairing chromatography (RP-HPIPC) using a volatile ion-pairing reagent, dibutyl ammonium acetate (*see* **Fig.3** and **Note 13**).

1. Equilibrate the column in starting buffer (30% methanol, 10 mM dibutylamine, 50 mM ammonium acetate).
2. Inject the sample (sample diluted in loading buffer).
3. Elute the phenylsemicarbazide derivatives with a linear gradient of methanol (30%–90% over 300 min) at a flow rate of 0.7 ml/min.
4. Monitor eluted phenylsemicarbazide tagged fragments at 242 nm and collect individual peaks at the detector outlet.
5. Wash the column with 90% methanol.

3.2.5. Mass Spectrometry Analysis of Phenyl Semicarbazide Heparin Fragments

ESI mass spectra were obtained using a Q-tof Micro instrument (Micromass) equipped with a picoview Nano-electrospray source (*see* **Fig. 4**). Ionization was achieved using a spray voltage of 900 V, sample voltage 25 V, extraction cone at 0.7 V and collision energy at 3 V. Under these optimised conditions, ions were dissolved

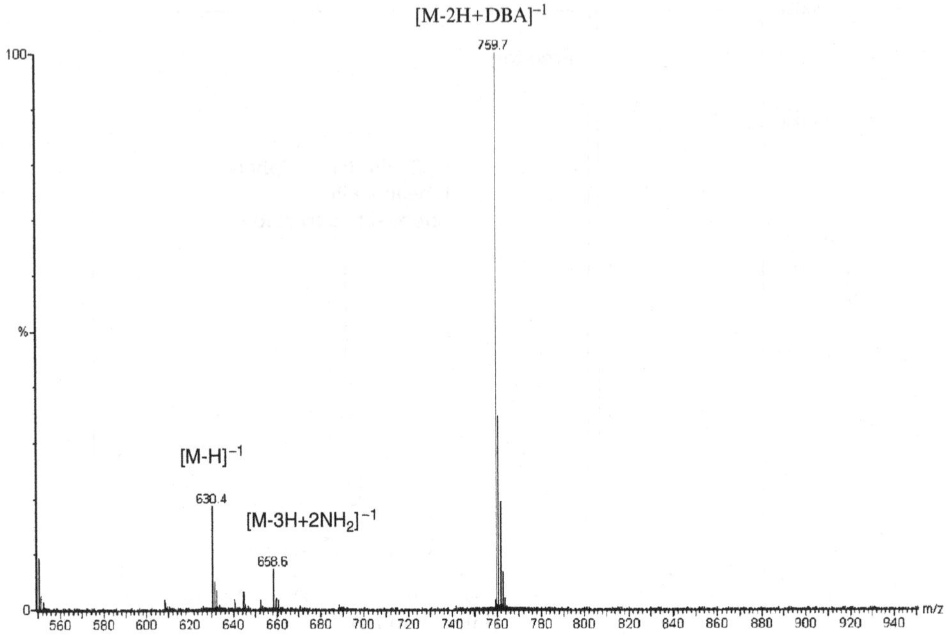

Fig. 4. Electrospray ionization mass spectrum of disaccharide with two sulphates plus tag (phenylsemicarbazide) (major peak eluting at 72 min from **Fig.1**).

without inducing losses of sulphate groups. Phenyl semicarbazide tagged fragments are used at a final concentration of 0.5–4 pmol/µl in 50% methanol.

4. Notes

1. Epoxide formation can occur between C_2 and C_3 of the 2-O sulfated iduronic acid residue in strong alkaline solutions or mildly alkaline conditions if heated (>pH 8, >50C), while conversion of GlcNAc to ManNAc has also been reported in mild basic conditions. De-N-sulfation can also occur in highly acidic environments (pH<4).
2. HPLC grade water was used in the preparation of all solutions (Fluka).
3. The reducing agent solution should be made fresh prior to use.
4. Although it is possible to use a dual inlet HPLC system, it is more convenient to use a four inlet setup in order to negate solution changing for the column regeneration 1.2 M sodium chloride, 300 mM sodium hydroxide wash step.

5. During the method development process it was found that the polymer based Dionex Propac PA1 column (originally engineered for protein separations) gave excellent separation of the eight disaccharides. Silica based SAX columns suffer from the binding of free fluorescent tag to the column, necessitating arduous and time consuming clean-up steps.
6. Sodium hydroxide solutions react with carbon dioxide present in the air to form carbonate salts (particularly found at the base of sodium hydroxide solution containers). These carbonate salts are strong counter ions when used with the Propac PA1 column, preventing the binding of low sulfated saccharides to the column upon sample loading and causing the elution of bound saccharides in an unresolved manner. It is prudent to minimise exposure to air (preferably under an inert gas) and refrain from the use of dregs in sodium hydroxide solutions).
7. Disaccharide analysis of Δ-disaccharides has previously been carried out using less sensitive ultraviolet detection. This is facilitated by means of the C=C bond, which is a chromaphore ($\lambda_{abs} = 232$ nm), introduced by bacterial lyase cleavage. Again, a Propac PA1 column can be used for separation and detection *(3)*.
8. Digestion is possible chemically using nitrous acid *(18)*. This method possesses the advantage of preserving the identity of the uronic acid residue, but suffers from two important drawbacks. Firstly, cleavage does not occur between an *N*-acetylated glucosamine residue and the subsequent uronic acid, and secondly the chromaphore introduced by lyase digestion (C=C bond) is not present. It should be noted that the reducing end glucosamines are converted to anhydromannitol during sample preparation (post reduction).
9. Conjugation relies on an unprotonated amine group being present on the fluorophore, while the reaction with the carbohydrate reducing end carbonyl is acid catalysed. In order to compensate for these contradictory requirements, an excess of free tag is used in an acidic environment. Higher yield reactions can be carried out with heating or stronger acids, but the possibility of de-N-sulfation needs to be considered. The reaction is thought to proceed more efficiently in an anhydrous reaction solution, therefore DMSO and glacial acetic acid storage needs to be considered appropriately.
10. Elution times are highly consistent and reproducible. The column possesses good stability and longevity (when used with ultraviolet detection, in excess of 750 runs were carried out over a 4-year period) although the cost is high compared to that of its silica based counterparts.

11. Nitrous acid cleavage is dependent on pH and can cut selectively at un-substituted *N*-glucosamine at pH 2.5–4. Chemical digestion with nitrous acid generates fragments of different sizes in which the reducing end glucosamine residue is converted to 2,5-anhydromannose *(19)*. The strong reactivity of the aldehyde of 2,5 anhydromannose allows these fragments to react more efficiently with nucleophiles such as hydrazides/amines generating very stable conjugates.
12. The hydrophobic phenylsemicarbazide tag in addition to improving the detection of nitrous acid fragments, provides good separation for both enzyme and nitrous acid generated fragments. Derivatization proceeds via reductive amination, involving the reducing end carbonyl of the heparin fragments and the semicarbazide amine group present within the chromophore. The resulting conjugates are stable at room temperature at 4°C and also at –20°C for several months.
13. RP-HPIPC methodology offers advantages over strong anion exchange chromatography (SAX). The use of ammonium acetate and methanol allows a resolution comparable to that achieved on a SAX column. However, no desalting step is needed as the products eluted from the column could be analysed by mass spectrometry after a drying step.

Acknowledgements

The authors would like to thank Dr Scott Guimond, Marc Prescott, Audrey Dumax-Vorzet and Rowena Sison for technical assistance. The work of the authors is funded by the Wellcome Trust Medical Research Council, Biotechnology & Biological Sciences Research Council, European Union and the Human Frontiers Science Program.

References

1. Turnbull, J., Powell, A., Guimond, S. (2001) Heparan sulfate: decoding a dynamic multifunctional cell regulator. *Trends Cell Biol.* 11, 75–82.
2. Radoff, S., Danishefsky, I. (1984) Location on heparin of the oligosaccharide section essential for anticoagulant activity. *J. Biol. Chem.* 259, 166–172.
3. Skidmore, M. A., Turnbull, J. E. (2005) Sequencing and separation of heparan sulfate, in Chemistry and Biology of Heparin and Heparan sulfate (Garg, H. G., Linhardt, R. J., Hales, C. A. ed.), Elsevier B.V., New York, NY, pp. 179–202.
4. Lindahl, U., Kusche-Gullberg, M., Kjellen, L. (1998) Regulated diversity of heparan sulfate. *J. Biol. Chem.* 273, 24979–24982.
5. Drummond, K. J., Yates, E. A., Turnbull, J. E. (2001) Electrophoretic sequencing of heparin/heparan sulfate oligosaccharides using a highly sensitive fluorescent end label. *Proteomics* 1, 304–310.
6. Isbell, H. S., Frush, H. L., Wade, C. W. R., Hunter, C. E. (1969) Transformations of sugars in alkaline solutions. *Carbohydr. Res.* 9, 163–175.
7. Jaseja, M., Rej, R. N., Sauriol, F. Perlin, A. S. (1989) Novel region- and stereoselective

modifications of heparin in alkaline solution. Nuclear magnetic resonance spectroscopic evidence. *Can. J. Chem.* 67, 1449–1457

8. Yamada, S., Watanabe, M., Sugahara, K. (1998) Conversion of N-sulfated glucosamine to N-sulfated mannosamine in an unsaturated heparin disaccharide by non-enzymatic, base-catalyzed C-2 epimerization during enzymatic oligosaccharide preparation. *Carbohydr. Res.* 309, 261–268
9. Rudd, T., Skidmore, M. A., Yates, E. A. (2005) Surface-Based Studies of Heparin/Heparan Sulfate–Protein Interactions: Considerations for Surface Immobilisation of HS/Heparin Saccharides and Monitoring their Interactions with Binding Proteins, in Chemistry and Biology of Heparin and Heparan sulfate (Garg, H. G., Linhardt, R. J., Hales, C. A. ed.), Elsevier B.V., New York, NY, pp. 347–368.
10. Lamari, N.L., Kuhn, R., Karamanos, N.K. (2003) Derivatization of carbohydrates for chromatographic, electrophoretic and mass spectrometric analysis. *J. Chromatogr.* 793, 15–36.
11. Ramsay, S.L., Freeman, C., Grace, P.B., Redmond, J.W. MacLeod, J.K. (2001) Mild tagging procedures for the structural analysis of glycans. *Carbohydr. Res.* 333, 59–71.
12. Kariya, Y., Hermann, J., Suzuki, K., Isomura, T., Ishihara, M. (1998) Disaccharide analysis of heparin and heparan sulfate using deaminative cleavage with nitrous acid and subsequent labelling with paranitrophenyl hydrazide. *J. Biochem.* 123, 240–246.
13. Militsopoulou, M., Lamari, F.N., Hjerpe, A., Karamanos, N. K. (2002) Determination of twelve heparin and heparan sulfate-derived disaccharides as 2-aminoacridone derivatives by capillary zone electrophoresis using ultraviolet and laser-induced fluorescence detection. *Electrophoresis* 23, 1104–1109.
14. Skidmore, MA; Guimond, SE; Dumax-Vorzet, AF; Atrih, A; Yates, EA, Turnbull, JE. (2006) High sensitivity separation and detection of heparan sulphate disaccharides. *J. Chrom. A.* 1135, 52–56.
15. Turnbull, J. E. (2001) Analytical and preparative strong anion-exchange HPLC of heparan sulfate and heparin saccharides. *Methods Mol. Biol.* 171, 141–147
16. Desai, U. R., Wang, H. M., Linhardt, R. J. (1993) Specificity studies on the heparin lyases from *Flavobacterium heparinum. Biochemistry* 32(32), 8140–8145.
17. Yoshida, K., Miyauchi, S., Kikuchi, H., Tawada, A., Tokuyasu, K. (1989) Analysis of unsaturated disaccharides from glycosaminoglycuronan by HPLC. *Anal. Biochem.* 177, 327–332.
18. Turnbull, J. E., Gallagher, J. T. (1988) Oligosaccharide mapping of heparan sulphate by polyacrylamide-gradient-gel electrophoresis and electrotransfer to nylon membrane. *Biochem. J.* 251(2), 597–608.
19. Bienkowski, M. J. Conrad, H. E. (1984) Structural characterisation of the oligosaccharides formed by depolymerisation of heparin with nitrous acid. *J. Biol. Chem.* 260, 356–365.

Chapter 13

Small-Scale Enzymatic Digestion of Glycoproteins and Proteoglycans for Analysis of Oligosaccharides by LC-MS and FACE Gel Electrophoresis

Ruby P. Estrella, John M. Whitelock, Rebecca H. Roubin, Nicolle H. Packer, and Niclas G. Karlsson

Summary

Structural characterization of oligosaccharides from proteoglycans and other glycoproteins is greatly enhanced through the use of mass spectrometry and gel electrophoresis. Sample preparation for these sensitive techniques often requires enzymatic treatments to produce oligosaccharide sequences for subsequent analysis. This chapter describes several small-scale methods for in-gel, on-blot, and in-solution enzymatic digestions in preparation for graphitized carbon liquid chromatography-mass spectrometry (LC-MS) analysis, with specific applications indicated for glycosaminoglycans (GAGs) and N-linked oligosaccharides. In addition, accompanying procedures for oligosaccharide reduction by sodium borohydride, sample desalting via carbon microcolumn, desialylation by sialidase enzyme treatment, and small-scale oligosaccharide species fractionation are included. Fluorophore-assisted carbohydrate electrophoresis (FACE) is another useful method to isolate derivatized oligosaccharides. Overall, the modularity of these techniques provides ease and flexibility for use in conjunction with mass spectrometric and electrophoretic tools for glycomic research studies.

Key words: Glycosaminoglycans, N-linked oligosaccharides, Fluorophore-assisted carbohydrate electrophoresis, In-gel enzyme digestion, On-blot enzyme digestion, Graphitized carbon LC-MS.

1. Introduction

Miniaturization of sample preparation is fundamental for sensitive oligosaccharide detection by mass spectrometry. In proteomics, it was recognized early that the sensitivity of mass spectrometry was not the limiting factor for successful protein identification and characterization. Scaling down was

Nicolle H. Packer and Niclas G. Karlsson (eds.), *Methods in Molecular Biology, Glycomics: Methods and Protocols, vol. 534*

DOI: 10.1007/978-1-59745-022-5_13

necessary prior to analysis, which involved incorporation of electrophoretic isolation and subsequent enzymatic digestion. As methods development in glycomics parallel those in proteomics, there are many emerging glycomic methods which are adaptations from gel and membrane enzyme digest approaches used in many proteomic laboratories today. For example, on-blot membrane digests have been successfully implemented for studying N-linked and O-linked oligosaccharides by subsequent mass spectrometry *(1, 2)*. There have also been successful in-gel digests for both types of oligosaccharides *(3–5)*. Here, we present small-scale enzymatic digestion methods that have been utilized in order to release oligosaccharides from glycoproteins for downstream analysis using graphitized carbon LC-MS. While selective release of N-linked oligosaccharides by PNGase F is the most established method, we have also extended the application in this chapter for methods to study glycosaminoglycans.

Glycosaminoglycans (GAGs) are a special class of carbohydrates that are characteristically attached to proteoglycans. They are long, linear and sulfated, and often comprise alternating hexuronic acid and *N*-acetylhexosamine subunits. Useful enzymes for this type of analysis are described in **Table 1**; however, these protocols are not limited to using these enzymes alone, as other enzymes may also be used for oligosaccharide degradation and release studies. In this chapter we present specific methods describing the application of chondroitinase ABC (which cleaves the β1–4 linkage between GalNAc and HexA resulting in glycosaminoglycan disaccharides released from the proteoglycans) and PNGase F (which cleaves the linkage between the reducing terminal GlcNAc of all types of mammalian N-linked glycoproteins and asparagine on the protein core). Individual researchers can easily substitute the described methods with other, appropriate exoglycosidase/endoglycosidases of choice for their specific applications *(6, 7, 10–12, 14–17)*.

We also include in this chapter the description of useful derivatization techniques, such as the reduction of oligosaccharides, in order to improve the chromatographic resolution, and sample desalting techniques for preparing the released oligosaccharides for mass spectrometry. Small-scale desialylation has also been shown to be a useful tool for analysis of oligosaccharides, as well as fractionation of oligosaccharides into neutral and acidic species for subsequent analysis *(18, 19)*. Fluorophore-assisted carbohydrate electrophoresis (FACE) is another useful method for the characterization and quantification of derivatized carbohydrates from enzyme-digested glycoproteins and proteoglycans. It is possible to apply an in-gel extraction to fluorophore-labeled oligosaccharides in a FACE gel for subsequent mass spectrometric analysis *(20)*.

Table 1
Endoglycosidases for N-linked Glycoproteins and Proteoglycans used in miniaturized sample preparation

Enzyme	EC number	Cleavage specificity	Supplier	Recommended buffer/concentration	References
PNGase F (*Flavobacterium meningosepticum*, recombinant in *E. coli*)	EC 3.5.1.52	Linkage between 1st GlcNAc and asparagine on protein core of all types of N-linked glycoproteins (asparagine becomes aspartic acid)	Roche	100 mM sodium phosphate buffer, 25 mM EDTA, pH 7.2	Wilson et al. *(1)* Karlsson et al. *(6)*
Chondroitinase ABC (protease free from *Proteus vulgaris*)	EC 4.2.2.4	β1–4 Linkage between GalNAc and HexA (for CSA, CSB (DS), CSC, chondroitin, HA)	Seikagaku	NH_4 acetate pH 8.0 (1 mU/10 µg proteoglycan, or 10 mU/mL)	Estrella et al. *(7)*
Chondro-4-sulfatase (*Proteus vulgaris*)	EC 3.2.6.9	Linkage between GalNAc4S and sulfate group at the C4 position	Seikagaku	NH_4 acetate in 0.025% (w/v) BSA, pH 7.4 (100mU/mL)	Calabro et al. *(8)*
Chondro-6-sulfatase (*Proteus vulgaris*)	EC 3.2.6.10	Linkage between GalNAc6S and sulfate group at the C6 position	Seikagaku	NH_4 acetate in 0.025% (w/v) BSA, pH 7.4 (100 mU/mL)	Calabro et al. *(8)*
Chondroitinase ACII Arthro (*Arthrobacter aurescens*)	EC 4.2.2.5	β1–4 Linkage between GalNAc and HexA (for CSA, CSC, chondroitin, HA, not CSB)	Seikagaku	NH_4 acetate in 0.025% (w/v) BSA, pH 6.2 (100 mU/mL)	West et al. *(9)*
Chondroitinase B (*Flavobacterium heparinum*)	EC 4.2.2	β1–4 Linkage between GalNAc and IdoA (for CSB only)	Seikagaku	0.33 µM Calcium acetate, pH 8.0 (100 mU/mL)	Zamfir et al. *(10)*
Keratanase I (*Pseudomonas* sp.)	EC 3.2.1.103	β1–4 Linkage between Gal and GlcNAc	Seikagaku	NH_4 acetate, pH 7.4 (10 mU)	Karlsson et al. *(11)*
Keratanase II (*Bacillus* sp.)	EC 3.2.1	β1–3 Linkage between GlcNAc and Gal	Seikagaku	NH_4 acetate, pH 7.4 (5 mU)	Zhang et al. *(12)* Plaas et al. *(13)*

(continued)

Table 1 (continued)

Enzyme	EC number	Cleavage specificity	Supplier	Recommended buffer/concentration	References
Heparinase I (*Flavobacterium heparinum*)	EC 4.2.2.7	α1–4 Linkage between GlcNAc and Ido2S of heparin and heparin-like HS (GlcNAc can be NS-6S, NS, 6S or 0S)	Seikagaku	PBS, pH 7.0 (25 mU/mL)	Knox et al. *(14)* Whitelock et al. *(15)*
Heparitinase (heparinase III) (*Flavobacterium heparinum*)	EC 4.2.2.8	α1–4 Linkage between GlcNAc and GlcA of HS and HS-like heparin(GlcNAc can be 6S or 0S)	Seikagaku	PBS, pH 7.0 (25 mU/mL)	Karlsson et al. *(11)* Knox et al. *(14)* Whitelock et al. *(15)*
Heparitinase II (*Flavobacterium heparinum*)	No EC number	α1–4 Linkage between GlcNAc and GlcA2S of HS and heparin (GlcNAc can be NS-6S, NS, 6S or 0S)	Seikagaku	PBS, pH 7.0 (25 mU/mL)	Knox et al. *(14)*
Hyaluronidase SD (*Streptococcus dysgalactiae*)	EC 4.2.2	β1–4 Linkage between GlcNac and GlcA (HA specific, but can cleave chondroitin)	Seikagaku	NH_4 Acetate, pH 7.5 (5 mU)	Saad et al. *(16)*
Hyaluronidase (bovine testes)	EC 3.2.1.35	β1–4 linkage between GlcNAc and GlcA (can cleave CS)	Sigma (H3506)	PBS in 0.01% (w/v) BSA, pH 4.5–6.0 (200 U/mL)	Zaia et al. *(17)* Karlsson et al. *(11)*
Hyaluronidase (*Streptomyces hyalurolyticus*)	EC 4.2.2.1	β1–4 linkage between GlcNAc and GlcA (HA Specific)	Sigma (H1136)	PBS in 0.01% (w/v) BSA, pH 4.5–6.0 (200 U/mL)	

EC Enzyme commission; *HexA* Hexuronic acid; *GlcA* Glucuronic acid; *IdoA* Iduronic acid; *Gal* Galactose; *GalNAc* *N*-Acetylgalactosamine; *GlcNAc* *N*-Acetylglucosamine; *xS* Sulfate group SO_3^- at position x, *CSA* Chondroitin-4-sulfate; *CSB (DS)* Chondroitin sulfate B, or Dermatan sulfate; *CSC* Chondroitin-6-sulfate; *HA* Hyaluronic acid; *PBS* Phosphate buffer saline; *BSA* Bovine serum albumin

The protocols presented here are thus very modular, and different combinations can be used to design individualized analyses of released oligosaccharides from glycoproteins and proteoglycans.

Graphitized carbon has been shown to be a versatile liquid chromatographic separation medium for separating various carbohydrates ranging from 2 to 20 monosaccharide units and for isomer separation of closely related oligosaccharides *(11, 21–25)*. Together with mass spectrometry, it has been shown to be a powerful tool for glycomic discovery and characterization of oligosaccharides. Hence, there is a great advantage in developing methods suitable for LC-MS using graphitized carbon, but the methods described here may also apply to other separation and analysis methods.

2. Materials

2.1. Gel and Electroblot Isolation of High Molecular Mass Glycoproteins and Proteoglycans

1. *1D AgPAGE Sample Buffer.* 30% (v/v) glycerol, 0.375 M Tris–HCl, 0.005% (w/v) bromophenol blue, 0.1% (w/v) sodium dodecyl sulfate (SDS).
2. Dithiothreitol (DTT) (10 mM).
3. Iodoacetamide acid (IAA) (25 mM).
4. *AgPAGE Running Buffer* (1 L volume). 1 mM ethylenediaminetetraacetic acid (EDTA), 0.192 M boric acid, 0.1% (w/v) SDS, buffered to pH 7.6 with ~0.5 M Tris (tris(hydroxymethyl)methylamine).
5. Agarose (1% (w/v))/polyacrylamide gel (0–6% (w/v)) (AgPAGE).
6. Electrophoresis apparatus.
7. Coomassie Brilliant Blue G250.
8. *Intensifying solution.* 10% (w/v) ammonium sulfate.
9. Electrophoresis apparatus for semi-dry protein to membrane transfer.
10. Polyvinylidene fluoride (PVDF), Immobilon PSQ membrane (Millipore).
11. Blotting Paper, cut to gel size (Bio-Rad).
12. *Cathode Buffer.* 25 mM Tris-HCl (Sigma-Aldrich), 40 mM α-amino-*n*-hexanoic acid (Sigma-Aldrich), 0.01% (w/v) SDS.
13. *Anode Buffer.* 25 mM Tris-HCl (Sigma-Aldrich), 20% (v/v) methanol.
14. *Ion Trap.* 0.3 M Tris-HCl (Sigma-Aldrich).
15. Alcian Blue 8GX (Sigma-Aldrich).
16. Methanol.

2.2. Oligosaccharide Release Methods

2.2.1. In-Gel Release of N-Linked Oligosaccharides Using PNGase F

1. Clean scalpel or blade.
2. Petri dish.
3. 96-Well plate and adhesive cover tape.
4. Acetonitrile (50% (v/v))/100 mM ammonium bicarbonate.
5. Speedivac rotating vacuum evaporator (Savant).
6. Incubator or oven at 37°C.
7. PNGase F (*Flavobacterium meningosepticum* recombinant in *E. coli*) (Roche).

2.2.2. On-Blot Release of N-Linked Oligosaccharides Using PNGase F

1. *Direct Blue 71* (Sigma-Aldrich) *solution.* 4 mL of 0.1% (w/v) in 40% (v/v) ethanol/10% (v/v) acetic acid.
2. *Destaining solution for Direct Blue.* 40% (v/v) ethanol/10% (v/v) acetic acid.
3. *Alcian Blue 8GX* (Sigma-Aldrich). 0.125% (w/v) Alcian Blue in 25% (v/v) ethanol and 10% (v/v) acetic acid.
4. *Destaining solution for Alcian Blue.* 100% (v/v) methanol.
5. Polyvinyl pyrolidone (PVP) (Sigma-Aldrich, PVP40).

2.2.3. In-Solution Release of Glycosaminoglycans Using Chondroitinase ABC

1. *Positive control.* Bovine articular cartilage aggrecan (Sigma-Aldrich, A1960) dissolved in MilliQ H_2O at 2 mg/mL stock solution.
2. Proteoglycan samples of similar concentration.
3. Chondroitinase ABC protease free (*Proteus vulgaris*) (Seikagaku 100332) (*see* **Note 1**).
4. *Enzyme Buffer.* 0.1 M ammonium acetate (Sigma-Aldrich), pH 8.0 (*see* **Note 2**).
5. Eppendorf microcentrifuge tubes, 1.5 mL.
6. Incubator or oven at 37°C.
7. Microcentrifuge.
8. Boiling water bath or hot plate.

2.2.4. In-Gel Release of Glycosaminoglycans Using Chondroitinase ABC

1. Clean scalpel or blade.
2. Petri dish, clean transparency, or perspex.
3. 96-Well plate and adhesive cover tape.
4. Acetonitrile (50% (v/v))/50 mM ammonium bicarbonate.
5. Acetonitrile (100% (v/v)).
6. Chondroitinase ABC protease free (*Proteus vulgaris*) (Seikagaku 100332) (*see* **Note 1**).
7. *Enzyme Buffer.* 0.1 M ammonium acetate (Sigma-Aldrich), pH 8.0 (*see* **Note 2**).
8. Speedivac rotating vacuum evaporator (Savant).

9. Incubator or oven at 37°C.

2.2.5. On-Blot Release of Glycosaminoglycans Using Chondroitinase ABC

1. *Alcian Blue 8GX* (Sigma-Aldrich). 0.125% (w/v) Alcian Blue in 25% (v/v) ethanol and 10% (v/v) acetic acid.
2. *Destaining solution for Alcian Blue.* 100% (v/v) methanol.
3. PVP (Sigma-Aldrich, PVP40).
4. Chondroitinase ABC protease free (*Proteus vulgaris*) (Seikagaku 100332) (*see* **Note 1**).
5. *Enzyme Buffer.* 0.1 M ammonium acetate (Sigma-Aldrich), pH 8.0 (*see* **Note 2**).

2.3. Reduction of Oligosaccharides Using Sodium Borohydride

1. *Reducing solution.* 1.0 M sodium borohydride ($NaBH_4$, Lancaster, 5859) in 100 mM sodium hydroxide (Sigma-Aldrich, S8045) (*see* **Note 3**).
2. Oven at 50°C.
3. Glacial acetic acid.
4. AG50W-X8 resin (H^+ form) (BioRad, Hercules CA).
5. C-18 ZipTip (Millipore).
6. Methanol.
7. Milli-Q water.
8. HCl (1 M).
9. Speedivac rotating vacuum evaporator (Savant).

2.4. Carbon Microcolumn Desalting Protocol (26)

1. Alltech SPE Carbograph column or equivalent carbon column.
2. C-18 ZipTip (Millipore).
3. Eppendorf microcentrifuge tubes.
4. *Activation solution.* 90% (v/v) acetonitrile in 0.5% (v/v) trifluoroacetic acid (TFA).
5. *Wash solution.* 0.5% (v/v) TFA in Milli-Q water.
6. *Elution Solution.* 40% (v/v) acetonitrile in 0.5% (v/v) TFA.
7. Speedivac rotating vacuum evaporator (Savant).

2.5. Separation of Oligosaccharides into Neutral and Acidic Species

1. DEAE-Sephadex (GE Bioscience or Sigma-Aldrich).
2. ZipTip (Millipore).
3. Pyridinium acetate (1 M, pH 5.4).

2.6. In-Solution Desialylation of Oligosaccharides

1. Sodium phosphate buffer (25 mM), pH 6.0.
2. Sialidase *Arthrobacter ureafaciens* (Glyko, Novato, CA, for 3- and 6-linked sialic acid) or *Streptococcus pneumoniae* (Glyko, Novato, CA, for 3-linked sialic acid).

2.7. FACE

2.7.1. Fluorophore labeling:

1. 2-Aminoacridone (AMAC, Fluka) (0.1 M in 85% (v/v) dimethylsulfoxide (DMSO) /15% (v/v) acetic acid).

2. Acetic Acid.
3. DMSO (Sigma).
4. Sodium cyanoborohydride (Aldrich) (1 M in DMSO) (*see* **Note 4**).

2.7.2. FACE gel preparation:

1. 2× Resolving gel stock (50% (w/v) T/7.5% (w/v) C = 46.25% (w/v) acrylamide/3.75% (w/v) bis) For 10 mL (two gels):
   (a) 5 mL 2× resolving gel stock (50% (w/v) T/7.5% (w/v) C = 46.25% (w/v) acrylamide/3.75% (w/v) bis).
   (b) 1.25 mL 1.5 M Tris–borate, pH 8.8.
   (c) 1.25 mL 1.5 M Tris-HCl, pH 8.8.
   (d) 1 mL 25%(v/v) glycerol.
   (e) 1.5 mL water.
   (f) 50 μL 10% (w/v) ammonium persulfate (APS).
   (g) 5 μL *N,N,N,N*-tetramethylethylenediamine (TEMED).
2. 10× Stacking gel stock (50% (w/v) T/15% (w/v) C = 42.5% (w/v) acrylamide/7.5% (w/v) bis). For 5 mL (two gels):
   (a) 0.5 mL 10× stacking gel stock (50% (w/v) T/15% (w/v) C = 42.5% (w/v) acrylamide/7.5% (w/v) bis).
   (b) 1.2 mL 1.5 M Tris–HCl pH 6.8.
   (c) 0.88 mL 25% (v/v) Poly(ethylene glycol) (PEG) 8000.
   (d) 2.42 mL water.
   (e) 50 μL 10% (w/v) APS.
   (f) 10 μL TEMED.
3. 1.5 M Tris–borate, pH 8.8.
4. 1.5 M Tris, pH 8.8.
5. 1.5 M Tris–HCl, pH 6.8.
6. 25% (v/v) glycerol (BDH AnalaR 87%).
7. 25% (w/v) PEG 8000.
8. 10% (w/v) APS – freshly made.
9. TEMED.
10. TBE running buffer:
    (a) 16.2 g Tris-HCl.
    (b) 8.25 g boric acid.
    (c) 1.35 g EDTA.
    (d) 1.5 L water.
11. Glyko Gel (Prozyme, CA) apparatus:
    (a) 1.0-mm-thick glass plates of approximately 10 × 10 cm.

(b) 0.5 mm spacers (*see* **Note 5**).

(c) 0.5 mm Combs with eight wells.

12. Glyko FACE gel electrophoresis tank and power pack.
13. BioRad Gel Doc or equivalent system for gel visualization.

2.8. Graphitized Carbon LC-MS of Oligosaccharides

1. *ESI-LC-MS system.* Agilent Technologies 1100 Series LC/MSD Trap XCT Plus or Thermo LCQ Deca-XP Ion Trap mass spectrometer or an equivalent mass spectrometer.
2. *Graphitized Carbon Column.* SGE ProteCol Hypercarb 250A, 300 μm ID × 100 mm, 5 μm.
3. *Mobile phase A1.* 10 mM ammonium bicarbonate.
4. *Mobile phase B1.* 10 mM ammonium bicarbonate in 80% (v/v) acetonitrile.
5. Suitable control standards (reduced as per **Subheading 3.3**):

   (a) *GAGs.* Chondroitin sulfate disaccharides di-0S, di-4S, di-6S; heparan sulfate disaccharides I-A, II-A, III-A, IV-A, I-S, II-S, III-S (Sigma-Aldrich).

   (b) *N-linked Oligosaccharides.* Fetuin from fetal calf serum (Sigma-Aldrich).

3. Methods

3.1. Gel Separations

Samples can be separated either by standard 1D or 2D SDS-PAGE electrophoresis or, for high molecular weight glycoproteins and proteoglycans, by SDS-AgPAGE composite gels (prepared as described in Karlsson et al. in **Subheading A.1.2** of this book) and carried out as follows:

1. Load 10 μg/10 μL per lane on to 1D SDS AgPAGE gels. Set the current to 30 mA per gel: run time approximately 2–3 h until dye front reaches the very bottom of gel.
2. After electrophoresis, gels are stained with Coomassie Brilliant Blue G250 overnight on a platform rocker at room temperature. Alternatively, the proteoglycans and glycoproteins can be blotted to PVDF membranes using the semi-dry transfer method as described by Schulz et al. *(2)*. After transfer, the blots are stained with nondestructive protein stains (e.g., direct blue) or Alcian Blue (stains negatively charged oligosaccharides) (*see* **Note 6**).
3. The next day, gels are destained with the intensifying solution (10% (w/v) ammonium sulfate) until a good contrast between the protein bands and the background is achieved. In **Fig. 1**,

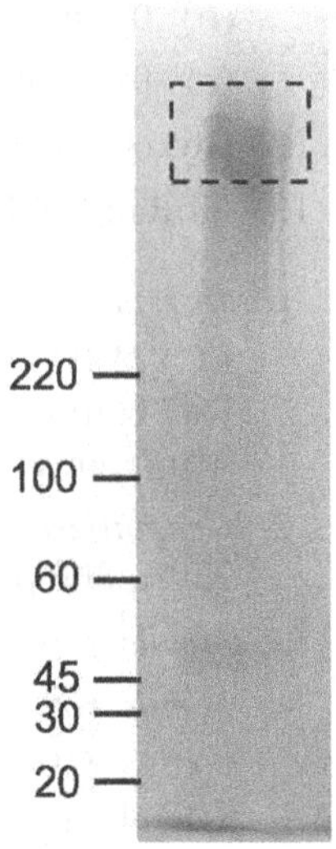

Fig. 1. Coomassie Blue–stained Aggrecan (10 μg) in SDS-AgPAGE gel. The box represents the area of band excision for digestion of the resultant disaccharides shown in Fig. 4.

the fully glycosylated heterogeneous aggrecan band appears greater than 220 kDa.

3.2. Oligosaccharide Release Methods

3.2.1. In-Gel Release of N-Linked Oligosaccharides Using PNGase F

1. Run the gels and stain proteins with Coomassie Brilliant Blue G250.
2. Cut the gel spots using a scalpel, and place the pieces in separate wells of a 96-well microtier plate. Larger pieces will need to be cut up into smaller pieces to make it easier for the enzyme to penetrate the gel.
3. Destain the Coomassie-stained spots using 100 μL of 50% (v/v) acetonitrile in 100 mM ammonium bicarbonate. Do this either manually by shaking the gel pieces in the destaining solution for approximately 20 min before changing the destaining solution. Pipette most of the destaining solution and repeat 2–3 times until the gel pieces are completely clear of color.
4. Place an adhesive cover sheet on the plate and, using a scalpel, cut small holes in the sheet corresponding to the wells containing the gel pieces. Place the plate in the Speedivac and dry at 35°C (vacuum drying) or at 37°C in a heated oven.
5. Add 5 μL of PNGase F (0.5 U/μL) and incubate for 3–12 h at 37°C. Larger spots may require the addition of water, so that the gel pieces will still be surrounded by liquid after they have swelled to their maximum capacity (*see* **Note 7**).
6. Recover the release oligosaccharides after sonication of the samples (in the 96-well plate) for approximately 5 min, and

remove the oligosaccharide-containing fluid surrounding the gel pieces.

7. Collect supernatant with a pipette into fresh tubes. Then, extract the gel pieces twice more with 50 μL of water, and combine the extracts with the initial extract.
8. Desalt the N-linked oligosaccharides from the wells using the carbon microcolumn desalting protocol, or perform a reduction and subsequent cation exchange desalting before applying the oligosaccharides to LC-MS (*see* **Note 8**).

3.2.2. On-Blot Release of N-Linked Oligosaccharides Using PNGase F

1. Run the gels and blot proteins onto PVDF membranes.
2. Visualize the proteins using nondestructive stains such as Direct Blue or Alcian Blue.
3. Fill the wells of a microtiter plate with 100 μL of 1% (w/v) PVP solution in 50% (v/v) methanol to wet and block the membrane.
4. Cut the PVDF membrane bands or spots from the membrane and place them in separate wells of the microtiter plate.
5. Shake the spots for 5 min in the 1% (w/v) PVP solution and wash three times for 5 min with water.
6. Add 5 μL of PNGase F (0.5 U/μL), incubate for 15 min at 37°C, and add 10 μL of water and incubate overnight at 37°C. Add 100 μL of water to surround sample wells and seal the plate to create a humidified environment and reduce evaporation.
7. Sonicate the samples (in the 96-well plate) for approximately 5 min.
8. Collect samples from the wells and desalt using the carbon microcolumn desalting protocol, or alternatively reduce the N-linked oligosaccharides and desalt using the cation exchange resin method.

3.2.3. In-Solution Release of Glycosaminoglycans Using Chondroitinase ABC

1. Digest 30 μg of isolated proteoglycans in solution with 1 μL of 1 U/mL chondroitinase ABC (final concentration 10 mU/mL) and 85 μL of 0.10 M ammonium acetate (pH 8.0) in a 1.5-mL Eppendorf tube. Digest overnight in an incubator at 37°C overnight (at least for 16 h).
2. Stop digestion by boiling the tube in a water bath or putting the tube on a hot plate for 10 min. Continue to the reduction step (**Subheading 3.3**) to prepare alditols of the released oligosaccharides.

3.2.4. In-Gel Release of Glycosaminoglycans Using Chondroitinase ABC

1. For gel excision, place the gel in a clean Petri dish, plastic transparent sheet, or Perspex, and use a clean scalpel and methanol for cleaning in between band excising. Cut out the

whole band, and then cut the band further into smaller pieces (about 1 mm^2 in size), which will ensure enzyme access to all available sites on the proteoglycan. Each cut-up set of gel pieces should be put in separate wells of the 96-well microtiter plate (*see* **Note 9**).

2. Destain the gel pieces with 150 μL 50% (v/v) acetonitrile/50 mM ammonium bicarbonate three times at 20 min each, and wash at 37°C until no blue color is detected (gel pieces should turn opaque). Decant, and then wash one more time with 85 μL 100% (v/v) acetonitrile. Dry the gel pieces in an oven (37°C) until the pieces shrink to a fraction of their original size.
3. Prepare chondroitinase ABC at 100 mU/mL 0.1 M ammonium acetate buffer, pH 8.0. Add 10 μL of the enzyme to each well, and keep at 37°C for 10 min to allow the enzyme solution to get absorbed into the gel pieces (the gel will grow back to almost the original size and become semitransparent). Then, add another 30 μL 0.1 M ammonium acetate buffer, pH 8.0, so that the gel pieces do not dry out (final concentration 25 mU/mL). Add water to the surrounding wells and cover the microtiter well plate with an adhesive cover sheet so that the incubating wells do not dry out. Incubate overnight at 37°C.
4. Next day, stop enzyme digestion by boiling the plate or putting it on a hot plate for 10 min. Check that there is supernatant buffer in wells; otherwise, supplement with additional ammonium acetate buffer (total volume of supernatant should be approximately 10–30 μL). Sonicate the wells for 15 min so that the excised oligosaccharides and peptides are extracted from the gel pieces. Collect the supernatant with a pipette into fresh tubes. Then, extract the gel pieces twice more with 50 μL of water and combine.
5. Speedivac the samples to dryness. Reconstitute in water prior to reduction (**Subheading 3.3**).

3.2.5. On-Blot Release of Glycosaminoglycans Using Chondroitinase ABC

1. Run the gels and blot proteins onto PVDF membranes.
2. Stain the proteoglycans on PVDF blots with Alcian Blue for 10 min, and then destain with 100% (v/v) methanol for 20 min or until good contrast is achieved.
3. Dry the membrane in between two sheets of dry blotting paper. Membranes can be kept dry until use, but remember to wet the membrane with methanol before digestion to make the membrane hydrophilic and receptive to enzyme.
4. Prepare 1% (w/v) PVP in 50% (v/v) methanol.

5. Excise the proteoglycan band (usually 1 cm) with a scalpel, and then cut it into 1 mm pieces. Put each cut-up band into separate wells of the 96-well microtiter plate.
6. Incubate the band with 100 μL 1% (w/v) PVP in 50% (v/v) methanol for 10 min at room temperature on a platform rocker. (PVP is a blocking agent to prevent enzyme adsorption on to PVDF).
7. Decant the PVP solution and then wash three times in 200 μL MilliQ H_2O, with 5 min for each wash.
8. Prepare chondroitinase ABC to 1 U/mL (1 mU/10uL) in 0.1 M ammonium acetate buffer, pH 8.0. Add 10 μL (or enough volume to cover) to each membrane well, and then keep at 37°C for 10 min to allow the enzyme solution to be absorbed. Add another 40 μL of buffer to keep the moisture. Incubate at 37°C overnight (final concentration 20 mU/mL).
9. Next day, remove and keep the supernatant.
10. Wash the membrane three times with 50 μL of water and combine with the supernatant. Dry the samples down in the Speedivac. Proceed with reduction of oligosaccharides (**Subheading 3.3**).

3.3. Reduction of Oligosaccharides Using Sodium Borohydride

1. Prepare fresh 1 M sodium borohydride in 100 mM sodium hydroxide.
2. Add 20 μL or an equal volume to the oligosaccharide sample (previously released by PNGase F or chondroitinase ABC) and incubate overnight at 50°C.
3. Next day, quench the reaction by adding 1 μL glacial acetic acid. Some effervescence might been seen upon addition if a reasonable amount of oligosaccharides is successfully reduced.
4. *Prepare the cation exchange microcolumn.* Add 25 μL of AG50 W-X8 resin on top of a C-18 Zip Tip and place column in an Eppendorf tube.
5. Wash the column three times with 50 μL of 1 M HCl (each wash is a 10-s, quick centrifuge spin). Then, wash the column three times with 50 μL of 100% (v/v) methanol. At this point, discard the wash eluate. To wash out the methanol, wash the column three times with 50 μL of MilliQ H_2O. Transfer the columns to fresh Eppendorf tubes to collect the oligosaccharides.
6. Add the reduced oligosaccharides to the top of the column and centrifuge briefly. Add 25 μL of MilliQ H_2O to sample wells or tubes to wash out the residual oligosaccharides and transfer to the loaded column and centrifuge. Repeat one more time. Elute the oligosaccharides by adding another 30 μL of MilliQ H_2O to the loaded column and centrifuge.

7. Dry the collected samples on the Speedivac for approximately 1 h. At this point, the white residue of the borate will be visible at the bottom of the tube. To remove the methyl borate, add 50 μL of methanol and evaporate in the Speedivac. Repeat four more times until no white residue is seen.
8. Dissolve the dry samples in 0.1% (v/v) formic acid or Milli Q H_2O prior to LC-MS analysis (*see* **Note 10**).

3.4. Carbon Microcolumn Desalting Protocol

1. Remove the carbon packing from the Alltech SPE Carbograph columns (*see* **Note 11**).
2. Wash and store the carbon in 50% (v/v) methanol.
3. Add 5 μL of carbon suspension on top of a C-18 ZipTip and place in an Eppendorf tube.
4. Activate the carbon with 0.5% (v/v) TFA in 90% (v/v) acetonitrile by adding 25 μL to the carbon column; then do a quick, 10-s centrifuge spin and repeat two more times (*see* **Note 12**).
5. Wash the column three times with 25 μL of 0.5% (v/v) TFA, using a quick, 10-s centrifuge spin per each wash.
6. Load the sample onto the column, and do a quick, 10-s centrifuge spin. Repeat **step 5**, then transfer column to a fresh Eppendorf tube to collect sample.
7. Elute the sugars three times with 25 μL of 40% (v/v) acetonitrile in 0.5% (v/v) TFA.
8. Speedivac the samples to dryness and then reconstitute in 0.1% (v/v) formic acid or Milli Q H_2O prior to LC-MS analysis.

3.5. Separation of Oligosaccharides into Neutral and Acidic Species (if Required)

1. *Prepare the DEAE-Sephadex anion exchange microcolumn (Ac^--form).* Add 50 μL of DEAE-Sephadex (GE Bioscience) on top of a ZipTip (Millipore) and place in an Eppendorf tube. Apply the oligosaccharide solution (e.g., a mixture of acidic and neutral N-linked oligosaccharides) to the top of the column and then centrifuge briefly. Add 60 μL of water, then centrifuge, and repeat twice more to collect the neutral oligosaccharides.
2. Place the microcolumn in a new Eppendorf tube.
3. *Collect the acidic oligosaccharides.* Add 60 μL of pyridinium acetate (1 M, pH 5.4) and centrifuge. Repeat two more times.
4. Speedivac the samples to dryness, and then reconstitute in 0.1% (v/v) formic acid or Milli Q H_2O prior to LC-MS analysis.

3.6. In-Solution Desialylation of Oligosaccharides (if Required)

1. Dry down the desalted oligosaccharides (reducing or alditols, unfractionated or fractionated) in a Speedivac.
2. Dissolve the samples in 18 μL of 25 mM sodium phosphate buffer, pH 6.0, and add the appropriate sialidases: 2 μL

(approximately 10 mU) of *Arthrobacter ureafaciens* sialidase (removes 3- and 6-linked sialic acid) or 2 μL (approximately 10 mU) of *Streptococcus pneumoniae* (removes 3-linked sialic acid).

3. Incubate at 37°C for 16 h.
4. Desalt the samples using the carbon microcolumn desalting protocol (**Subheading 3.3**).

3.7. FACE (Fig. 2)

3.7.1. Preparation of the Fluorescent Derivative by Reductive Amination of Reducing Oligosaccharides (8)

1. Prepare 0.1 M AMAC in 85% (v/v) DMSO/15% (v/v) acetic acid.
2. Place 50 nmol of disaccharide reducing equivalents based on 1,9-dimethylmethylene blue (DMMB) assay (for sulfated GAG concentration determination) *(27)* in 0.1 M ammonium acetate for labeling in an 1.5 mL Eppendorf tube, and then Speedivac or lyophilize to dryness (no heat setting) (*see* **Note 13**).
3. Add 5 μL 0.1 M AMAC (the pellet will turn yellow and become easier to resuspend) and vortex briefly.
4. Add 5 μL 1 M sodium cyanoborohydride in DMSO (vortex before use) and vortex briefly.
5. Incubate the samples at 37°C overnight (approximately 16 h).

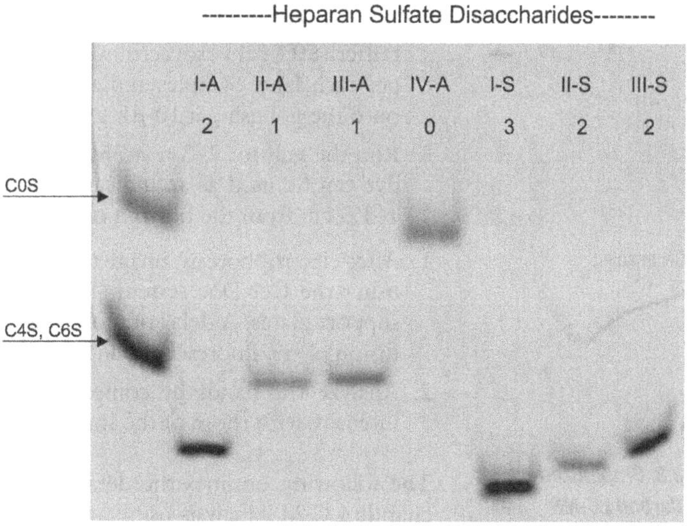

Fig. 2. FACE analysis of AMAC-labeled chondroitin sulfate disaccharides di-0S, di-6S, and di-4S, and heparan sulfate (HS) disaccharides I-A, II-A, III-A, IV-A, I-S, II-S, and III-S. The number of sulfate groups is indicated underneath each of the HS disaccharide headings, showing all the possible combinations of the sulfated HS. These could potentially be subjected to in-gel extraction for mass spectrometric confirmation and characterization.

3.7.2. Preparation of FACE Gels

1. Clean the gel apparatus with ethanol (clean with a tooth brush and detergent after use).
2. Put a few droplets of water on the plate to allow adhesion of spacers to the glass plates. Prevent movement/sliding.
3. Attach the plates to the gel-casting apparatus.
4. Check the plates for leaks with water.
5. Mark the plates for the stacking gel – about ½ cm below the comb.

Resolving Gel

1. Mix 10% (w/v) APS and TEMED to the resolving gel components just before pouring the resolving gel onto the glass plates.
2. Using a transfer pipette, slowly pour the resolving gel.
3. Before the gel sets, overlay it with 1 mL MilliQ H_2O. Give the gel approximately 20 min to set.

Stacking Gel

1. Remove the overlaid water from the glass plates.
2. Mix 10% (w/v) APS and TEMED into the stacking gel mixture just before pouring the stacking gel on top of the resolving gel.
3. Insert the comb before the gels set. Give the gels approximately 20–25 min to set.

Running Gel

1. After making the gels, prepare the FACE gel apparatus in a cold room (4°C) and fill it with prechilled (4°C) TBE buffer. Pre-electrophorese the gel for 40 min at 200 V per gel.
2. *Prepare the samples.* 5 μL (2.5 nmol) sample plus 5 μL loading buffer (50% (v/v) glycerol with/without bromophenol blue) per each lane. Samples and appropriate standards are loaded onto the gel using a 10-μL glass syringe (*see* **Note 14**).
3. Run the gels for 2–3 h at 350–400 V. (If available, a tracking dye can be used as an indication of migration approximately 1–1.5 cm from the bottom of the gel.)

Gel Imaging

1. After electrophoresis, image the gels immediately by UV light using the Gel Doc system (BioRad) while still in their glass support plates. A delay of more than 20 min may result in diffusion of the fluorescent bands.
2. Analyze the bands by comparing their migration and pixel intensity with those of the standards run on the same gel.

3.8. Graphitized Carbon LC-MS of N-Linked Oligosaccharides and Glycosaminoglycans Digested with Chondroitinase ABC (Figs. 3 and 4)

The following outlines the details and conditions for oligosaccharide LC-MS analysis using an Agilent 1100 Series LC/MSD Trap XCT Plus or another suitable mass spectrometer. Note that ion signal optimization may be required.

1. Column details and reparation
   (a) *Column.* SGE ProteCol Hypercarb 250A, 300 μm ID × 100 mm, 5 μm

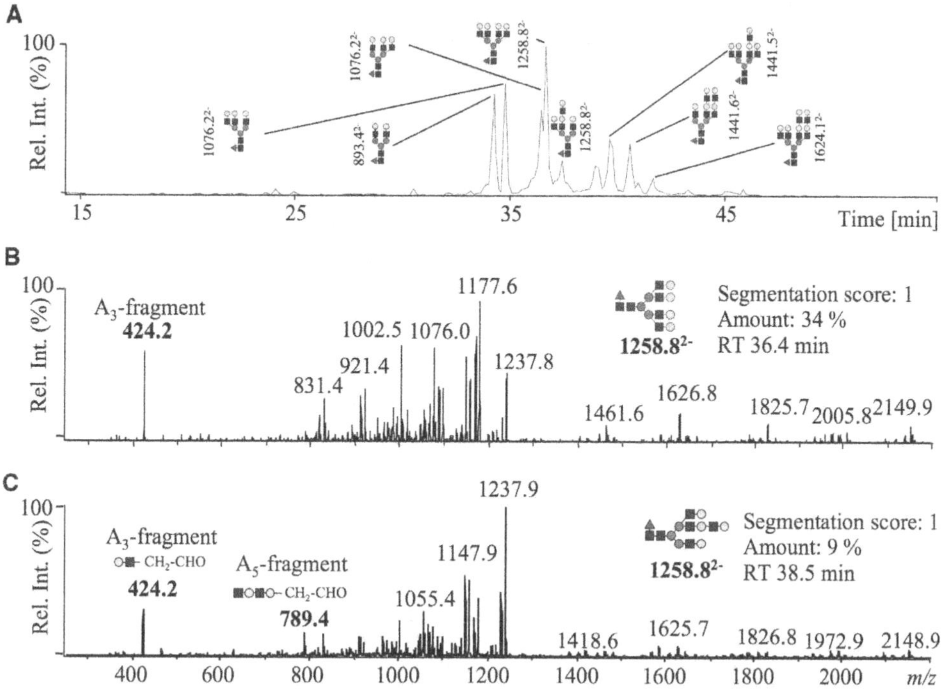

Fig. 3. (**A**) Base peak chromatogram of N-linked desialylated oligosaccharides from recombinant erythropoietin (10 μg). (**B**)and (**C**) The corresponding MS^2 spectra from components at 36.4 and 38.5 min. The spectra were searched with GlycosidIQ (Proteome Systems Ltd, Sydney, Australia) for best match *(28)*, and some key fragments for identifying branching pattern are labeled using the Domon and Costello nomenclature *(29)*.

(b) *Mobile phases.* A1: 10 mM ammonium bicarbonate, B1: 10 mM ammonium bicarbonate in 80% (v/v) acetonitrile (ACN)

(c) *Column equilibration.* With flow rate at 7 μL /min

(d) Start with 100% A1 for 30 min, then 50% A1/50% B1 for next 30 min to clean out the column, then equilibrate column back to starting conditions with 100% A1 for 30 min (*see* **Note 15**).

2. LC conditions for separation of N-linked oligosaccharides

(a) Gradient:

0–6 min at 0% B1
6–35 min gradual increase to 25% B1
35–37 min sharp increase to 100% B1
37–39 min steady at 100% B1
39–39.01 min back to 0% B1
Stop time at 55 min

(b) Flow rate: 7 μL/min

(c) Sample Injection Volume: 5 μL

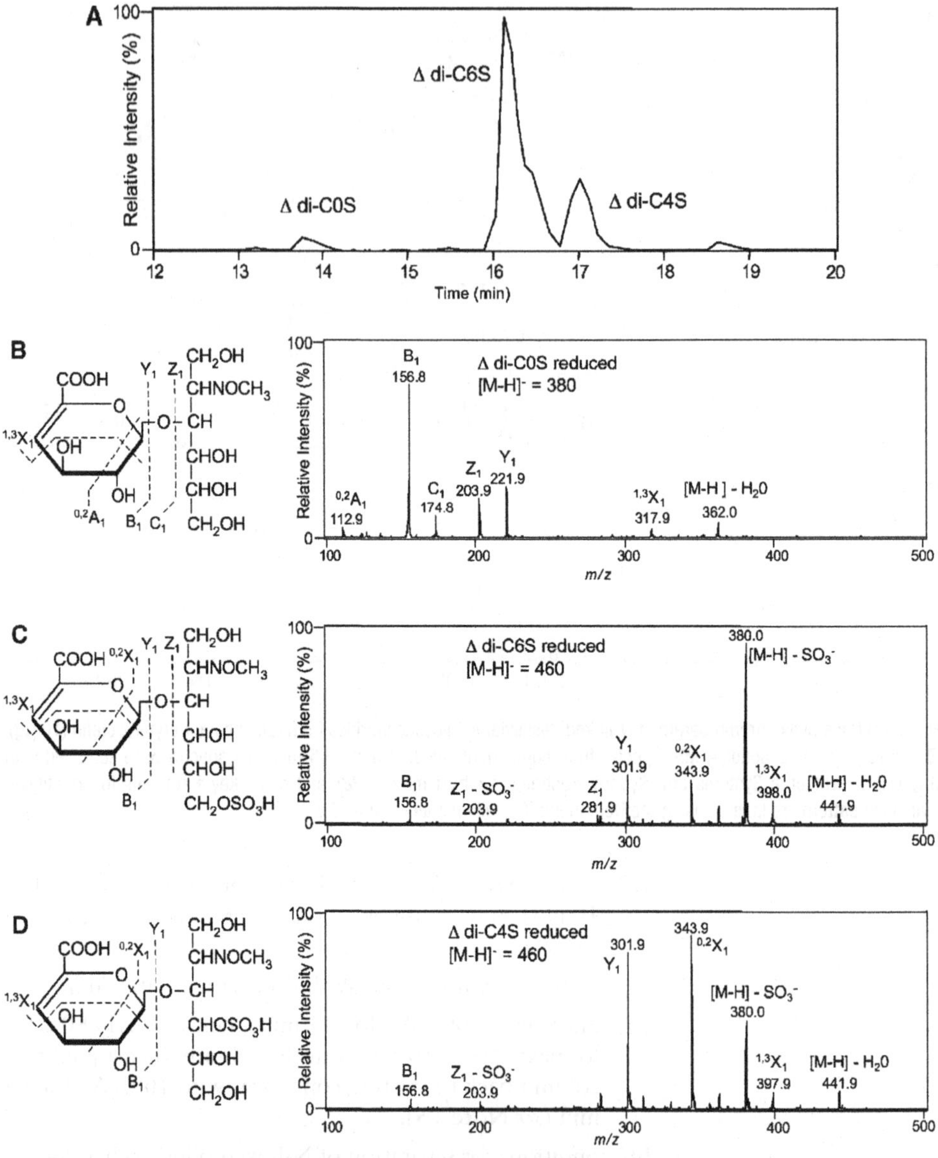

Fig. 4. (**A**) Base Peak chromatogram of Aggrecan (10 μg) disaccharides released from SDS-AgPAGE gel with chondroitinase ABC. (**B–D**) The corresponding MS^2 spectra and proposed fragmentation. Annotation of the spectra and fragments are according to the Domon and Costello nomenclature *(29)*.

3. LC Conditions for separation of GAG oligosaccharides
   (a) Gradient:
   0–2 min at 0% B1
   2–30 min gradual increase to 35% B1
   30–35 min sharp increase to 90% B1
   35–45 min steady at 90% B1

45–48 min sharp decrease to 0% B1
48–50 min back to 0% B1
Stop time at 75 min

(b) Flow rate: 7μL/min

(c) Sample Injection Volume: 5 μL

4. Typical MS conditions for oligosaccharide analysis

(a) Ionization mode: Negative electrospray

(b) Drying gas flow: 7 L/min

(c) Drying gas temperature: 320.0°C

(d) Nebulizer gas: 20.0 psi

(e) Skimmer: -30.0 V

(f) Capillary exit: -180.0 V

(g) Trap drive: 59.0 V

(h) ICC: On

(i) Maximum accumulation time: 300 ms

(j) Target: 100,000 (MS/MS)

(k) Scan range: 250–2,200 for N-linked oligosaccharides and 170–800 for digested GAG oligosaccharides

4. Notes

1. There are two types of CS ABC enzymes from Seikagaku; the protease-free version (Seikagaku 100332) only has endolyase activity giving a hexasaccharide linkage region oligosaccharide still attached to the protein backbone. The other preparation (Seikagaku 100330) contains both endolyase and an exolyase activity capable of digesting CS tetrasaccharides *(30)*.
2. Ammonium acetate buffer was selected because it is volatile and thus compatible with the subsequent electrospray mass spectrometry. Other buffers may also be used because of critical phosphate sensitivity and pH for optimal enzyme activity according to the enzyme manufacturer's instructions; however, one needs to ensure that the sample is fully desalted prior to injection into the electrospray source of the mass spectrometer.
3. This solution needs to be prepared fresh, just before use. Also, take care when handling both the sodium borohydride (it is corrosive and moisture sensitive) and sodium hydroxide.
4. Prepare sodium cyanoborohydride in a fume hood to avoid toxic inhalation of cyanoborohydride. The resultant solution is stable for at least 6 months at -80°C.

5. Thin (0.5 mm) spacers are used to facilitate cooling during the gel run.
6. It is our experience that electroblotting to Immobilon PSQ (pore size 0.2 μm) is the best for subsequent blot staining with Alcian Blue, whereas Immobilon P (pore size 0.45 μm) is the best for subsequent western blotting. With the smaller pore size, Alcian Blue is easier to destain and obtain better contrast, while with the larger pore size the blots were too dark to distinguish the stained glycoproteins/proteoglycans.
7. Oligosaccharides could be released by PNGase F and desialylated at the same time by the addition of sialidases (e.g. 1 μL (5 mU) of *Arthrobacter ureafaciens* sialidase (3- and 6-linked sialic acid released) or *Streptococcus pneumoniae* (3-linked sialic acid) (Glyko, Novato, CA).
8. Unreduced oligosaccharides in their native solution state show unfavorable separation on the graphitized carbon column because of splitting of the component peaks into their α and β anomers. Reduced oligosaccharides show a single peak per component.
9. If not ready for excision, gels can be stored in MilliQ H_2O in plastic bags at 4°C until use. Take care not to store for too long (months), because in that case the Coomassie Blue staining will be lost.
10. Samples may contain peptide fragments and residual salts after reduction, which will interact with the online graphitized carbon capillary column at the front end of the LC-MS and will decrease the potential chromatographic performance of the column. Additional desalting through a carbon microcolumn may be required.
11. Prepare the carbon/methanol suspension in a fume hood to avoid dangerous inhalation of the carbon powder. Removing the carbon retaining frits to access the material from the carbon SPE columns can be easily done using a small screw as a cork screw. The resultant suspension is stable at room temperature for many months.
12. Reagent volumes of the activation solution, wash solution, and elution should be equal to sample volume; e.g. if have 50 μL sample volume, then use 50 μL wash solution, etc.
13. Try not to dry completely, as dried sample will be difficult to resuspend and will degrade.
14. As the width of the gel is 0.5 mm, gel-loading tips are too wide for this gel. To load, use a 10-μL glass syringe (no more than 6 μL), filling from the bottom of the well. Rinse the syringe well with water between samples.

15. Typical pressure after equilibration at 100% A1 (flat baseline) is 70–80 mbar (for a new column this pressure increases slightly over time and use).

Acknowledgments

We thank Dr Neil Davies for the preparation of the FACE gel used in **Fig. 2**.

References

1. Wilson, N. L., Schulz, B. L., Karlsson, N. G., and Packer, N. H. (2002) Sequential analysis of N- and O-linked glycosylation of 2D-PAGE separated glycoproteins. *J. Proteome. Res.* 1, 521–529.
2. Schulz, B. L., Packer, N. H., and Karlsson, N. G. (2002) Small-scale analysis of O-linked oligosaccharides from glycoproteins and mucins separated by gel electrophoresis. *Anal. Chem.* 74, 6088–6097.
3. Kuster, B., Wheeler, S. F., Hunter, A. P., Dwek, R. A., and Harvey, D. J. (1997) Sequencing of N-linked oligosaccharides directly from protein gels: in-gel deglycosylation followed by matrix-assisted laser desorption/ionization mass spectrometry and normal-phase high-performance liquid chromatography. *Anal. Biochem.* 250, 82–101.
4. Papac, D. I., Briggs, J. B., Chin, E. T., and Jones, A. J. (1998) A high-throughput microscale method to release N-linked oligosaccharides from glycoproteins for matrix-assisted laser desorption/ionization time-of-flight mass spectrometric analysis. *Glycobiology* 8, 445–454.
5. Taylor, A. M., Holst, O., and Thomas-Oates, J. (2006) Mass spectrometric profiling of O-linked glycans released directly from glycoproteins in gels using in-gel reductive beta-elimination. *Proteomics* 6, 2936–2946.
6. Karlsson, N. G., Wilson, N. L., Wirth, H. J., Dawes, P., Joshi, H., and Packer, N. H. (2004) Negative ion graphitised carbon nano-liquid chromatography/mass spectrometry increases sensitivity for glycoprotein oligosaccharide analysis. *Rapid Commun. Mass Spectrom.* 18, 2282–2292.
7. Estrella, R. P., Whitelock, J. M., Packer, N. H., and Karlsson, N. G. (2007) Graphitized carbon LC-MS characterization of the chondroitin sulfate oligosaccharides of aggrecan. *Anal. Chem.* 79, 3597–3606.
8. Calabro, A., Benavides, M., Tammi, M., Hascall, V. C., and Midura, R. J. (2000) Microanalysis of enzyme digests of hyaluronan and chondroitin/dermatan sulfate by fluorophore-assisted carbohydrate electrophoresis (FACE). *Glycobiology* 10, 273–281.
9. West, L. A., Roughley, P., Nelson, F. R., Plaas, A. H. (1999) Sulphation heterogeneity in the trisaccharide (GalNAcSbeta1, 4GlcAbeta1,3GalNAcS) isolated from the non-reducing terminal of human aggrecan chondroitin sulphate. *Biochem. J.* 15, 223–229.
10. Zamfir, A., Seidler, D. G., Kresse, H., and Peter-Katalinic, J. (2003) Structural investigation of chondroitin/dermatan sulfate oligosaccharides from human skin fibroblast decorin. *Glycobiology* 13, 733–742.
11. Karlsson, N. G., Schulz, B. L., Packer, N. H., and Whitelock, J. M. (2005) Use of graphitised carbon negative ion LC-MS to analyse enzymatically digested glycosaminoglycans. *J. Chromatogr. B Analyt. Technol. Biomed. Life Sci.* 824, 139–147.
12. Zhang, Y., Kariya, Y., Conrad, A. H., Tasheva, E. S., and Conrad, G. W. (2005) Analysis of keratan sulfate oligosaccharides by electrospray ionization tandem mass spectrometry. *Anal. Chem.* 77, 902–910.
13. Plaas, A. H., West, L. A., Midura, R. J. (2001) Keratan sulfate disaccharide composition determined by FACE analysis of keratanase II and endo-beta-galactosidase digestion products. *Glycobiology.* 11, 779–790.
14. Knox, S., Merry, C., Stringer, S., Melrose, J., and Whitelock, J. (2002) Not all perlecans are created equal: interactions with fibroblast growth factor (FGF) 2 and FGF receptors. *J. Biol. Chem.* 277, 14657–65.

15. Whitelock, J. M., Murdoch, A. D., Iozzo, R. V., and Underwood, P. A. (1996) The degradation of human endothelial cell-derived perlecan and release of bound basic fibroblast growth factor by stromelysin, collagenase, plasmin, and heparanases. *J. Biol. Chem.* 271, 10079–10086.

16. Saad, O. M., Myers, R. A., Castleton, D. L., and Leary, J. A. (2005) Analysis of hyaluronan content in chondroitin sulfate preparations by using selective enzymatic digestion and electrospray ionization mass spectrometry. *Anal. Biochem.* 344, 232–239.

17. Zaia, J., and Costello, C. E. (2001) Compositional analysis of glycosaminoglycans by electrospray mass spectrometry. *Anal. Chem.* 73, 233–239.

18. Karlsson, N. G., Schulz, B. L., and Packer, N. H. (2004) Structural determination of neutral O-linked oligosaccharide alditols by negative ion LC-electrospray-MSn. *J. Am. Soc. Mass Spectrom.* 15, 659–672.

19. Karlsson, N. G., Karlsson, H., and Hansson, G. C. (1995) Strategy for the investigation of O-linked oligosaccharides from mucins based on the separation into neutral, sialic acid- and sulfate-containing species. *Glycoconjugate J.* 12, 69–76.

20. Robbe, C., Capon, C., Flahaut, C., and Michalski, J. C. (2003) Microscale analysis of mucin-type *O*-glycans by a coordinated fluorophore-assisted carbohydrate electrophoresis and mass spectrometry approach. *Electrophoresis* 24, 611–621.

21. Davies, M. J., Smith, K. D., Carruthers, R. A., Chai, W., Lawson, A. M., and Hounsell, E. F. (1993) Use of a porous graphitised carbon column for the high-performance liquid chromatography of oligosaccharides, alditols and glycopeptides with subsequent mass spectrometry analysis. *J. Chromatogr.* 646, 317–326.

22. Koizumi, K. (1996) High-performance liquid chromatographic separation of carbohydrates on graphitized carbon columns. *J. Chromatogr. A* 720, 119–126.

23. Kawasaki, N., Ohta, M., Hyuga, S., Hashimoto, O., and Hayakawa, T. (1999) Analysis of carbohydrate heterogeneity in a glycoprotein using liquid chromatography/mass spectrometry and liquid chromatography with tandem mass spectrometry. *Anal. Biochem.* 269, 297–303.

24. Ninonuevo, M., An, H., Yin, H., Killeen, K., Grimm, R., Ward, R., German, B., and Lebrilla, C. (2005) Nanoliquid chromatography-mass spectrometry of oligosaccharides employing graphitized carbon chromatography on microchip with a high-accuracy mass analyzer. *Electrophoresis* 26, 3641–3649.

25. Thomsson, K. A., Karlsson, N. G., and Hansson, G. C. (1999) Liquid chromatography-electrospray mass spectrometry as a tool for the analysis of sulfated oligosaccharides from mucin glycoproteins. *J. Chromatogr. A* 854, 131–139.

26. Packer, N. H., Lawson, M. A., Jardine, D. R., and Redmond, J. W. (1998) A general approach to desalting oligosaccharides released from glycoproteins. *Glycoconjugate J.* 15, 737–747.

27. Farndale, R. W., Sayers, C. A., and Barrett, A. J. (1982) A direct spectrophotometric microassay for sulfated glycosaminoglycans in cartilage cultures. *Connective Tissue Res.* 9, 247–248.

28. Joshi, H. J., Harrison, M. J., Schulz, B. L., Cooper, C. A., Packer, N. H., and Karlsson, N. G. (2004) Development of a mass fingerprinting tool for automated interpretation of oligosaccharide fragmentation data. *Proteomics* 4, 1650–1664.

29. Domon, B., and Costello, C. E. (1988) A systematic nomeclature for carbohydarate fragmentations in FAB-MS/MS spectra of glyconjugates. *Glycoconjugate J.* 5, 397–409.

30. Huckerby, T. N., Lauder, R. M., and Nieduszynski, I. A. (1998) Structure determination for octasaccharides derived from the carbohydrate-protein linkage region of chondroitin sulphate chains in the proteoglycan aggrecan from bovine articular cartilage. *Eur. J. Biochem.* 258, 669–676.

Subsection B

Glycopeptides

Chapter 14

Enrichment Strategies for Glycopeptides

Shigeyasu Ito, Ko Hayama, and Jun Hirabayashi

Summary

In order to understand glycoprotein functionality, information on the structure of both the core proteins and the glycan moieties is necessary. From a practical viewpoint, glycopeptides rather than whole glycoproteins are the general targets for structural analysis, which is primarily carried out by employing mass spectrometry (MS). Using the "glycoproteomics" concept, several techniques have recently been developed to allow the preparation of a series of reference glycopeptides. In this chapter, we describe two selective capturing methods for glycopeptides, i.e., lectin-affinity chromatography and polysaccharide hydrophilic affinity physicochemical chromatography. The combined use of these methods effectively removes non-glycosylated peptides, the inclusion of which substantially interferes with glycopeptide ionization in MS analysis.

Key words: Glycopeptides, Lectin, High-performance liquid chromatography, Hydrophilic affinity chromatography, Mass spectrometry.

1. Introduction

Glycosylation is undoubtedly one of the major post-translational modifications that occur in eukaryotic proteins. Although their biological functions largely remain unclear, more than half of eukaryotic proteins are glycosylated *(1)*. Increasing lines of evidence clearly indicate that protein glycosylation plays a critical role in protein properties, including solubility, stability, folding, and function. In the light of these considerations, it seems probable that glycosylation is involved in diverse biological phenomena, such as fertilization, development, differentiation, morphogenesis, tumorigenesis, metastasis, and infection. In this context, a

Nicolle H. Packer and Niclas G. Karlsson (eds.), *Methods in Molecular Biology, Glycomics: Methods and Protocols, vol. 534*

DOI: 10.1007/978-1-59745-022-5_14

detailed structural analysis of glycoproteins is an essential element of any serious approach to determine the molecular basis of eukaryotic protein activity. From an analytical viewpoint, various methods based on different separation principles are available for glycans and their derivatives, such as NMR *(2, 3)*, lectin-affinity chromatography *(4, 5)*, multidimensional liquid chromatography *(6)*, and capillary electrophoresis *(7, 8)*. In general, these methods require prior liberation of glycans attached to core proteins by either chemical or enzymatic treatments. Since such procedures result in loss of information regarding both glycosylation sites and core proteins, a full and proper understanding of glycoprotein structure cannot be obtained. To elucidate glycoprotein functions, information on both liberated glycans and protein moieties is required. The concept of "glycoproteomics" has recently been introduced, which involves analysis of glycoproteins/glycopeptides rather than the separate glycans and proteins/peptides *(9)*.

Mass spectrometry (MS) has recently emerged as an extremely sensitive, accurate, and high-throughput tool, indispensable for proteomics. The situation is also true for glycoproteomics, even though the latter approach poses many more difficulties. Nevertheless, advances in electrospray ionization mass spectrometry coupled with reversed-phase high-performance liquid chromatography now allow the systematic and high-throughput analysis of glycopeptides *(10–12)*. Unfortunately, the inclusion of non-glycopeptides in proteolytic digests leads to substantial decreases in sensitivity for glycopeptide analysis, due to considerable ion suppression. Appropriate pretreatments to remove interfering materials such as non-glycosylated peptides are essential for efficient analysis.

Glycans and glycopeptides are in general more hydrophilic than non-glycosylated peptides. For structure-dependent separation of non-derivatized glycans, porous graphitized carbon has often been used *(13–15)*. This stationary phase is not suitable for selective capturing of glycopeptides from proteolytic digests, as it shows even stronger binding to non-glycosyated peptides. In this chapter, two affinity procedures to enrich glycopeptides are described: lectin-affinity chromatography *(16, 17)* and hydrophilic (homophilic) affinity chromatography *(18, 19)*. Lectin-affinity chromatography is often used to prepare a series of glycopeptides by taking advantage of the relatively broad specificity of lectins (as opposed to antibodies). The hydrophilic affinity isolation method is in principle more physicochemically based on homophilic hydrogen bonding between the hydroxyl groups of chromatography resins (polysaccharides) and glycopeptides. The latter method was originally developed to isolate fluorescently labeled (i.e., pyridylaminated) glycans *(20)*, but recently its performance was re-evaluated as a useful enrichment device targeting glycopeptides *(19)*. Though the method is extremely simple, it is effective for the removal of non-glycosylated peptides. However, the method is not necessarily perfect, as inclusion of some non-glycosylated peptides and even leakage of short glycopeptides are sometimes observed. Combined use of these two methods is therefore desirable, all other conditions permitting.

2. Materials

2.1. Denaturation and Enzymatic Digestion of Glycoproteins

1. *Denaturation reagent.* 7 M guanidine–HCl, 0.5 M Tris–HCl, pH 8.5, 10 mM ethylenediaminetetraacetic acid (EDTA), 0.01% (w/v) dithiothreitol.
2. *Iodoacetamide.* Freshly prepared at a concentration of 20 mg/mL for each experiment.
3. *Ammonium bicarbonate.* Freshly prepared at a concentration of 50 mM for each experiment.
4. *Ultrafiltration device.* Ultrafree-0.5 centrifugal filter device, Biomax-5 membrane (Millipore, USA).
5. *Lysylendopeptidase.* Dissolved at 10 AU/mL in 50 mM acetic acid and stored at –30°C.
6. Trypsin (autolysis-free, modified trypsin, a product of Promega): dissolved at 0.2 mg/mL in 50 mM acetic acid, and stored at –30°C.

2.2. Lectin-Affinity Chromatography

1. *Equilibration buffer.* 0.1 M Tris–HCl, 1.5 M NaCl, 1 mM $CaCl_2$, 1 mM $MgCl_2$.
2. *Elution buffer.* 0.2 M lactose dissolved in the above equilibration buffer.
3. *Lectin-conjugated agarose.* Packed in an appropriate empty column (e.g., 150 mm × 4.5 mm ID). Various resins of lectin-agarose are commercially available. In this study, RCA-120 agarose (4.3 mg/mL, Seikagaku Corporation, Tokyo, Japan) is used to capture a series of β-galactoside-containing glycopeptides.

2.3. C18 Cartridge Treatment

1. Sep-Pak Light C18 Cartridge (Waters, Massachusetts, USA) connected to a 10-mL syringe.
2. *Washing solvent.* 0.1% (v/v) trifluoroacetic acid.
3. *Elution solvent.* 100% acetonitrile.

2.4. Further Purification of Glycopeptides by Hydrophilic Affinity Chromatography

1. *Equilibration solvent.* 1-Butanol/ethanol/H_2O (5:1:1, v/v) containing 1 mM $MnCl_2$.
2. *Washing solvent.* 1-Butanol/ethanol/H_2O (5:1:1, v/v).
3. *Elution solvent.* Ethanol/H_2O (1:1, v/v).
4. *Sepharose 4B* (Amersham Biosiciences, NJ, USA). Used as an affinity adsorbent for hydrophilic affinity isolation method. Cellulose Cartridge Column Pack (TaKaRa, Kyoto, Japan) can also be used for the same purpose with almost equal performance. In the latter case, the cartridge is disassembled and cellulose resin is used to pack an online column.

3. Methods

3.1. Denaturation and Digestion of Glycoprotein

As usual for proteomic analysis, protein samples are denatured and *S*-alkylated before proteolytic digestion. Hereafter, a standard procedure is described for *S*-carbamoylmethylation and trypsinization.

1. One milligram (dry weight) of glycoprotein is dissolved in 500 μL of denaturation reagent, and the reaction is allowed to proceed for 3 h at 37°C.
2. In the absence of light (by shielding the reaction tube), 125 μL of iodoacetamide solution is added, and the *S*-alkylation reaction is allowed to proceed for 3 h at 37°C.
3. The reduced and *S*-carbamoylmethylated glycoprotein is ultrafiltrated using a Biomax-5 membrane against 50 mM ammonium bicarbonate at 9,000 *g* until the concentration of guanidine–4HCl becomes less than 1 M (*see* **Note 1**).
4. To the above solution, 10 μL of lysylendopeptidase solution is added, and the proteolysis reaction is performed for 16 h (over night) at 37°C.
5. The enzyme is inactivated by boiling for 10 min.

Hereafter, optional procedures:

6. In some cases, further digestion is necessary to generate smaller fragments, because lysylendopeptidase may generate fragments which are too large (i.e., 100 amino acid residues) and consequently not suitable for MS analysis. For this purpose, e.g., 100 μL of trypsin solution is added to the above solution, and the reaction is allowed to proceed for 16 h (over night) at 37°C.
7. Trypsin is inactivated by boiling for 10 min, and the resultant digest is dried using a conventional concentrator.

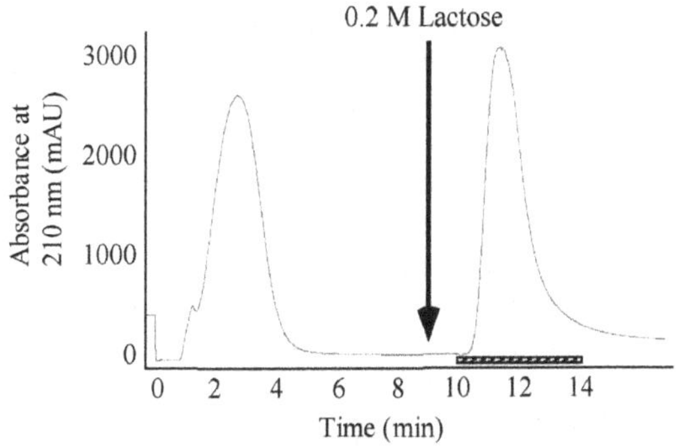

Fig. 1. Lectin-affinity chromatography for the capture of glycopeptides of asialofetuin (ASF). A proteolytic digest of ASF is applied to a column of RCA-120 agarose according to the protocol described in **Subheading 3.2**. The glycopeptides thus retained are specifically eluted with a buffer containing 0.2 M lactose. Fractions spanning retention times 10–14 min are collected as the "glycopeptide" fraction.

3.2. Lectin-Affinity Chromatography

1. As an instrument for purification of glycopeptides by lectin-affinity chromatography, an ÄKTAprime system (Amersham Biosiciences, NJ, USA) can be used, equipped with a column packed with RCA-120-agarose, at 1.0 mL/min flow rate (*see* **Note 2**).
2. The digested glycoprotein is dissolved in the equilibration buffer and loaded into the above lectin column through an injector (*see* **Note 3**).
3. Glycopeptides adsorbed on the column are eluted with a buffer containing a specific hapten sugar (0.2 M lactose in this case) after extensive washing of the column (i.e., until the UV signal due to flow-through peptides returns to the baseline level). An example of the chromatogram is shown in **Fig. 1**.

3.3. Removal of Hapten Sugar

The above-eluted glycopeptide fraction contains a considerable amount of hapten sugar used for elution (0.2 M lactose in this case). However, the sugar must be removed before the following hydrophilic affinity chromatography. For this purpose, a conventional Sep-Pak C18 solid-phase extraction cartridge is utilized.

1. The cartridge is washed with 10 mL of elution solvent.
2. The cartridge is washed with 10 mL of washing solvent.
3. Glycopeptides obtained in the above lectin-affinity separation procedure are loaded on the cartridge.
4. The cartridge is washed with 10 mL of washing solvent.
5. Glycopeptides are eluted with 3 mL of elution solvent and are dried using a concentrator (*see* **Note 4**).

3.4. Further Purification of Glycopeptides by Hydrophilic Affinity Chromatography

A glycopeptide fraction obtained solely by the above lectin-affinity chromatography often contains non-glycosylated peptides. This is particularly the case when lectins, for which the affinity to glycopeptides is relatively low, are used, because researchers may reduce the volume of washing solution to avoid leakage of such "weak" glycopeptides from the column. Alternatively, non-glycosylated peptides may interact with glycopeptides by some specific or nonspecific forces.

The principle of hydrophilic affinity chromatography is based on the homophilic hydrogen bonding that is formed between the polysaccharide resin and the target glycopeptides under less hydrophilic conditions. For this purpose, a relatively high concentration of 1-butanol (50%, v/v) is used to achieve firm binding of the glycopeptides to the resin (under less hydrophilic conditions). Glycopeptides are eluted by increasing the water concentration (50%, v/v).

1. To an appropriate size (e.g., 1.5 mL) of microtube, a polysaccharide resin, either cellulose (15 mg) or agarose (e.g., Sepharose 4B, 20 μL), is added (*see* **Note 5**).
2. The resin in the microtube is washed once with the elution solvent (1 mL). After a brief centrifugation (9,000 *g*, 5 min), the supernatant is removed carefully.
3. The resultant resin is similarly washed with 1 mL of the washing solvent.
4. Glycopeptides derived as described in **Subheading 3.2** are dissolved in 500 μL of equilibration solvent and mixed with the above resin.
5. After gently shaking for 45 min, the microtube is centrifuged and the derived supernatant removed.
6. The resin is washed three times with 1 mL of eluting solvent by centrifugation as above.

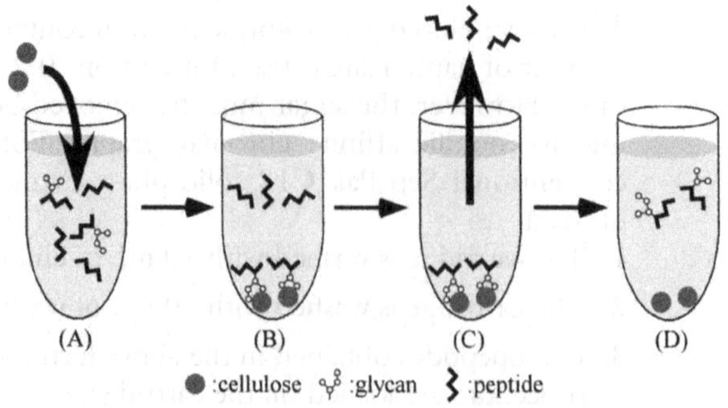

Fig. 2. A schematic drawing of hydrophilic affinity chromatography. (**A**) To an appropriately sized microtube, cellulose resin and the proteolytic digest of ASF are added, and the binding reaction is performed by gentle shaking of the microtube. (**B**) By centrifugation of the microtube, cellulose–glycopeptide complexes are separated from non-glycosylated peptides. (**C**) Washing of the cellulose–glycopeptide complexes three times removes non-glycosylated peptides. (**D**) Adsorbed glycopeptides are recovered by adding an elution solvent (ethanol/H_2O, 1:1, v/v).

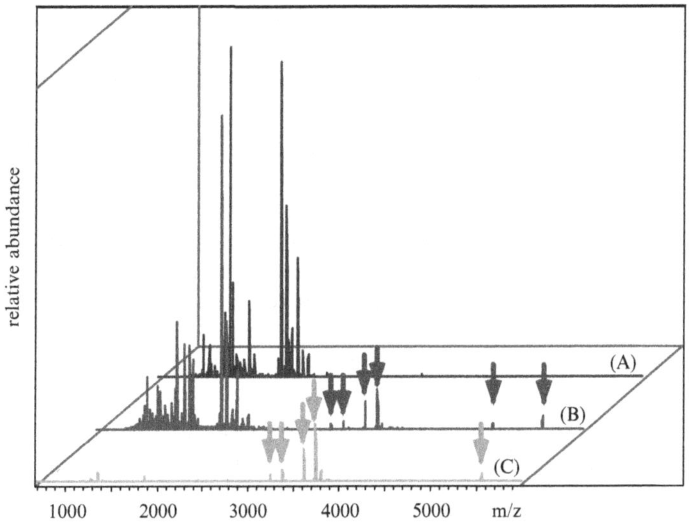

Fig. 3. Matrix-assisted laser desorption/ionization time-of-flight mass spectrometry (MALDI-TOF-MS) spectra showing an enrichment process for glycopeptides derived from ASF. (**A**) Tryptic digest of ASF. No glycopeptides can be observed, possibly because of strong ion suppression by non-glycosylated peptides. (**B**) After purification by lectin-affinity chromatography. Glycopeptides become visible in the spectrum (*arrows*), while a considerable amount of non-glycosylated peptides still remain in the fraction. (**C**) After further purification by hydrophilic affinity chromatography. Note that the signals due to non-glycosylated peptides (*m/z* < 3,000) are not completely eliminated. Signals corresponding to the glycopeptides are indicated with *arrows*.

7. Glycopeptides bound to the resin are eluted with 700 μL of eluting solvent by increasing the concentration of H_2O (up to 50%, v/v). After centrifugation of the microtube, the supernatant containing the glycopeptides is carefully recovered (**Fig. 2**). A representative mass spectrum of the asialofetuin glycopeptides so obtained is shown in **Fig. 3**.

4. Notes

1. Since the efficiency of the subsequent protease digestion largely depends on the concentration of the substrates (glycoproteins), ultrafiltration, rather than dialysis, is used to exchange buffers and also to reduce the reaction volume.
2. Other chromatography systems, e.g., other ÄKTA series and conventional HPLC systems, can be substituted whenever appropriate.
3. After proteolysis of glycoproteins, the resultant glycopeptides often show lowered affinity to lectins because of loss of

"multivalency" with respect to glycans. In order to avoid substantial leakage of glycopeptides having weaker affinity for the lectin, the sample volume should be reduced as much as possible.

4. Glycopeptides obtained by lectin-affinity chromatography must be "cleaned up" before the following hydrophilic affinity chromatography, as they contain a considerable amount of hapten sugar (in this case 0.2 M lactose). Since hydrophilic affinity chromatography is based on hydrogen bonds formed between carbohydrate moieties of glycopeptides and resin polysaccharides, such contamination by sugars in the elution buffer is undesirable and must be removed. For this purpose, a reversed-phase C18 cartridge is used.
5. For optimal performance, a suitable ratio of resin volume (cellulose or Sepharose) to the amount of glycopeptide should be maintained. If an excess volume of polysaccharide is used, free peptides will contaminate the eluted fraction, whereas if too small a volume of resin is used, a significant amount of glycopeptides will leak from the polysaccharide column.

Acknowledgments

This work was supported in part by New Energy and Industrial Technology Development Organization (NEDO) in Japan.

References

1. Apweiler, R., Hermjakob, H., and Sharon, N. (1999) On the frequency of protein glycosylation, as deduced from analysis of the SWISS-PROT database. *Biochim. Biophys. Acta* 1473, 4–8.
2. Yamaguchi, Y., Kato, K., Shindo, M., Aoki, S., Furusho, K., Koga, K., Takahashi, N., Arata, Y., and Shimada, I. (1998) Dynamics of the carbohydrate chains attached to the Fc portion of immunoglobulin G as studied by NMR spectroscopy assisted by selective 13C labeling of the glycans. *J. Biomol. NMR* 12, 385–394.
3. Deisenhofer, J. (1981) Crystallographic refinement and atomic models of a human Fc fragment and its complex with fragment B of protein A from *Staphylococcus aureus* at 2.9- and 2.8-A resolution. *Biochemistry* 20, 2361–2370.
4. Nakamura, S., Yagi, F., Totani, K., Ito, Y., and Hirabayashi, J. (2005) Comparative analysis of carbohydrate-binding properties of two tandem repeat-type Jacalin-related lectins, *Castanea crenata* agglutinin and *Cycas revolute* leaf lectin. *FEBS J.* 272, 2784–2799.
5. Hirabayashi, J., Hashidate, T., Arata, Y., Nishi, N., Nakamura, T., Hirashima, M., Urashima, T., Oka, T., Futai, M., Muller, E.W., Yagi, F., and Kasai, K. (2002) Oligosaccharide specificity of galectin: a search by frontal affinity chromatography. *Biochim. Biophys. Acta* 1572, 232–254.
6. Natsuka, S., and Hase, S. (1998) Analysis of N and O-glycans by pyridylamination. *Methods Mol. Biol.* 76, 101–113.
7. Sei, K., Nakano, M., Kinoshita, M., Masuko, T., and Kakehi, K. (2002) Collection of a1-acid glycoprotein molecular species by capillary electrophoresis and the analysis of their molecular masses and carbohydrate chains: basic studies on the analysis of glycoprotein glycoforms. *J. Chromatogr. A* 958, 273–281.

8. Kakehi, K., Kinoshita, M., Kawakami, D., Tanaka, J., Sei, K., Endo, K., Oda, Y., Iwaki, M., and Masuko, T. (2001) Capillary electrophoresis of sialic acid-containing glycoprotein: effect of the heterogeneity of carbohydrate chains on glycoform separation using an α1-acid glycoprotein as a model. *Anal. Chem.* 73, 2640–2647.
9. Hirabayashi, J. (2004) Lectin-based structural glycomics: glycoproteomics and glycan profiling. *Glycoconj. J.* 21, 35–40.
10. Huddleston, M.J., Bean, M.F., and Carr, S.A. (1993) Collisional fragmentation of glycopeptides by electrospray ionization LC/MS and LC/MS/MS: Methods for selective detection of glycopeptides in protein digests. *Anal. Chem.* 65, 877–884.
11. Carr, S.A., Huddleston, M.J., and Bean, M.F. (1993) Selective identification and differentiation of *N*- and *O*-linked oligosaccharides in glycoproteins by liquid chromatography-mass spectrometry. *Protein Sci.* 2, 183–196.
12. Harazono, A., Kawasaki, N., Itoh, N., Hashii, N., Ishii-Watabe, A., Kawanishi, T., and Hayakawa, T. (2006) Site-specific *N*-glycosylation analysis of human plasma ceruloplasmin using liquid chromatography with electrospray ionization tandem mass spectrometry. *Anal. Biochem.* 348, 259–268.
13. Koizumi, K. (1996) High-performance liquid chromatographic separation of carbohydrates on graphitized carbon columns. *J. Chromatogr. A* 720, 119–126.
14. Larsen, R.M., Hojrup, P., and Roepstorff, P. (2005) Characterization of gel-separated glycoproteins using two-step proteolytic digestion combined with sequential microcolumns and mass spectrometry. *Mol. Cell Proteomics* 4, 107–119.
15. Davis, M.J., Smith, K.D., Carruthers, R.A., Chai, W., Lawson, A.M., and Hounsell, E.F. (1993) Use of a porous graphitised carbon column for the high-performance liquid chromatography of oligosaccharides, alditols and glycopeptides with subsequent mass spectrometry analysis. *J. Chromatogr.* 646, 317–326.
16. Qiu, R., and Regnier, F.E. (2005) Comparative glycoproteomics of *N*-linked complex-type glycoforms containing sialic acid in human serum. *Anal. Chem.* 77, 7225–7231.
17. Kaji, H., Saito, H., Yamauchi, Y., Shinkawa, T., Taoka, M., Hirabayashi, J., Kasai, K., Takahashi, N., and Isobe, T. (2003) Lectin affinity capture, isotope-coded tagging and mass spectrometry to identify *N*-linked glycoproteins. *Nat. Biotech.* 21, 667–672.
18. Tajiri, M., Yoshida, S., and Wada, Y. (2005) Differential analysis of site-specific glycans on plasma and cellular fibronectins: application of a hydrophilic affinity method for glycopeptide enrichment. *Glycobiology* 15, 1332–1340.
19. Wada, Y., Tajiri, M., and Yoshida, S. (2004) Hydrophilic affinity isolation and MALDI multiple-stage tandem mass spectrometry of glycopeptides for glycoproteomics. *Anal. Chem.* 76, 6560–6565.
20. Shimizu, Y., Nakata, M., Kuroda, Y., Tsutsumi, F., Kojima, N., and Mizuochi, T. (2001) Rapid and simple preparation of *N*-linked oligosaccharides by cellulose-column chromatography. *Carbohydr. Res.* 332, 381–388.

Chapter 15

Introductory Glycosylation Analysis Using SDS-PAGE and Peptide Mass Fingerprinting

Nicole Wilson, Raina Simpson, and Catherine Cooper-Liddell

Summary

Glycosylation is extremely complex, with the potential for a protein to have oligosaccharides attached at multiple sites, and for each site of glycosylation to have multiple structures attached to it. Structural information on the oligosaccharides bound to either asparagine residues (N-linked) or serine and threonine residues (O-linked) requires sensitive, specialised, and complex techniques and equipment. We show here, however, that a large amount of information regarding the glycosylation of glycoproteins can be obtained with common protein techniques such as 1D SDS-PAGE and peptide mass fingerprinting (PMF). Enzymatic deglycosylation in combination with SDS-PAGE and PMF analysis can determine the relative percentage of N-linked carbohydrate on the glycosylated protein, as well as attachment sites of the oligosaccharides.

Key words: SDS-PAGE, Glycosidase, PNGase F, PMF, Peptide mass fingerprinting, Deglycosylation.

1. Introduction

Most proteins undergo post-translational modification, in particular glycosylation, which can alter their physical and chemical properties (e.g., MW, pI, folding, stability, activity, antigenicity, and function). The presence or absence of glycosylation may be significant to both the activity and longevity of the protein in a biological system *(1–3)*.

An efficient way to determine if a protein is glycosylated is to observe a molecular weight difference between the glycosylated and de-glycosylated protein by SDS-PAGE. This can be achieved using enzymes that remove the oligosaccharides while leaving the protein backbone intact. The shift in molecular weight can be

Nicolle H. Packer and Niclas G. Karlsson (eds.), *Methods in Molecular Biology, Glycomics: Methods and Protocols, vol. 534*

DOI: 10.1007/978-1-59745-022-5_15

used to determine whether the increased apparent mass is due to the oligosaccharides. Furthermore, using a combination of enzyme treatments to selectively remove the N-linked oligosaccharides vs. removal of both the N- and some of the O-linked oligosaccharides enables the determination of the type/s of glycosylation present on the protein.

Peptides mass fingerprinting (PMF) methods *(4, 5)* such as MALDI-MS are not usually able to detect glycopeptides using general methods, due to the large mass of the N-linked glycopeptides, as well as to their heterogeneity. Removing the oligosaccharides allows the chance of seeing the de-glycosylated peptides, and can also lead to critical information being obtained on the sites of glycosylation.

Removing the N-linked oligosaccharides using Peptide-*N*-glycosidase F (PNGase F), cleaves the GlcNAc-Asn linkage *(6)*, converting the asparagine (Asn) to an aspartic acid (Asp). The difference in the molecular weight between these two amino acids (+1 Da) will alter the mass of the peptide produced when the protein is digested using trypsin. The change in mass of the peptide enables the determination of whether the site is fully or partially glycosylated or unglycosylated.

Similarly, extra peptides that contain serine or threonine residues visible after removal of O-linked oligosaccharides, can indicate probable sites of O-linked glycosylation. As there is no enzyme that can release all O-linked oligosaccharide structures we use a cocktail of glycosidases that will remove many of the substituents of the mammalian mucin-type O-linked oligosaccharides.

2. Materials

Due to the use of mass spectrometry, which is a very sensitive technique, precautions such as wearing gloves and using high quality water at all times, are recommended.

2.1. Enzymatic Deglycosylation

1. Protein denaturation solution: 5 mM CHAPS, 30 mM 2-mercaptoethanol, and 10 mM Tris HCl (pH 7–7.2).
2. PNGase F (*N*-Glycosidase F, Roche Diagnostics).
3. Sialidase A (Neuraminidase, Roche Diagnostics).
4. β*(1–4)*Galactosidase (Prozyme – Glyko).
5. β-*N*-Acetylglucosaminidase (*N*-acetyl-β-D-glucosaminidase, Prozyme – Glyko).
6. O-Glycanase (Prozyme – Glyko).

2.2. SDS-PAGE

1. SDS-PAGE sample buffer: 10% Glycerol, 1% SDS, 10 mM DTT.
2. 4–20% Tris HCl gels (BioRad).
3. Protein stain such as Coomassie Brilliant Blue G250 (Sigma) or a fluorescent stain (Deep Purple – GE Healthcare, or Flamingo Pink – BioRad).

2.3. Trypsin Digestion

1. Trypsin (20 mg/ml in 50 mM NH_4HCO_3) (sequencing grade trypsin – Promega) (*see* **Note 1**).
2. C_{18} resin tip (available from Eppendorf or Millipore in 10 μl size).

2.4. MALDI-MS/PMF

1. MALDI target plate.
2. 3 μg/ml α-cyano-4-hydroxycinnamic acid in 70% acetonitrile, 0.1% TFA (make fresh).

3. Methods

3.1. Enzymatic Deglycosylation

This method is designed to deglycosylate small quantities of sample (<10 μg). Ideally, the sample buffer should be 100 mM NH_4HCO_3 pH 8.5 or 100 mM Tris HCl pH 8.5 (*see* **Note 2**).

1. Divide the glycoproteins into three equal aliquots (~3–5 μg each) in 0.5 ml Eppendorf tubes, for a 'control', 'de-*N*-glycosylated' and a 'combined de-*N*- and de-O-glycosylated' sample.
2. Reduce sample volume to 5 μl using a vacuum centrifuge.
3. To each aliquot, add 5 μl of denaturation solution (*see* **Note 3**).
4. Seal the tubes and leave at room temperature for 30 min.
5. Add the enzymes as shown in **Table 1** to the appropriate samples, seal and vortex the tubes before incubating the samples for 12 h at 37°C.

3.2. SDS-PAGE (Fig. 1)

These instructions assume the use of precast one-dimensional SDS-PAGE gels. We have used the Invitrogen Surelock, and the BioRad Mini or Criterion systems.

1. Add 5 μl of SDS-PAGE sample buffer (*see* **Note 5**).
2. Boil samples for 5 min at 100°C.
3. Allow the samples to cool and load each sample in a separate lane of a 4–20% Tris HCl 1D gel (*see* **Note 6**).
4. Load 5 μl of a broad range molecular weight marker in a separate lane next to the samples. We use precision plus markers from BioRad.

Table 1
Enzymes used for deglycosylation

Sample	Enzymes
Control (glycosylated protein)	None
De-N-glycosylated protein (N-linked oligosaccharides released)	1 μl PNGase F
De-N- and de-O-glycosylated protein (N-linked and selective O-linked oligosaccharides released) (*see* **Note 4**)	1 μl PNGase F
	1 μl Sialidase A (Neuraminidase)
	1 μl β*(1-4)*Galactosidase
	1 μl β-*N*-Acetylglucosaminidase
	1 μl *O*-Glycanase

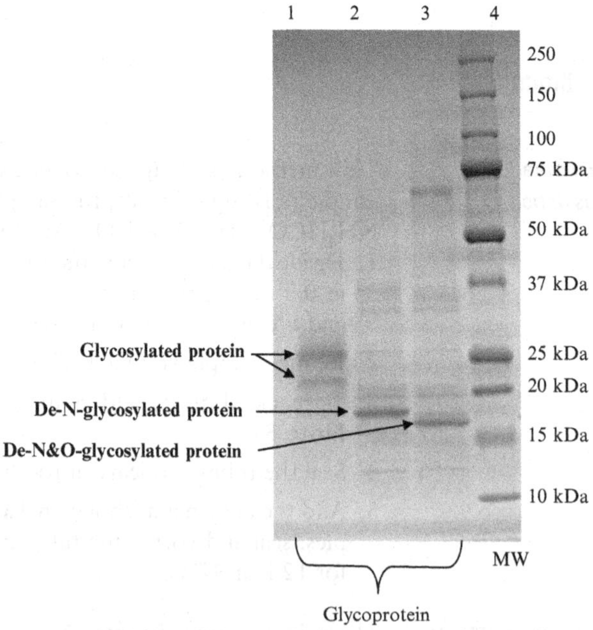

Fig. 1. 1D-PAGE of a glycoprotein, de-N-glycosylated, and de-N- and de-O-glycosylated (in lanes 1, 2, and 3, respectively). The samples were run on a 4–20% linear 1D-PAGE with BioRad Precision Plus protein standards for molecular weight determination.

5. Run the gel at a constant 30 mA per gel or at 200 V for approximately 1 hr, until dye front runs off the bottom of the gel (*see* **Note 7**).
6. Remove the gel from the cassette, fix and stain the gel using a protein stain. For a visual stain we use Coomassie brilliant blue

G250 (BioRad) as per manufacturer's instructions or for more sensitive staining we use deep purple (GE Healthcare) as per manufacturer's instructions.

7. Image the stained gels on an appropriate imaging system as per manufacturer's instructions. For example we use a flat bed transmission scanner to image Coomassie or other visually based stained gels, and the LAS4000 (FUJI) or Typhoon (GE Healthcare) to image fluorescently stained gels.

3.3. Tryptic Digestion

1. Identify and cut bands of interest (*see* **Note 8**).
2. Place gel pieces in either separate wells of a 96-well plate or into separate Eppendorf tubes (*see* **Note 9**).
3. De-stain gel plugs using 50% acetonitrile in 50 mM NH_4HCO_3 with gentle agitation for 20 min, or until the stain has been completely removed from the gel plug (*see* **Note 10**).
4. Remove de-staining solution using a pipette and dry the gel plugs completely by vacuum centrifugation or in an oven (5 min at 100°C).
5. Cool samples on ice and add 3 µl of trypsin (20 mg/ml in 50 mM NH_4HCO_3) to each gel plug (*see* **Note 11**). Keep samples on ice for a further 15 min.
6. Add 20 µl of either H_2O or 50 mM NH_4HCO_3 to the gel plugs. Seal the plate or tubes and incubate for 12 h at 37°C or for 1–2 h at 50°C.
7. Ensure that the gel plug is surrounded by solution. If the gel plug has absorbed all of the liquid add 5–10 µl of H_2O or 50 mM NH_4HCO_3 to the plug. Sonicate the samples for 15 min to ensure that all of the digested peptides are in solution (*see* **Note 12**).
8. Remove any non-protein contaminants and concentrate the tryptic digests on C_{18} resin tips using the following steps.
   (a) Activate C_{18} tips with 2 × 10 µl 90% acetonitrile, 0.1% TFA.
   (b) Equilibrate C_{18} tips with 2 × 10 µl 0.1% TFA.
   (c) Pipette samples onto C_{18} tips. Pipette samples up and down at least ten times.
   (d) Wash C_{18} tips with 2 × 10 µl 0.1% TFA.
9. Elute digested peptides onto a MALDI target plate (*see* **Note 13**) using 3 µg/ml α-cyano-4-hydroxycinnamic acid in 70% acetonitrile, 0.1% TFA. Allow samples to dry.

3.4. Peptide Mass Fingerprinting

We typically use a MALDI TOF/TOF mass spectrometer (Applied Biosystems 4700 Proteomics Analyser) in positive reflectron mode with automated MS/MS performed on the most intense peaks.

3.5. Peptide and Glycosylation Site Identification (Fig. 2)

1. Take the peptide peak lists produced by PMF of the untreated, de-N-glycosylated, and the de-N- and de-O-glycosylated protein and submit each to a peptide matching program such as Mascot (Matrix Science www.matrixscience.com) or FindMod (ExPASy Proteomics Tools http://au.expasy.org/tools/) (*see* **Note 14**).
2. If the protein of interest is known, or once it is identified using Mascot or a similar program, use the FindMod tool to look for sites of glycosylation. For N-linked sites, search the protein sequence for asparagine (Asn, N) residues, and where it exists in the Asn-Xaa-Ser/Thr/Cys (N-X-S/T/C) where Xaa is not proline motif, change the asparagine to an aspartic acid (N to D which occurs when the N-linked oligosaccharides are removed using PNGase F). Paste this modified sequence into FindMod.
3. Match the peptide mass list from the untreated, de-N-glycosylated and de-N- and de-O-glycosylated protein bands against this modified protein sequence.
4. Compare the peptide matches against the normal and modified sequences. If new matches appear against the modified sequence for the de-N-glycosylated sample, it can be concluded that de-N-glycosylation of the protein prior to SDS-PAGE has removed N-linked oligosaccharides, modifying the protein backbone from an Asn (N) to an Asp (D). Thus the site is N-linked glycosylated. (*see* **Note 15**).

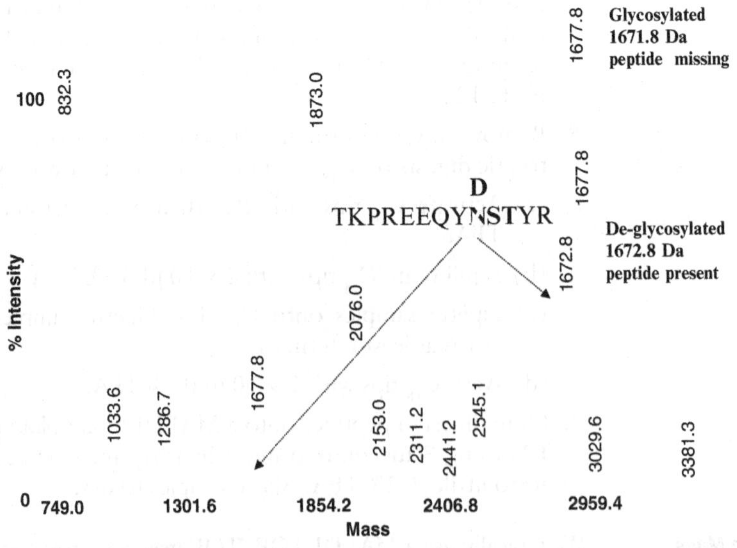

Fig. 2. MALDI-MS spectra of a protein tryptic digestion containing a single N-linked site, both before and after removal of the N-linked oligosaccharides using enzymatic de-glycosylation (PNGase F, Roche Diagnostics).

Example:
Protein amino acid sequence
TKPREEQYNSTYR $m/z = 1{,}671.8$
When de-N-glycosylated will become
TKPREEQYDSTYR $m/z = 1{,}672.8$

5. Similarly look for new peptide matches between the de-N- and the de-N- and de-O-glycosylated protein. If new peptide matches containing serine or threonine residues are found, this indicates that the peptide may have been O-glycosylated.

4. Notes

1. Trypsin can be made up, aliquoted and stored at -20°C for 3 months. Ensure the trypsin is thawed immediately before use to reduce the incidence of autolysis, which may cause peptide masses to be quenched in the MS spectrum.
2. A high salt buffer may decrease the enzyme efficiency.
3. This solution denatures the protein sample to increase the accessibility of the sugar attachment sites for the enzymes.
4. There is no enzyme that can release all O-linked oligosaccharide structures. *O*-glycanase is the only known enzyme capable of cleaving the linkage between an oligosaccharide and serine or threonine. However, this enzyme can only release the disaccharide Gal(β*1-3)*GalNAc from a serine or threonine linkage *(7, 8)*. It cannot release oligosaccharides with substituents on the disaccharide, such as sialic acid, galactose or *N*-acetylglucosamine. Thus we use a cocktail of glycosidases that will first remove these substituents on the O-linked oligosaccharides, often, but not universally, leaving the core disaccharide, which can then be released using O-glycosidase.
5. Bromophenol blue (0.001%) can be added to the sample so that it is easy to visualise when the proteins have migrated through the gel.
6. The type of gel used may vary depending on the theoretical weight of the target protein. See manufacturer's instructions to determine the appropriate gel for any particular protein. Generally the 4–20% 1D gel will give good resolution of protein between 5 and 200 kDa.
7. Gel run time will depend on the type of gel used. The 1 h time limit is best used as a guide only.
8. Make sure that if whole bands are cut that they are cut into approximately 1–1.5 mm^3 gel pieces to improve digestion.

9. Ensure that the plate or tubes used can withstand temperatures up to 100°C. A 96-well plate is our preference due to the ease of use for later clean-up steps.
10. This destaining solution is used for Coomassie-stained gel plugs. For deep purple (GE Healthcare) or flamingo pink (BioRad) stained gels, the plugs do not require de-staining, but wash in 50 mM NH_4HCO_3 for 10 min.
11. The average gel plug size is approximately 1–2 mm^3. For plugs larger than this the amount of trypsin added should be adjusted. For a 1D band (1 × 5 mm) a volume of 9 µl of trypsin would be required.
12. If the plate format is used, do not attempt to recover the sample solution that has condensed on the adhesive cover as this may contaminate the sample with polymer.
13. The MALDI plate should be thoroughly cleaned by sonication in methanol for 15 min, and then dried completely prior to use.
14. FindMod is a good tool to use where the protein is known. This program allows you to search the experimental masses against the known protein sequence.
15. It is possible for a site to be only partially glycosylated, and therefore the peptide peak list should always be searched against both the theoretical sequence, as well as N D modified sequence. It is also important to search the untreated peptide peak list against the N D modified sequence to eliminate false hits.

References

1. Kukuruzinska, M.A. and Lennon, K. (1998) Protein N-glycosylation: molecular genetics and functional significance. *Crit Rev Oral Biol Med* 9, 415–448
2. Hang, H.C. and Bertozzi, C.R. (2005) The chemistry and biology of mucin-type O-linked glycosylation. *Bioorg Med Chem* 13, 5021k–5034k
3. Helenius, A. and Aebi, M. (2004) Roles of N-linked glycans in the endoplasmic reticulum. *Annu Rev Biochem* 73, 1019–1049
4. Jonsson, A.P. (2001) Mass spectrometry for protein and peptide characterisation. *Cell Mol Life Sci* 58, 868–884
5. Thiede, B., Hohenwarter, W., Krah, A., Matotow, J., Schmid, M., Schmidt, F., and Jungblut, P.R. (2005) Peptide Mass Fingerprinting. *Methods* 35, 237–247
6. Takahashi, N. (1977) Demonstration of a new amidase acting on glycopeptides. *Biochem Biophys Res Commun* 76, 1194–1201
7. Umemoto, J., Bhavandan, V.P., and Davidson, E.A. (1977) Purification and properties of an Endo-alpha-N-acetyl-d-galactosaminidase from Diplococcus pneumoniae. *J Biol Chem* 252, 8609–8614
8. Iwase, H. and Hotta, K. (1993) Release of O-linked glycoprotein glycans by endo-alpha-N-acetylgalactosaminidase. *Methods Mol Biol* 14, 151–159

Chapter 16

Characterization of N-Linked Glycosylation on Recombinant Glycoproteins Produced in *Pichia pastoris* Using ESI-MS and MALDI-TOF

Bing Gong, Michael Cukan, Richard Fisher, Huijuan Li, Terrance A. Stadheim, and Tillman Gerngross

Summary

The production of recombinant therapeutic glycoproteins is an active area of research and drug development. Typically, improvements in therapeutic glycoprotein efficacy have focused on engineering additional N-glycosylation sites into the primary amino acid sequence or attempting to control a particular glycoform profile on a protein through process improvements. Recently, a number of alternative expression systems have appeared that are challenging the dominance of mammalian cell culture. Our laboratory has focused on the re-engineering of the secretory pathway in the yeast *Pichia pastoris* to perform glycosylation reactions that mimic processing of *N*-glycans in humans. We have demonstrated that human antibodies with specific human *N*-glycan structures can be produced in glycoengineered lines of *Pichia pastoris* and that antibody-mediated effector functions can be optimized by generating specific glycoforms. In this chapter we provide detailed protocols for the analysis of glycosylation on intact glycoproteins by MALDI-TOF and site specific *N*-glycan occupancy on digested glycoprotein using ESI-MS.

Key words: N-linked glycosylation, ESI-MS, MALDI-TOF, Site-specific occupancy.

1. Introduction

The production of recombinant therapeutic glycoproteins requires carefully controlled processes that allow for the production of the most efficacious product while maximizing product uniformity. Biotherapeutics are subject to several post-translational modifications (PTMs) including N-linked glycosylation. N-linked glycosylation occurs through a conserved pathway, which starts in the endoplasmic reticulum (ER) with the transfer of a core oligosaccharide

Nicolle H. Packer and Niclas G. Karlsson (eds.), *Methods in Molecular Biology, Glycomics: Methods and Protocols, vol. 534*

DOI: 10.1007/978-1-59745-022-5_16

to an asparagine residue in the sequence Asn-X-Ser/Thr within the polypeptide chain *(1)*. This modification can affect protein structure and function, as well as impact serum stability. Thus, analytical techniques that efficiently monitor the micro- and macroheterogeneity of glycans on a glycoprotein are essential.

Since mammalian expression systems produce mainly human glycans, they are currently the dominant platform for therapeutic glycoprotein production. However, major advances in the bioengineering of glycosylation pathways has allowed for the production of glycoproteins with functionally more desirable glycoforms *(2–19)*. The analysis of N-linked glycosylation on therapeutic proteins requires accurate, fast and reliable methods. Here, we describe methods for determining *N*-glycan composition and site-specific occupancy by analyzing intact and protease-digested glycoproteins, respectively.

2. Materials

2.1. Matrix Assisted Laser Desorption Ionization-Time of Flight (MALDI-TOF) Mass Spectrometry

1. Centrifugal concentrator with 10,000 Da molecular weight cutoff columns (Sartorius AG, Goettingen, Germany).
2. *Ammonium bicarbonate (Sigma, St Louis, MO)*. 10 mM solution in HPLC grade water (EMD Biosciences, San Diego, CA), adjust pH to 7.8 with HCl.
3. *α-Cyano-4-hydroxycinnamic acid (CHCA) (Sigma)*. Dissolve until saturation in 50% acetonitrile (EMD Biosciences), 0.1% formic acid (Sigma).
4. *Sinapinic acid (Sigma)*. Dissolve until saturation in 50% acetonitrile, 0.1% trifluoroacetic acid (TFA, 99% spectrophotometric grade) (Sigma).
5. Stainless steel MALDI sample plate (Applied Biosystems, Foster City, CA).
6. Voyager DE linear MALDI-TOF (Applied Biosystems).
7. Symbiot I system, optional (Applied Biosystems).

2.2. Nano-Electrospray Ionization Mass Spectrometry (Nano-ESI-MS)

1. HPLC grade water and acetonitrile containing 0.1% (v/v) formic acid.
2. *Commercial lyophilized monoclonal IgG1 antibody (mAb)*. Dissolve in HPLC grade water at 3 mg/ml, store at 4°C.
3. *Dithiothreitol (DTT) (Sigma)*. Prepare 1 M solution in HPLC grade water and store 50 μl aliquots at −20°C.
4. PNGase F (New England Biolabs, Ipswich, MA).
5. LTQ mass spectrometer (Thermo Finnigan, San Jose, CA).

6. Surveyor HPLC system (Thermo Finnigan).
7. Jupiter 5u C4 300 Å column (150 × 2.0 mm) (Phenomenex, Torrance, CA).
8. Advion Triversa Nanomate (Ithaca, NY).
9. 96-well sample plate (Abgene, Epsom, UK)

2.3. Trypsin Digest of Protein

1. *Guanidine hydrochloride (Sigma).* 8 M solution in HPLC grade water, adjust pH to 7.8 with NaOH.
2. *Dithiothreitol (DTT) (Sigma).* Prepare 1 M solution in HPLC grade water and store 50 ml aliquots at –20°C.
3. *Alkylating reagent (10×).* 400 mM (freshly prepared) from recrystallized iodoacetic acid (Sigma) in 1 M Tris–base (Sigma); neutralize with ammonium hydroxide (EMD Biosciences).
4. *Recrystallization of IAA.* Saturate iodoacetic acid in boiling hexane, then cool solution in an ice bucket. IAA will form crystals. Allow to dry. Transfer IAA crystals to a sterile 50 ml conical tube. Wrap with aluminum foil to protect from light. Store at –20°C.
5. Centrifugal concentrator with 50,000 Da molecular weight cutoff columns (Sartorius AG).
6. *Digestion buffer.* 100 mM Tris–HCl in water, pH 7.4.
7. *Trypsin (Sigma).* Dissolve to a concentration of 1 mg/ml in 10 mM HCl (Fisher, Hampton, NH).
8. *PNGase F (New England Biolabs).* Provided at a concentration of 600 U/μl.

2.4. Reversed Phase HPLC

1. Jupiter 4u Proteo 90 Å column (250 × 4.6 mm, Phenomenex).
2. *Solvent A.* Prepare a 0.1% TFA (Sigma) solution in HPLC grade water.
3. *Solvent B.* Prepare a 0.08% TFA solution in HPLC grade acetonitrile.
4. Deep-well 96-well plate (Axygen Scientific, Union City, CA).
5. CentriVap concentrator (Labconco, Kansas City, MO)

3. Methods

Intact glycoproteins are commonly analyzed using MALDI-TOF and ESI-MS. The ability to resolve individual glycoforms depends on the mass of the glycoprotein, the extent of glycosylation and also the resolution of the instrument (*see* **Fig. 1**). Here we describe the use of matrix-assisted laser desorption ionization-time of flight (MALDI-TOF) mass spectrometry to detect the N-glycosylation present on a small glycoprotein (Kringle-3-His6) with a single

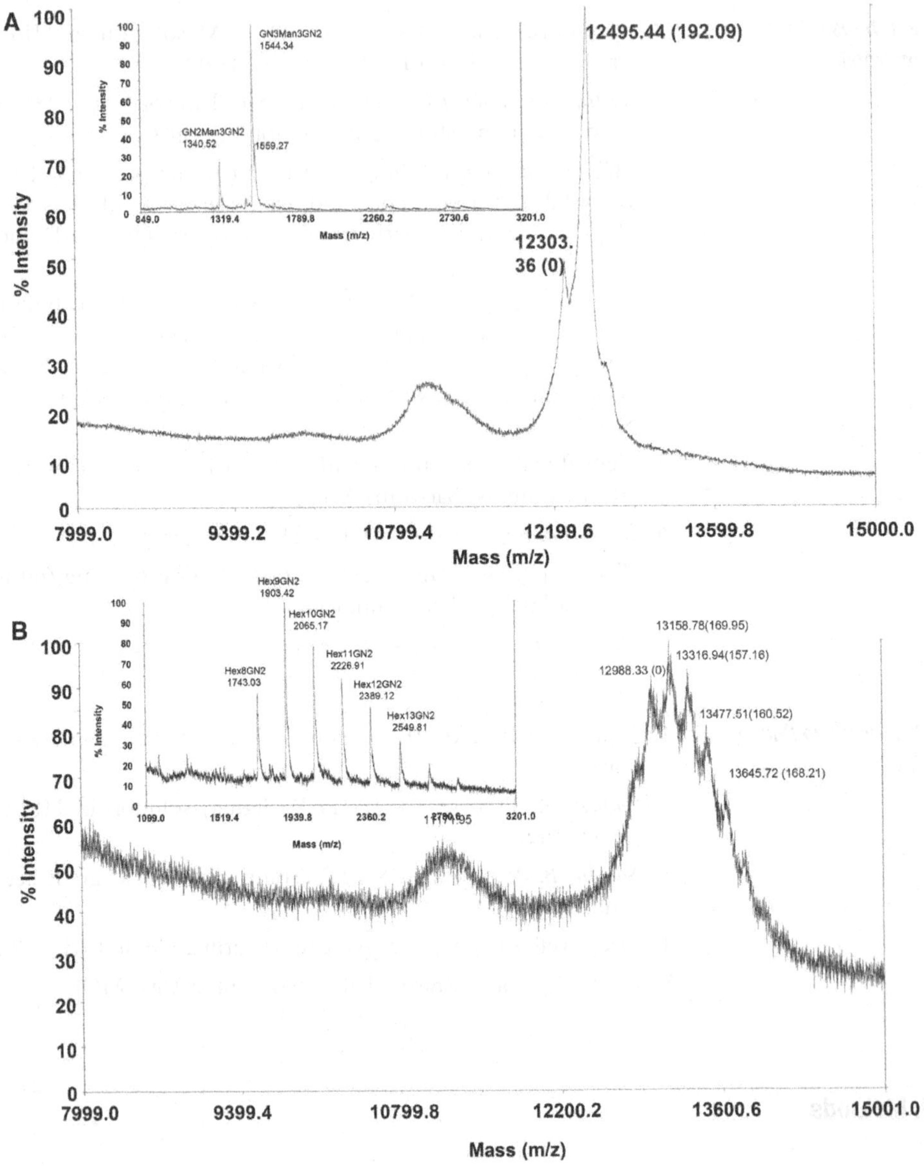

Fig. 1. MALDI-TOF of intact glycoproteins. His-tagged K3 purified from a glycoengineered *P. pastoris* strain with high N-glycan uniformity (**A**). Insert shows a MALDI-TOF profile of glycans released from the protein by PNGase F digest *(20)*. His-tagged K3 purified from a wild-type *P. pastoris* strain with heterogeneous N-glycans (**B**). Insert shows a MALDI-TOF profile of glycans released from the protein by PNGase F digest.

glycosylation site and electrospray ionization-mass spectrometry (ESI-MS) to measure N-glycosylation on a commercial recombinant monoclonal antibody.

The determination of N-linked occupancy is achieved by cleavage of the glycoprotein such that each N-linked site is isolated as a separate glycopeptide. Efficient proteolysis is achieved by reduction and alkylation of the glycoprotein followed by digestion with trypsin. The digested material is divided into two equal fractions, one without further treatment and another digested with PNGase F. Peptides from both samples are separated by HPLC followed by ESI-MS/MS analysis. The peak areas corresponding to the deglycosylated peptides are then used to calculate site-specific occupancy.

3.1. MALDI-TOF Analysis of Intact Glycoprotein

1. Low salt containing protein samples should be prepared by buffer exchanging into 10 mM NH_4HCO_3 using a centrifugal concentrator with 10,000 Da molecular weight cutoff columns (Vivascience) at least three times.
2. Prepare a matrix solution by dissolving 10 mg of sinapinic acid in 1 ml of 50% acetonitrile, 0.1% TFA, vortex to mix well. Spin at 9,300 × *g* for 1 min before use.
3. Spot 0.5 μl to a stainless steel MALDI plate followed by applying an equal amount of matrix solution (*see* **Note 1**). Allow the sample to dry. Automatic spotting is accomplished using a Symbiot I system.
4. Determine protein molecular mass using a Voyager DE linear MALDI-TOF mass spectrometer using delayed extraction. Ions are generated by irradiation with a pulsed nitrogen laser (337 nm) with a 2 ns pulse time. The instrument is operated in the delayed extraction mode with a 300 ns delay and an accelerating voltage of 25 kV. The grid voltage is 92.00%, the guide wire voltage is 0.05%, and the internal pressure is less than 5×10^{-7} Torr. Spectra are generated from the sum of 100–200 laser pulses and acquired with a 500 MHz digitizer in the positive ion mode (**Fig. 1**).

3.2. ESI-MS of Intact Glycoprotein

1. Commercial mAb (300 μg) is dissolved in 100 μl 25 mM NH_4HCO_3, pH 7.8.
2. Divide the above sample into two equal aliquots in microcentrifuge tubes, one for a PNGase F reaction (Sample 1) and the other for non PNGase F-treated sample (Sample 2).
3. Add PNGase F (50 units) to Sample 1 and incubate overnight at 37°C.
4. Aliquot one-half of the PNGaseF-digested sample into a fresh microcentrifuge tube (Sample 3). Add 1 M DTT to a final concentration of 10 mM, and incubate the sample at 37°C for 30 min.

5. Aliquot the remaining one-half of the non PNGase-treated sample into a fresh microcentrifuge tube (Sample 4). Add 1 M DTT to a final concentration of 10 mM and incubate the sample at 37°C for 30 min.
6. Add formic acid to a final concentration of 0.1% (v/v) to both the PNGase F treated and non-PNGase F treated samples.
7. Calibrate a Thermo Finnigan LTQ linear ion-trap mass spectrometer with PPG 2700 and tune manually using an Advion Triversa Nanomate as the nanospray source (*see* **Note 2**). Listed below are the instrument parameters optimized in our setup:

Ionization mode:	Positive ESI ionization
Spray voltage (on Nanomate):	1.5 kV
Capillary temperature:	200°C (reduced mAb)/250°C (intact mAb)
Capillary voltage:	40 V
Tube lens voltage:	175 V

8. Inject 10 μl of sample onto the Phenomenex Jupiter C4 column and perform, a 20-min HPLC run at 70°C according to the following gradient:

Time (min)	Flow (μl/min)	A (%)	B (%)
0	600	97	3
5	600	97	3
15	600	3	97
17	600	3	97
17.1	600	97	3
20	600	97	3

Using a postcolumn splitter, a 300 nl/min flow was split to the mass spectrometer with the remainder of the flow going to waste.

9. Data analysis was achieved by averaging the spectra of the protein peak on the total ion current (TIC) chromatogram and then processed using ProMass deconvolution software (Novatia, Monmouth Junction, NJ, *see* **Note 3**, **Figs. 2** and **3**).

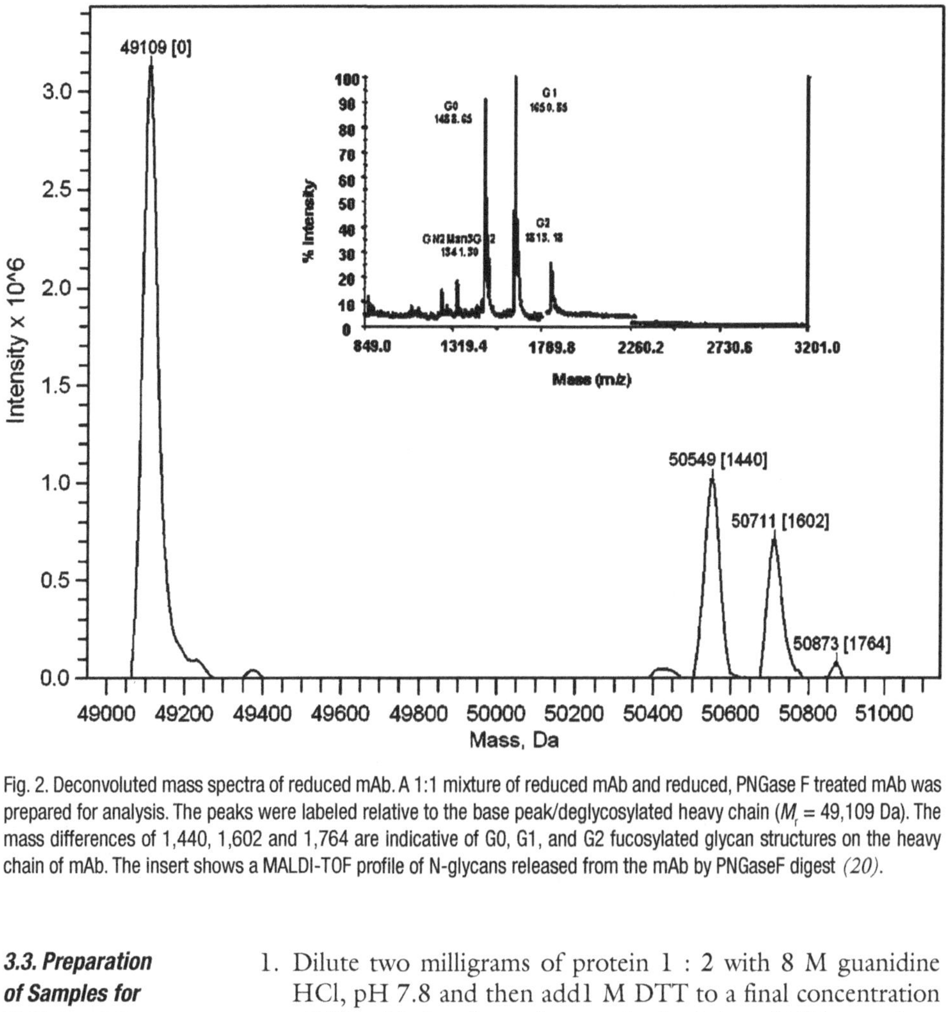

Fig. 2. Deconvoluted mass spectra of reduced mAb. A 1:1 mixture of reduced mAb and reduced, PNGase F treated mAb was prepared for analysis. The peaks were labeled relative to the base peak/deglycosylated heavy chain (M_r = 49,109 Da). The mass differences of 1,440, 1,602 and 1,764 are indicative of G0, G1, and G2 fucosylated glycan structures on the heavy chain of mAb. The insert shows a MALDI-TOF profile of N-glycans released from the mAb by PNGaseF digest *(20)*.

3.3. Preparation of Samples for N-Glycosylation Occupancy Analysis

1. Dilute two milligrams of protein 1 : 2 with 8 M guanidine HCl, pH 7.8 and then add1 M DTT to a final concentration of 10 mM. Incubate the sample for 1 h at 37°C to reduce disulfide bonds.
2. Allow the sample to cool to room temperature and then alkylate free sulfhydryl groups with freshly-prepared 40 mM IAA in the dark for 1 h at room temperature.
3. Buffer-exchange the sample and concentrate to ~1 mg/ml in digestion buffer with a centrifugal concentrator (*see* **Note 4**).
4. Divide the sample into equal aliquots; add 5.2 µg of trypsin to each sample and incubate overnight at 37°C.
5. Boil both aliquots for 5 min to terminate the trypsin digestion.
6. Deglycosylate one aliquot by adding 3,000 units of PNGase F and incubating for 30 min at 37°C; incubate the second aliquot without PNGase F for 30 min at 37°C.

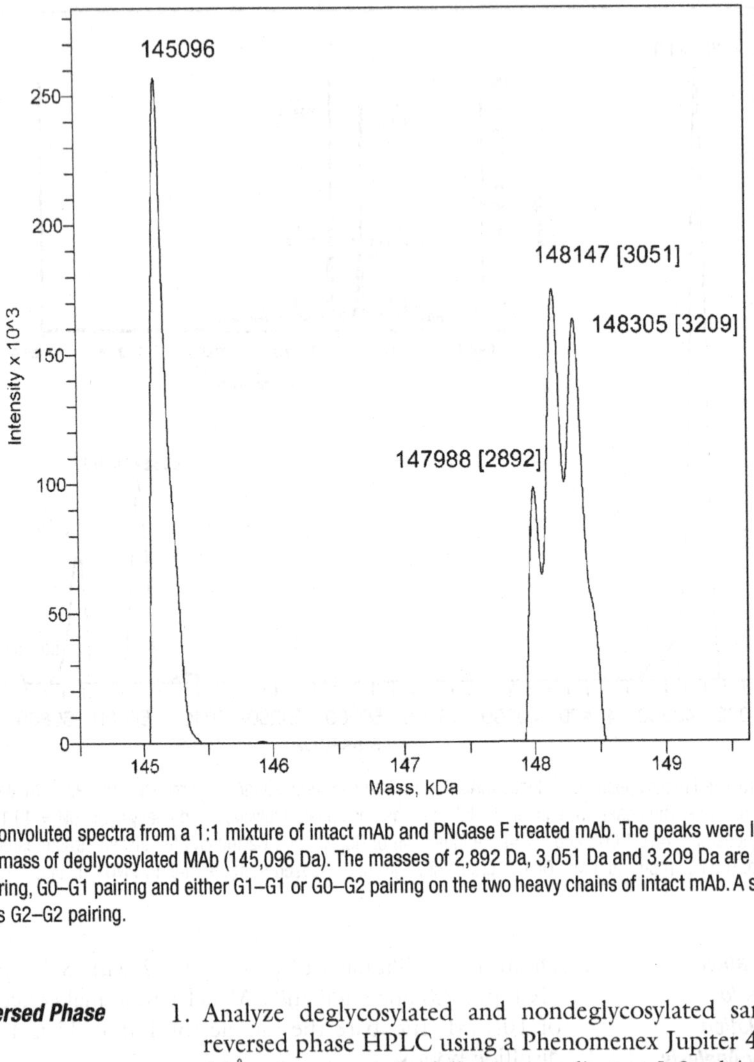

Fig. 3. Deconvoluted spectra from a 1:1 mixture of intact mAb and PNGase F treated mAb. The peaks were labeled relative to the mass of deglycosylated MAb (145,096 Da). The masses of 2,892 Da, 3,051 Da and 3,209 Da are indicative of G0–G0 pairing, G0–G1 pairing and either G1–G1 or G0–G2 pairing on the two heavy chains of intact mAb. A small portion of mAb has G2–G2 pairing.

3.4. Reversed Phase HPLC

1. Analyze deglycosylated and nondeglycosylated samples by reversed phase HPLC using a Phenomenex Jupiter 4u Proteo 90Å column and the following gradient at 1.0 ml/min:

Time (min)	Flow (μl/min)	A (%)	B (%)
0	1,000	95	5
3	1,000	95	5
90	1,000	55	45
95	1,000	95	5

Maintain the column temperature at 30°C using a column oven. A Hitachi diode array detector monitoring at 220 nm was used for detection. Collect 1 min fractions and dry overnight in

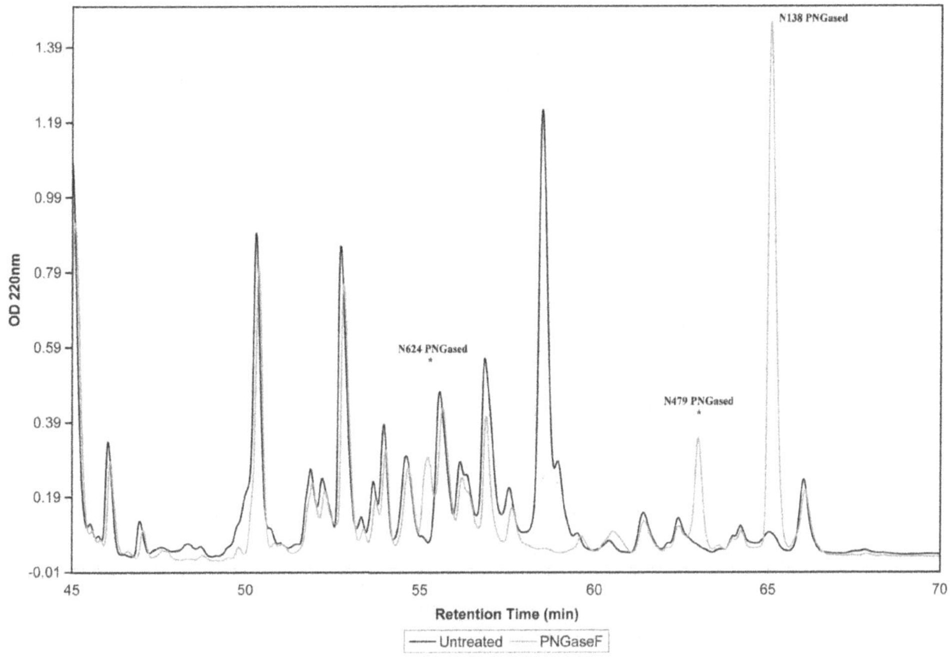

Fig. 4. Reverse phase peptide mapping of glycosylated and deglycosylated tryptic digest. A trypsin digest was separated on HPLC after boiling for 5 min and incubation for 30 min at 37°C with (*pink*) or without (*blue*) PNGase F. A direct comparison of chromatographs allows identification of fractions containing glycopeptides (*blue* peaks not present in deglycosylated sample) and deglycosylated peptides (*pink* peaks not present in untreated sample). Nanospray ESI identified deglycosylated peptides containing Asn624 (55 min), Asn479 (63 min), and Asn138 (65 min).

a CentriVap concentrator. Chromatograms of deglycosylated and nondeglycosylated samples are presented in **Fig. 4**.

2. Dissolve dried HPLC fractions in 50 μl 30% acetonitrile/0.1% formic acid and then analyze by MALDI-TOF mass spectrometry. Spot 0.5 μl sample onto a stainless-steel MALDI sample plate followed by the addition of 0.5 μl of saturated CHCA matrix. Allow sample to dry. Examine peptides by MALDI-TOF mass spectrometry. Ions are generated by irradiation with a pulsed nitrogen laser (337 nm) with a 2-ns pulse time. The instrument is operated in the delayed extraction mode with a 300 ns delay and an accelerating voltage of 25 kV. The grid voltage is 92.00%, the guide wire voltage is 0.05%, the internal pressure is $<5 \times 10^{-7}$ Torr, and the low mass gate is set at 600 Da. Spectra are generated from the sum of 100–200 laser pulses and acquired with a 500 MHz digitizer.
3. Fractions which appear on the HPLC trace after PNGase F digest (55, 56, 63, 64, 65, and 66 min) are transferred to a nanospray sample plate and sealed with adhesive foil. Set the LTQ mass spectrometer to data-dependent triple play mode

to analyze the top five most intense peaks for 2.9 min while the Nanomate sprays each fraction for 3 min in the infusion mode. Collected data are then searched against the protein database using SEQUEST software (Thermo Finnigan).

4. Notes

1. Several matrix-laser combinations have been tested successfully. For peptides and small molecular mass proteins (<10,000 Da), we have observed good results with α-cyno-4-hydroxycinnamic acid (CHCA), whereas higher molecular weight proteins are better ionized with sinapinic acid. The use of 3-amino-4 hydroxybenzoic acid and 2,5-dihydroxybenzoic acid (DHB) are recommended for the analysis of oligosaccharides.
2. Proper tuning of the instrument for the analyte of interest is critical for obtaining an optimal signal. On LTQ ion trap mass spectrometers, the most important parameters include spray voltage, capillary temperature, capillary voltage, tube lens voltage as well as sheath gas and auxiliary gas. LTQ was manually tuned by infusing mAb through a nano-ESI source. The auto gain control (AGC) should be properly set since high space-charge can affect the mass accuracy by distorting the analyte to a higher mass.
3. ProMass can also be integrated into Xcalibur software for automated data processing. For a reduced mAb sample, select the input m/z range from 600 to 2,000, select output mass range from 20,000 to 60,000 (this provides for the detection of heavy chain (~50 kDa) and light chain polypeptides (~25 kDa); for an intact non-reduced mAb sample, select the input m/z range from 1,500 to 3,500, select the output mass range from 20,000 to 160,000 (this provides for the detection of fully assembled mAb (150 kDa).
4. Protein can precipitate as the guanidine-HCl concentration decreases. This problem can be remedied by adding guanidine-HCl back to the sample until the protein redissolves and then resume the buffer exchange process.

References

1. Trombetta, E. S. (2003) The contribution of *N*-glycans and their processing in the endoplasmic reticulum to glycoprotein biosynthesis. *Glycobiology* 13, 77R–91R.

2. Rothman, R. J., Perussia, B., Herlyn, D., and Warren, L. (1989) Antibody-dependent cytotoxicity mediated by natural killer cells is enhanced by castanospermine-induced alterations of IgG glycosylation. *Mol Immunol* 26, 1113–23.
3. Garcia-Casado, G., Sanchez-Monge, R., Chrispeels, M. J., Armentia, A., Salcedo, G., and Gomez, L. (1996) Role of complex asparagine-linked glycans in the allergenicity of plant glycoproteins. *Glycobiology* 6, 471–7.
4. Liang, M., Guttieri, M., Lundkvist, A., and Schmaljohn, C. (1997) Baculovirus expression of a Human G2-Specific, neutralizing IgG monoclonal antibody to *Puumala* virus. *Virology* 235, 252–60.
5. Friedman, B., Vaddi, K., Preston, C., Mahon, E., Cataldo, J. R., and McPherson, J. M. (1999) A comparison of the pharmacological properties of carbohydrate remodeled recombinant and placental-derived β-Glucocerebrosidase: Implications for clinical efficacy in treatment of gaucher disease. *Blood* 93, 2807–16.
6. Umana, P., Jean-Mairet, J., Moudry, R., Amstutz, H., and Bailey, J. E. (1999) Engineered glycoforms of an antineuroblastoma IgG1 with optimized antibody-dependent cellular cytotoxic activity. *Nat Biotechnol* 17, 176–80.
7. Fukuta, K., Abe, R., Yokomatsu, T., Kono, N., Asanagi, M., Omae, F., Minowa, M. T., Takeuchi, M., and Makino, T. (2000) Remodeling of sugar chain structures of human interferon-γ. *Glycobiology* 10, 421–30.
8. Giddings, G., Allison, G., Brooks, D., and Carter, A. (2000) Transgenic plants as factories for biopharmaceuticals. *Nat Biotechnol* 18, 1151–5.
9. Houdebine, L. M. (2002) Antibody manufacture in transgenic animals and comparisons with other systems. *Curr Opin Biotechnol* 13, 625–9.
10. Shields, R. L., Lai, J., Keck, R., O'Connell, L. Y., Hong, K., Meng, Y. G., Weikert, S. H., and Presta, L. G. (2002) Lack of fucose on Human IgG1 *N*-linked oligosaccharide improves binding to Human FcγRIII and antibody-dependent cellular toxicity. *J Biol Chem* 277, 26733–40.
11. van Berkel, P. H., Welling, M. M., Geerts, M., van Veen, H. A., Ravensbergen, B., Salaheddine, M., Pauwels, E. K., Pieper, F., Nuijens, J. H., and Nibbering, P. H. (2002) Large scale production of recombinant human lactoferrin in the milk of transgenic cows. *Nat Biotechnol* 20, 484–7.
12. Choi, B. K., Bobrowicz, P., Davidson, R. C., Hamilton, S. R., Kung, D. H., Li, H., Miele, R. G., Nett, J. H., Wildt, S., and Gerngross, T. U. (2003) Use of combinatorial genetic libraries to humanize N-linked glycosylation in the yeast Pichia pastoris. *Proc Natl Acad Sci U S A* 100, 5022–7.
13. Hamilton, S. R., Bobrowicz, P., Bobrowicz, B., Davidson, R. C., Li, H., Mitchell, T., Nett, J. H., Rausch, S., Stadheim, T. A., Wischnewski, H., Wildt, S., and Gerngross, T. U. (2003) Production of complex human glycoproteins in yeast. *Science* 301, 1244–6.
14. Yuen, C. T., Storring, P. L., Tiplady, R. J., Izquierdo, M., Wait, R., Gee, C. K., Gerson, P., Lloyd, P., and Cremata, J. A. (2003) Relationships between the N-glycan structures and biological activities of recombinant human erythropoietins produced using different culture conditions and purification procedures. *Br J Haematol* 121, 511–26.
15. Bobrowicz, P., Davidson, R. C., Li, H., Potgieter, T. I., Nett, J. H., Hamilton, S. R., Stadheim, T. A., Miele, R. G., Bobrowicz, B., Mitchell, T., Rausch, S., Renfer, E., and Wildt, S. (2004) Engineering of an artificial glycosylation pathway blocked in core oligosaccharide assembly in the yeast *Pichia pastoris*: production of complex humanized glycoproteins with terminal galactose. *Glycobiology* 14, 757–66.
16. Gerngross, T. U. (2004) Advances in the production of human therapeutic proteins in yeasts and filamentous fungi. *Nat Biotechnol* 22, 1409–14.
17. Kost, T. A., Condreay, J. P., and Jarvis, D. L. (2005) Baculovirus as versatile vectors for protein expression in insect and mammalian cells. *Nat Biotechnol* 23, 567–75.
18. Zhu, L., van de Lavoir, M. C., Albanese, J., Beenhouwer, D. O., Cardarelli, P. M., Cuison, S., Deng, D. F., Deshpande, S., Diamond, J. H., Green, L., Halk, E. L., Heyer, B. S., Kay, R. M., Kerchner, A., Leighton, P. A., Mather, C. M., Morrison, S. L., Nikolov, Z. L., Passmore, D. B., Pradas-Monne, A., Preston, B. T., Rangan, V. S., Shi, M., Srinivasan, M., White, S. G., Winters-Digiacinto, P., Wong, S., Zhou, W., and Etches, R. J. (2005) Production of human monoclonal antibody in eggs of chimeric chickens. *Nat Biotechnol* 23, 1159–69.
19. Li, H., Sethuraman, N., Stadheim, T. A., Zha, D., Prinz, B., Ballew, N., Bobrowicz, P., Choi, B. K., Cook, W. J., Cukan, M., Houston-Cummings, N. R., Davidson, R., Gong, B., Hamilton, S. R., Hoopes, J. P., Jiang, Y., Kim, N., Mansfield, R., Nett, J. H., Rios, S., Strawbridge, R., Wildt, S., and Gerngross, T. U. (2006) Optimization of humanized IgGs in glycoengineered Pichia pastoris. *Nat Biotechnol* 24, 210–5.
20. Li, H., Miele, R.G., Mitchell, T., and Gerngross, T.U. (2007) N-linked glycan characterization of heterologous proteins. In Pichia Protocols, 2nd Ed, Humana Press, Totowa, NJ

Chapter 17

N-Glycosylation Site Analysis of Human Platelet Proteins by Hydrazide Affinity Capturing and LC-MS/MS

Urs Lewandrowski and Albert Sickmann

Summary

A starting point for many glycosylation analysis pathways is marked by the determination of the respective carbohydrate attachment sites to the polypeptide backbone of proteins. Several methods have been reported for this purpose in the past, commonly divided into a three-step approach (1) affinity purification of glycoproteins/-peptides, (2) processing/trimming of the glycopeptides and (3) elucidation of the glycan attachment site by mass spectrometry. For N-glycosylation site analysis the last two steps are usually similar, while methods differ in the affinity purification step. Here, we describe the oxidative derivatisation of carbohydrate moieties and covalent trapping of glycopeptides on hydrazide functionalized beads. This method is suitable for large scale analysis of glycoproteins as well as isolated glycoproteins and can be applied readily to a number of different samples. In the described protocol, the elucidation of N-glycosylation sites of human platelet proteins is demonstrated as an example.

Key words: Glycosylation, Platelet, Hydrazide affinity, Mass spectrometry, Affinity capture, Carbohydrate oxidation.

1. Introduction

Elucidation of glycosylation sites is of primary interest regarding the analysis of glycoproteins. While enhanced studies of carbohydrate moieties often comprise structural details of the glycan (sequence, branching, anomericity etc.), site analysis is a mandatory step for characterization of microheterogeneities. Direct analysis of glycopeptides by mass spectrometry often yields inferior results for structure and sequence of both parts – glycan and peptide – due to unequal fragmentation efficiencies of the constituents. Furthermore, this analysis is rendered too complex on the background of whole proteomes and hampered by ion suppression effects of

Nicolle H. Packer and Niclas G. Karlsson (eds.), *Methods in Molecular Biology, Glycomics: Methods and Protocols, vol. 534*

DOI: 10.1007/978-1-59745-022-5_17

glycopeptides in the presence of non-modified peptides. Therefore, the logical initial step in many glycoprotein studies is the determination of modification sites prior to the in depth analysis of the glycan part. For the former purpose, analysis can be focussed on the glycopeptides alone. Thereby, the complexity of large scale proteomic data is reduced to the glycopeptide level. By discarding the carbohydrate part and marking the site of modification in the process, further reduction down to the mere peptide level is achieved.

Many procedures for the isolation of glycopeptides or – proteins have been proposed in the past. Prominent among these are lectin based procedures, which use specific carbohydrate–protein interactions to recover individual glycopeptide species from complex peptide mixtures *(1)*. Although quite successful, this approach is nevertheless limited by the selectivity of the lectin and thereby to specific subsets of peptides. Furthermore, isolation of glycopeptides is based on a competitive and reversible interaction which is most often not compatible with stringent washing conditions, thereby resulting in the co-purification of non-modified peptides. Other modes of affinity purification e.g. take advantage of phenylboronic acid *(2, 3)* – carbohydrate interactions or aim at the enrichment of glycopeptides by their comparably high molecular weight using size exclusion chromatography *(4)*.

A further technique, hydrazide affinity purification, is based on a simple reaction used for decades e.g. for staining glycoproteins on SDS-PAGE gels *(5)*. The process of glycosylation site elucidation by hydrazide coupling is demonstrated in **Fig. 1**.

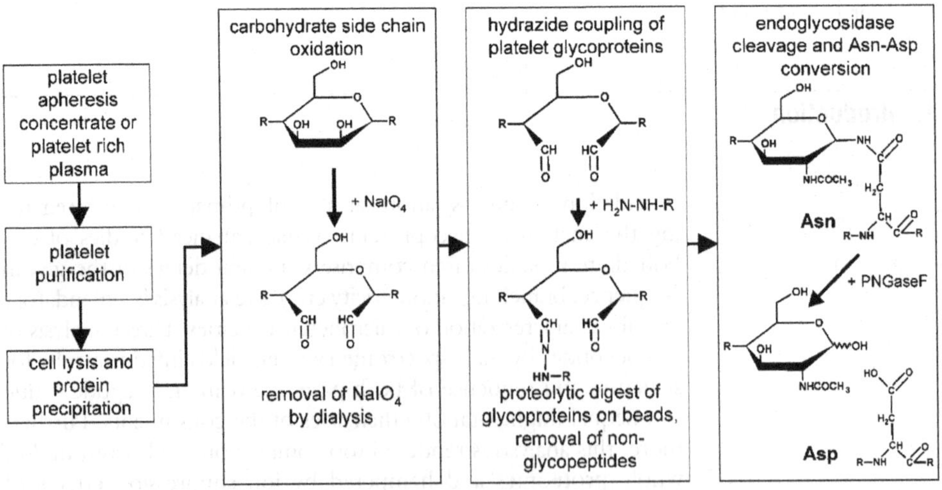

Fig. 1. Hydrazide affinity and N-glycosylation site analysis for human platelet proteins. Upon platelet isolation and protein preparation, glycan side chains are modified at their vicinal diol groups by sodium (meta-) periodate oxidation resulting in reactive aldehyde groups. These are coupled to hydrazide functionalized beads to selectively immobilize glycoproteins. Following proteolytic digest, respective glycopeptides are cleaved from the solid support by endoglycosidase treatment. Due to amidase activity of PNGaseF a characteristic conversion of Asn to Asp at the former glycan attachment site is induced which can furthermore be detected by mass spectrometry.

Initially, carbohydrate residues featuring vicinal *cis*-diol groups are oxidized by periodic acid resulting in two aldehyde groups. These groups are uncommon for peptides since no amino acid carries free aldehydes. Thereby, upon coupling of the aldehyde to a strong nucleophile, a method for specific enrichment of glycopeptides is given. In this context, hydrazide is most often used for derivatisation and coupling of oxidized carbohydrates to solid supports. The resulting Schiff base is sufficiently stable to allow stringent washing conditions. Subsequently, glycoproteins can be digested proteolytically on the resin while removing all non-glycopeptides by subsequent washing steps. Since glycopeptides are covalently bound to the support, glycosidase cleavages result in free peptides which can furthermore be sequenced by mass spectrometry *(6)*. Primarily, peptide:*N*-glycosidase (PNGaseF) *(7)* is used for the cleavage reaction; thereby, not only the glycan part is cleaved, but the enzyme also specifically converts glycosylated asparagine within the consensus sequence of NXS/T/C (X not proline) to aspartic acid resulting in a +1 amu mass shift. In itself, the combination of specific affinity, specialized enzymatic derivatisation and given consensus sequence is sufficient for the confirmation of glycosylation sites by mass spectrometric sequencing. However, other approaches have been performed using Endo H for cleaving the glycan moiety, leaving a single sugar residue attached to the peptide backbone. Nevertheless, this might result in side chain fragmentations and hampered peptide identification during mass spectrometric analysis. Problems arising from the differentiation between PNGaseF-based and other – artificial or natural – deamidation events of asparagine, can be avoided by endoglycosidase digests performed in $H_2^{18}O$ *(1)*. Thereby, the shift for the conversion from asparagine to aspartic acid is increased to + 3 amu and since no other amino acid or modification bears this mass, the confidence in modification site analysis is increased.

The application of the hydrazide affinity method to human platelets enabled us to screen for N-glycosylation sites on a number of known as well as unknown platelet proteins *(8)*. Moreover, the broad range of identified proteins showed no particular bias against certain subsets like membrane bound species. Therefore, glycosylation sites on low abundant membrane receptors could be determined; as an example, the mass spectrum of a formerly unknown glycosylation site of G6B, a putative novel plasma membrane receptor on platelets is depicted in **Fig. 3**. Other identified membrane proteins/sites included purinergic receptors such as P2X1 as well as common platelet proteins like multimerin, gpIIb/IIIa, gp IV, etc. On the background of these results, hydrazide affinity in combination with mass spectrometric sequencing enables detection of glycosylation sites on a range of soluble as well as membrane bound glycoproteins, covering, furthermore, high dynamic ranges and protein diversity.

2. Materials

2.1. Sample Preparation

2.1.1. Isolation of Platelets

- Resuspension buffer: 5 mM KCl, 145 mM NaCl, 14 mM Glucose, 1 mM $MgCl_2$, 10 mM citric acid, pH 6.4, stored at 4°C.
- 10 mL single use plastic pipettes.
- Liquid nitrogen.

2.1.2. Protein Purification

- Ethanol, p.A. (Merck KGaA, Darmstadt, Germany), cool to −25°C.
- 2.5% (w/v) SDS in 25 mM NH_4HCO_3, pH 7.8, prepare directly before use.

2.2. Oxidative Hydrazide Coupling

- 100 mM Na_2HPO_4, adjust pH to 7.0.
- Sodium (meta) periodate, $NaIO_4$, p.A., (Fluka, Steinheim, Germany; strong oxidation reagent), grind to fine powder before use.
- Dialysis membrane, 3500 MWCO, Spectra/por ® 3 (Carl Roth GmbH + Co. KG, Karlsruhe, Germany).
- CarboLink™ coupling gel (Perbio Science, Bonn, Germany).
- 50 mM NH_4HCO_3, prepare directly before use.
- 1 M urea in 50 mM NH_4HCO_3, prepare directly before use.
- Disposable 5 mL polypropylene columns (Perbio Science, Bonn, Germany).

2.3. Protein Processing

- Trypsin, modified, sequencing grade, 20 μg lyophilized/vial (Promega, Madison, WI, USA).
- 1 M urea in 50 mM NH_4HCO_3; 1 M NaCl in 50 mM NH_4HCO_3; 20 mM NH_4HCO_3; prepare all solutions directly before use.
- PNGaseF, 100 U, *Flavobacterium meningosepticum*, expressed in *E. coli*, lyophilized (Roche, Mannheim, Germany).
- 5% formic acid (use formic acid, 98–100% GR for analysis, Merck KGaA, Darmstadt, Germany).

2.4. Mass Spectrometric Sequencing

- High purity water, 18.2 MΩ (e.g. prepared by Purelab ultra, ELGA LabWater, Celle, Germany).
- Acetonitrile, gradient grade (avoid exposure to and inhalation of this solvent); formic acid, 98–100% GR for analysis; trifluoroacetic acid for protein sequence analysis (Merck KGaA, Darmstadt, Germany).
- ESI needles, PicoTip™ Emitter, 150 μm outer diameter, 20 μm inner diameter, 10 μm tip opening, non coated (FS150-20-10-N-20-C12, New Objective, Woburn, MA, USA).

- Famos™, Switchos™, Ultimate™-nano-HPLC system in inert version (Dionex, Idstein, Germany).
- Qtrap4000™ triple quadrupole linear ion trap mass spectrometer (Applied Biosystems, Darmstadt, Germany).

3. Methods

3.1. Sample Preparation

3.1.1. Isolation of Platelets

In order to obtain large amounts of platelets suitable for a series of analysis, apheresis derived platelet preparations are used for sample preparation. However, these concentrates, although leukocyte depleted, contain trace amounts of erythrocytes, leukocytes as well as a high content of plasma proteins. These contaminations have to be removed to avoid co-purification of non-platelet glycoproteins during hydrazide affinity.

1. Divide 250 mL apheresis derived platelet preparation evenly into six 50 mL reaction tubes and spin at 310 × *g* for 15 min at 21°C. Transfer the supernatants to fresh tubes and discard the pellets (*see* **Notes 1** and **2**).
2. **Step 1** is repeated once.
3. Spin at 385 × *g* for 20 min at 21°C. Discard the supernatant and dissolve each pellet in a total of 30 mL resuspension buffer by gentle suction with a 1,000 μL pipette.
4. Repeat **step 2** once.
5. Dissolve the six pellets in a total volume of 10 mL resuspension buffer and divide them into 1 mL aliquots within 2 mL reaction tubes.
6. Spin samples at 10,000 × *g* for 1 min and carefully remove supernatants with a pipette. Freeze samples immediately in liquid nitrogen and store at −80°C until further use.

3.1.2. Protein Purification

For removal of interfering cell contents, platelets are lysed and proteins precipitated prior to oxidative hydrazide coupling.

1. Lyse one aliquot of platelets (preparation 3.1, approximately 100 mg wet weight) by addition of 2 mL 2.5% SDS, 25 mM NH_4HCO_3.
2. Transfer the sample to a 50 mL reaction tube and add 18 mL ethanol (chilled to −25°C). Mix briefly by inverting the tube and incubate at −25°C for at least 3 h. Overnight incubation is possible as well (*see* **Note 3**).
3. Spin reaction tube at 4,000 × *g* for 15 min at 4°C. Remove the supernatant and withdraw residual ethanol with a pipette.

3.2. Oxidative Hydrazide Coupling of Glycoproteins

1. Precipitated proteins are dissolved in 25 mL 100 mM Na_2HPO_4, pH 7.0 (*see* **Note 4**).
2. Add 20 mg ground $NaIO_4$ to the sample and incubate at 21°C for 1 h in the dark. The grinding ensures rapid solubilisation of crystalline periodate.
3. Pre-wet and wash a 3500 MWCO dialysis membrane with 100 mM Na_2HPO_4, pH 7.0. Close one end of the dialysis membrane tube with a clip, insert the sample, withdraw air bubbles and subsequently close the tube with a second clip. The sample is dialysed against 1 L 100 mM Na_2HPO_4, pH 7.0 at 4°C for 3 h under gentle agitation. After buffer exchange, dialysis is continued against 2 L 100 mM Na_2HPO_4, pH 7.0 at 4°C for 16 h or overnight (*see* **Notes 5** and **6**).
4. Meanwhile, fill an empty disposable 5 mL column (single frit at the outlet) with a column volume of 100 mM Na_2HPO_4, pH 7.0 and add 1 mL hydrazide functionalized agarose beads (equal to 2 mL bead slurry).
5. The beads are washed three times with 5 mL 100 mM Na_2HPO_4, pH 7.0.
6. Add the washed beads to the dialysed sample and incubate at 21°C for at least 6 h under gentle agitation in a 50 mL reaction tube (*see* **Note 7**).
7. The beads are transferred to a disposable 5 mL column with a single frit at the outlet. Avoid any air bubbles to be trapped at the outlet. During the following steps the beads should be covered with sufficient liquid as not to let them dry.
8. Rinse surplus sample and wash the column three times with 5 mL 50 mM NH_4HCO_3.
9. Wash twice with 5 mL 1 M urea in 50 mM NH_4HCO_3.
10. Equilibrate the column five times with 5 mL 50 mM NH_4HCO_3.
11. Fill the column with 50 mM NH_4HCO_3 up to a total volume of 2 mL including the 1 mL beads.

3.3. Protein Processing

After covalent coupling of respective glycoproteins by their carbohydrate side chains to the hydrazide functionalized support, proteolytic digest and enzymatic derivatisation by endoglycosidase cleavage is used to generate modified (glyco-) peptides suitable for mass spectrometric sequencing of glycosylation sites. In general, several modifications of the main processing method are possible of which some are mentioned in **Subheading 4**. Here, only the most common form – using trypsin in combination with PNGaseF is described.

1. Dissolve 20 μg sequencing grade trypsin in 100 μL 50 mM NH_4HCO_3 and add the protease solution to the hydrazide

bead slurry in the column (**Subheading 3.2, step 11**) equilibrated with the same buffer. Close the column on both sides with the respective caps and seal with parafilm® (*see* **Note 8**).

2. The sample is incubated under gentle agitation at 37°C for 14 h. Ensure that beads do not settle at the bottom of the column but stay well dispersed.
3. Remove column caps, rinse the column from surplus sample volume and wash the column three times with 5 mL 50 mM NH_4HCO_3. Retain as few liquid in the column as possible throughout the washing steps, without letting the column run dry.
4. Rinse column twice with 5 mL 1 M urea in 50 mM NH_4HCO_3.
5. Wash the column with additional 1 M NaCl in 50 mM NH_4HCO_3.
6. Finally, beads are equilibrated five times with 5 mL 50 mM NH_4HCO_3.
7. Seal the bottom of the tube and add 500 μL 50 mM NH_4HCO_3 resulting in a total volume of 1.5 mL including the beads.
8. Dissolve 100 U PNGaseF in 200 μL H_2O and add 10 μL thereof to the equilibrated beads (*see* **Note 9**).
9. Incubate at 37°C for 14 h under gentle agitation. Ensure that beads are not settling at the bottom of the column but stay well dispersed.
10. The digestion buffer, containing released peptides, is removed and beads are washed twice with 1 mL 20 mM NH_4HCO_3 (compare **step 3**, *see* **Note 10**). The three volumes are combined, frozen and lyophilized by vacuum evaporation to approximately 400 μL. Mix 10 μL sample with 5 μL 5% formic acid and apply the sample to mass spectrometric sequencing.

3.4 Mass Spectrometric Sequencing

3.4.1. Nano-HPLC Separation of Peptides

In order to ensure sensitive and accurate determination of N-glycosylation sites, the modified peptide mixture gained by hydrazide affinity is separated by nano-scale HPLC prior analysis by mass spectrometry. The system described applies online pre-concentration and desalting steps as well as separation of peptides based on hydrophobic reversed-phase interaction. A column switching scheme is illustrated in **Fig. 2**. The HPLC system itself consists of a Famos™ (autosampler), Switchos™ (loading pump and switching module), and Ultimate™ (nano-HPLC pump) combination as an inert version in nano-LC configuration (all modules by Dionex, Idstein, Germany). Typically, the sample is loaded onto the precolumn (custom made, 100 μm ID × 360 μm OD, 2 cm length,

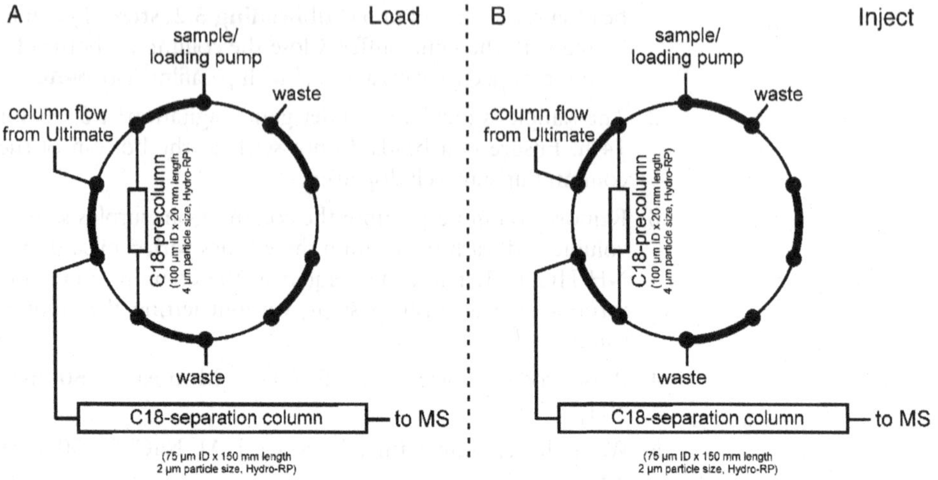

Fig. 2. Column switching scheme for online preconcentration of peptide samples and nano-HPLC – mass spectrometry. During the loading cycle of the sample (Load, **A**), peptides are concentrated and desalted on a short precolumn under high flow rates, while the flow-through is discarded to the waste. Upon valve switching (Inject mode, **B**), the precolumn is in-line with the separation column; peptides are separated by gradient elution under nano-flow conditions and direct sequencing by tandem mass spectrometry is performed.

4 µm particle size, 80 Å, Hydro-RP, Phenomenex, Aschaffenburg, Germany) for 5 min at 6 µL/min flow rate using 0.1% trifluoroacetic acid delivered by the Switchos™ pump. Thereafter, the precolumn is switched in-line with the separation column (custom made, 75 µm ID × 360 µm OD, 15 cm length, 2 µm particle size, 100 Å pore size, Hydro-RP, Phenomenex) maintained at a flow rate of 290 nL/min. Solvents for reversed – phase chromatography consist of 0.1% formic acid (solvent A) and 0.1% formic acid in 84% acetonitrile (solvent B), respectively (*see* **Note 11**). Gradient elution is performed as follows (1) 9% solvent B from 0 to 5 min, (2) linear slope from 9 to 50% solvent B during 90 min, (3) linear slope to 95% B from 95 to 100 min, (4) regeneration of the column at constant 95% B from 100 to 105 min, (5) linear slope from 95% B to 9% B from 105 to 107 min, and (6) equilibration to 9% B from 107 to 120 min.

3.4.2. Mass Spectrometry

After the complexity of the peptide mixture has been reduced by online coupled reversed-phase HPLC, peptides are ionised by electrospray mass spectrometry. Currently, we use a triple quadrupole linear ion trap (Qtrap4000™, Applied systems, Darmstadt, Germany) for this purpose, which is a compromise between sensitivity and scan rate on the one hand and mass accuracy on the other hand. For all measurements non coated electrospray needles with 150 µm outer diameter, 20 µm inner diameter and 10 µm tip opening, in combination with the nano-spray source supplied by Applied Biosystems are used.

Scanning is performed in triple play mode: A survey scan is followed by an enhanced resolution scan of the three most intensive ions and three subsequent MS/MS using the linear ion trap for readout (for example see **Fig. 3**). For survey scans the enhanced multiple charge mode is used, thereby focussing predominantly on doubly and triply charged precursor ions. For the enhanced multiple charge scan a scan rate of 4,000 amu/s covering a mass range of 380–1,500 *m/z* is chosen while summing three scans per spectrum. Furthermore, Q0 trapping is activated for higher sensitivity and ion trap fill time as well as Q3 empty time is set to 20 ms. Information dependent acquisition (IDA) is enabled by excluding former target ions within a mass window of 4 amu after two occurrences for 45 s. For the enhanced product ion scan mass range is between 115 and 1,500 *m/z* with a scan rate of 4,000 amu/s, while two scans are summed for each spectrum. Moreover, rolling collision energy is used for doubly (collision energy = $0.05 \times (m/z) + 5$)) and triply (collision energy = $0.044 \times (m/z) + 4$)) charged ions, respectively. Maximum ion trap fill time for enhanced product ion scans is set to 350 ms. For all scan modes, ion spray voltage is adjusted to 2,300 V, declustering potential to 60, and collision gas to "high", meanwhile, the interface heater is maintained at 150°C (*see* **Note 12**).

3.4.3. Database Search

The evaluation of mass spectrometric datasets is performed with the MASCOT™ search algorithm (version 2.1., MatrixScience, London, UK, www.matrixscience.com) and subsequent manual validation. For conversion of initial datafiles a Mascot plug-in to the Analyst 1.4.0 software is used (mascot.dll, version 1.6. b15, MatrixScience). Thereby, only peaks above a threshold of 0.1% of the highest peak intensity in every spectrum are transferred to a centroided peak list. No grouping of spectra is allowed and spectra must contain more than ten peaks in total. Important parameters for the Mascot algorithm itself are discussed in the following:

1. Database choice; for hydrazide affinity purification of platelets we employ the Swiss-Prot database at www.expasy.org, since it is well annotated and thereby reduces the probability of one hit wonders from unknown database entries. '*Homo sapiens*' is furthermore chosen as taxonomy. However, in regard to obtaining sites on unknown proteins, the NCBI non-redundant database (www.ncbi.nih.gov) is used for experiments with $H_2{}^{18}O$ (compare **Subheading 3.3, step 2**), since this method offers further proof of identifications by the unique mass shift of + 3 amu.
2. Carbamidomethylation is chosen as fixed modification, as well as a custom designed deamidation of asparagine (only Asn is chosen; monoisotopic mass is 115.0269 amu; average mass is 115.0874 amu) for variable modifications, respectively. In case

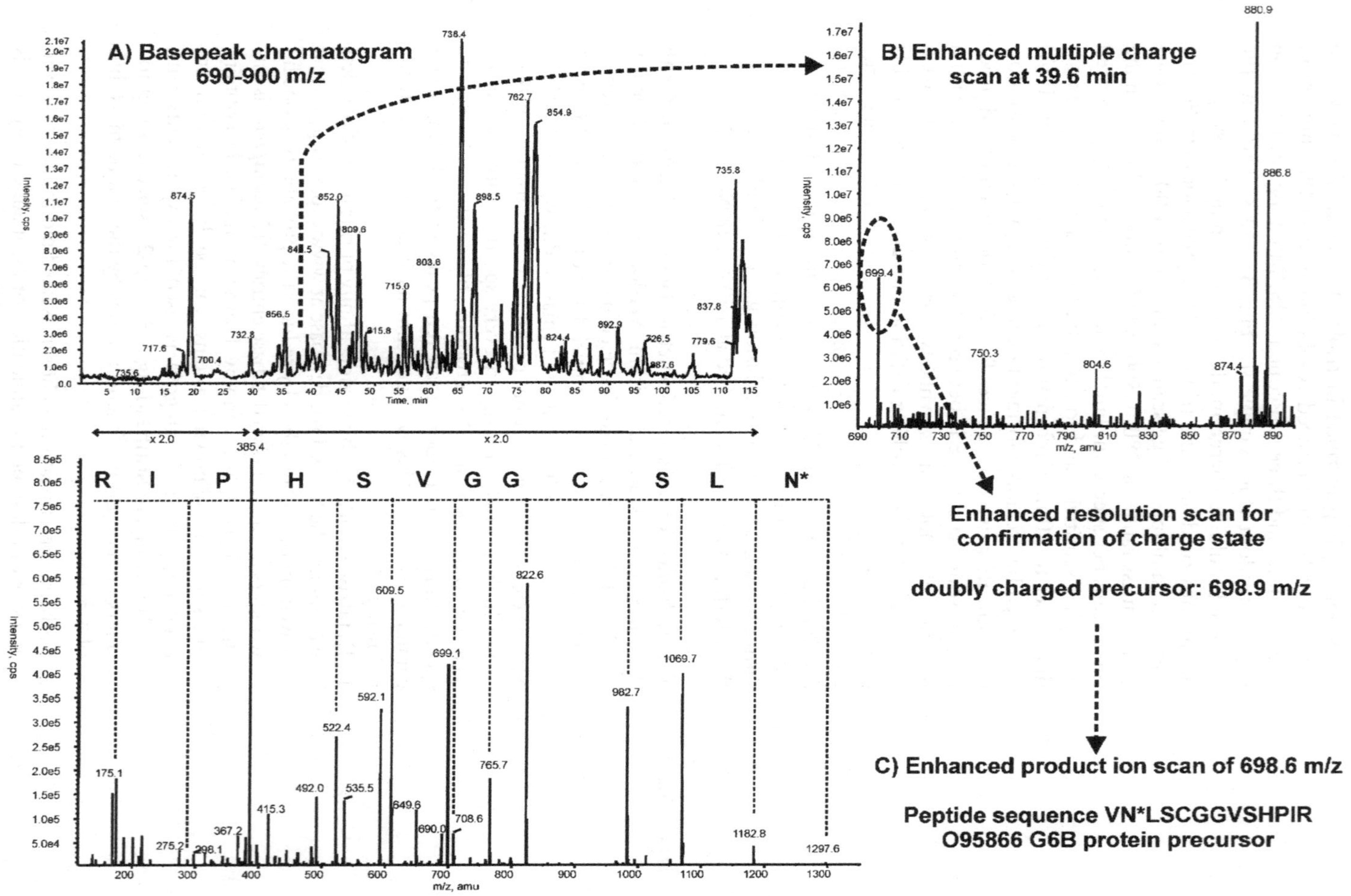

Fig. 3. Determination of G6B glycosylation site by mass spectrometry. Hydrazide affinity was performed with platelet proteins as stated in the methods section, while gas phase fractionation experiments were performed thereafter. (**A**) Base peak chromatogram of a 690–900 *m/z* enhanced multiple charged survey scan, (**B**) Single enhanced multiple charge scan at retention time 39.6 min during a 2-h gradient; the signal at *m/z* 699.4 was chosen for a subsequent enhanced resolution scan; (**C**) Enhanced product ion scan of 698.9; Mascot search resulted in peptide sequence VN*LSCGGVSHPIR of 095866 G6B protein precursor (N* marks the site of glycosylation).

PNGaseF digest was performed with ^{18}O-water, modification values are changed to a monoisotopic mass of 117.0550 amu and an average mass of 117.1103 amu.

3. Trypsin is specified as protease and a single miscleavage is allowed.
4. Peptide tolerance is set to 0.2 amu, which is within a reasonable range regarding the scan speed and resolution offered by the enhanced resolution scan for precursor mass determination. Moreover, MS/MS tolerance is set to 0.4 amu, since the higher scan speed of the enhanced product ion scan offers less resolution. Only doubly and triply charged ions are used for database search.

In general, spectra should be validated manually in addition to the scoring algorithms of Mascot™ or equivalent search engines. For the Qtrap4000™ y-ions usually comprise the highest abundant ion series, especially in high *m/z* regions. Completeness of ion series as well as peptide typical fragmentation intensities, e.g. with proline or glycine residues involved, in combination with immonium marker ions should be present in order to confirm the assignments.

4. Notes

1. Avoi d contact of intact platelets to glass labware, e.g. pipettes, centrifuge tubes, etc. Thereby, activation and subsequent clotting might occur, interfering with the cleanup procedure. Use plastic pipettes, tubes, etc. instead.
2. A typical platelet apheresis preparation contains 1–2 × 10^{11} cells. If no platelet concentrate is available, it may be replaced in the protocol by platelet rich plasma, which in turn can be isolated from citrated whole blood samples.
3. Ethanol precipitation was chosen because it is non-denaturing for proteins and in practice it is possible to solubilize proteins in aqueous or low amounts of detergent- containing solutions afterwards. Furthermore, SDS does not precipitate in ethanol at –25°C (*see* **Notes 4** and **5**).
4. Depending on the protein concentration, it may not be possible to dissolve proteins in 100 mM Na_2HPO_4. In this case, add 2.5 mL 5% SDS to the sample. The SDS seems not to have adverse effects on the downstream process of coupling but will dissolve proteins sufficiently. SDS can be replaced by other detergents; however, these have to be compatible with the subsequent trapping steps (e.g. no primary amines or other strong nucleophiles, no carbohydrate moieties). If the

replacing detergent is dialyzable, fresh detergent should be added after the dialysis step.

5. When SDS has been added to the sample (compare **Note 4**) it will precipitate overnight at 4°C in fine needle shaped crystals – it dissolves as soon as brought to room temperature again.
6. As an alternative to time consuming dialysis, gel filtration can be used to remove surplus sodium (meta)periodate from the sample. Therefore, commercially available products as e.g. D-Salt, 5 mL dextran desalting columns (Perbio Science) can be used. However, in our experience, certain subclasses of proteins may be lost and remain on the column during the process of gel filtration. Therefore, we recommend the use of dialysis in case of platelet proteins, although gel filtration may have advantages with other sample types, e.g. soluble plasma proteins.
7. Instead of agarose based hydrazide functionalized beads, 200 µL (25 mg/mL, 1 µm particle size) magnetic beads may be used (chemicell GmbH, Berlin, Germany) with equivalent results. These beads are silica-based and differ from each other in the attachment form of hydrazide linkers to silica beads. However, we found a mixture of equal amounts of the available variants (H1 to H5) the most convenient. Advantages of these beads can be seen in easier handling as well as increased stringency of washing conditions (e.g. by addition of organic solvents). Furthermore, volumes for elution can be kept rather low (100 µL) and thereby vacuum concentration and possible peptide loss can be avoided. For the washing steps, use 1.5 mL volumes in a 2.0 mL reaction tube and after the salt washing steps add steps with 50% methanol or acetonitrile in NH_4HCO_3 buffer. However, for trapping of the particles, a magnetic separator (e.g. Dynal MPC-S, Invitrogen, Karlsruhe, Germany) is recommended, since mere centrifugation is not convenient.
8. The use of different proteases such as chymotrypsin or Glu-C as well as respective combinations can replace the standard tryptic digest at this time point in the protocol, while buffer conditions may require subtle adjustments. However, the use of unspecific proteases should be avoided, since generally the complexity of the peptide mixture is increased to a high degree, while sensitivity is reduced.
9. PNGaseF stock solution should be freshly prepared before use. Nevertheless, upon storage at 4°C usage for up to a month is possible. At this step of the protocol, $H_2{}^{18}O$ labelling of glycosylation sites is possible. However, we recommend this procedure only in combination with magnetic beads (*see* **Note 7** since these can be easily equilibrated with ^{18}O-water, while working with the agarose beads may result in a mix

ture of ^{18}O- and ^{16}O-water. For labelling, withdraw the 50 mM NH_4HCO_3 from the magnetic beads and wash beads with 100 μL $H_2{}^{18}O$ (97% ^{18}O content, Cambridge Isotope Laboratories, Inc. Andover, MA, USA) to withdraw residual ^{16}O-water as far as possible. Add 100 μL fresh $H_2{}^{18}O$ followed by 1 μL 2 M NH_4HCO_3 (in $H_2{}^{16}O$) to the beads. Finally add 2 μl 1 U/μL PNGaseF (in $H_2{}^{16}O$) and incubate at 37°C for 4 h. Since residual trypsin may lead to slow exchange of ^{16}O to ^{18}O at the C-terminal carboxyl-group, do not extend this time span.

10. So far, no major advantage could be seen by eluting the peptides with high salt or organic solvent containing buffers. The former potentially obstructs the following of mass spectrometric analysis and the latter can cause clogging of the beads, depending on the concentration of organic solvent used. However, no significant gain was observed by applying these elution forms.
11. For unobstructed LC-MS analysis, high quality grade solvents are mandatory. Formic acid (store at 4°C), in particular, can degrade in a short time-span and should be replaced on a regular basis. Solvents should be filtered (0.2 μm PVDF filter, Millipore, Schwalbach, Germany) and degassed prior to use. The use of 0.2% formic acid containing solvents is possible to improve spray conditions, but stable conditions can already be achieved with 0.1% formic acid in solvent A and B.
12. The number of assigned glycopeptides can be further increased by gas phase fractionation. Therefore, the mass range of initial survey scans is limited and focussed to smaller sections per LC-MS run. The sample is applied in five subsequent 2-h gradients, while each gradient elution is scanned for a different mass range. Commonly, we differentiate five sections ranging between 380–500 *m/z*, 490–600 *m/z*, 590–700 *m/z*, 690–900 *m/z* and 890–1,500 *m/z*. Co-eluting peptides can therefore be sequenced while they might have been excluded by the simple triple play scanning technique covering the whole mass range between 380 and 1,500 *m/z* in a single scan event.

References

1. Kaji, H., et al. Lectin affinity capture, isotope-coded tagging and mass spectrometry to identify N-linked glycoproteins. *Nat Biotechnol* 21, 667–72 (2003).
2. Sparbier, K., Koch, S., Kessler, I., Wenzel, T. & Kostrzewa, M. Selective isolation of glycoproteins and glycopeptides for MALDI-TOF MS detection supported by magnetic particles. *J Biomol Tech* 16, 407–13 (2005).
3. Liu, X. C. & Scouten, W. H. Boronate affinity chromatography. *Methods Mol Biol* 147, 119–28 (2000).

4. Alvarez-Manilla, G., et al. Tools for glycoproteomic analysis: size exclusion chromatography facilitates identification of tryptic glycopeptides with N-linked glycosylation sites. *J Proteome Res* 5, 701–8 (2006).
5. Jay, G. D., Culp, D. J. & Jahnke, M. R. Silver staining of extensively glycosylated proteins on sodium dodecyl sulfate-polyacrylamide gels: enhancement by carbohydrate-binding dyes. *Anal Biochem* 185, 324–30 (1990).
6. Zhang, H., Li, X. J., Martin, D. B. & Aebersold, R. Identification and quantification of N-linked glycoproteins using hydrazide chemistry, stable isotope labeling and mass spectrometry. *Nat Biotechnol* 21, 660–6 (2003).
7. Tarentino, A. L., Gomez, C. M. & Plummer, T. H., Jr. Deglycosylation of asparagine-linked glycans by peptide:N-glycosidase F. *Biochemistry* 24, 4665–71 (1985).
8. Lewandrowski, U., Moebius, J., Walter, U. & Sickmann, A. Elucidation of N-glycosylation sites on human platelet proteins: A glycoproteomic approach. *Mol Cell Proteomics* 5, 226–233 (2006).

Chapter 18

LC/MS^n for Glycoprotein Analysis: N-Linked Glycosylation Analysis and Peptide Sequencing of Glycopeptides

Nana Kawasaki, Satsuki Itoh, and Teruhide Yamaguchi

Summary

Liquid chromatography/multiple-stage mass spectrometry (LC/MS^n) is an effective means for the site-specific glycosylation analysis of a limited quantity of glycoproteins, such as gel-separated proteins. Generally, a tryptic digest of the glycoprotein is separated by reversed-phase LC, and peptide sequencing and glycosylation analysis are achieved with on-line MS^n. In this chapter, a protocol for the LC/MS/MS/MS of a proteolytic digest of a gel-separated glycoprotein is described.

Key words: Glycosylation analysis, Peptide sequencing, LC/MS^n.

1. Introduction

A glycoprotein consists of many different glycoforms due to their carbohydrate heterogeneity at multiple glycosylation sites. Glycosylation analysis at each glycosylation site is important for gaining an understanding of the function of the glycoproteins. Mass spectrometric peptide mapping by liquid chromatography/mass spectrometry (LC/MS) is one of the effective means in use for the site-specific glycosylation analysis of glycoproteins *(1, 2)*. Glycosylation at each site can be elucidated by the MS of glycopeptides which are separated from the proteolytic digests by on-line LC *(3, 4)*.

LC/multiple-stage mass spectrometry (MS^n) has become a powerful tool for the site-specific glycosylation analysis of a limited quantity of glycoproteins, such as gel-separated proteins *(5)*. A mixture of cellular proteins is reduced and alkylated, and then separated by SDS-polyacrylamide gel electrophoresis (PAGE) or

Nicolle H. Packer and Niclas G. Karlsson (eds.), *Methods in Molecular Biology, Glycomics: Methods and Protocols, vol. 534*

DOI: 10.1007/978-1-59745-022-5_18

2D-electrophoresis. The gel band or spot that seems to contain a target glycoprotein is excised and crushed. The proteins are extracted from the gel in an intact form by shaking in 1% SDS and precipitation with acetone. The extracted proteins are digested with an appropriate enzyme which ideally provides glycopeptides with one attached sugar chain. The complex mixture of peptides is separated by reversed-phase LC, and glycosylation analysis and peptide sequencing are achieved by MS/MS and MS/MS/MS of the glycopeptides. The carbohydrate structures are fragmented and may be deduced from the MS/MS spectra whilst the peptide often remains with only a single GlcNAc attached. The peptide sequence of the glycopeptide can then be confirmed by further MS/MS/MS spectra of the [peptide + GlcNAc + H]$^+$ as a parent ion.

In this chapter we describe the protocol for the LC/MS/MS/MS of a gel-separated glycoprotein for the identification and analysis of the site-specific N-glycosylation. Thy-1, which is one of the GPI-anchored proteins and glycosylated at Asn23, 74, and 98, is used as a typical model glycoprotein *(6)*.

2. Materials

2.1. Reduction and Carboxymethylation

1. *Buffer.* 0.5 M Tris–HCl (pH 8.6) containing 8 M guanidine hydrochloride and 5 mM EDTA.
2. *Reducing reagent.* 2-mercaptoethanol. [Dithiothreitol, and tris(carboxyethyl)phosphin (TCEP) can be used instead.]
3. *Alkylation reagent.* Sodium iodoacetate. [Iodoacetamide can be used instead].
4. *Gel filtration column for desalting.* Sephadex G-25 column (e.g., PD-10 column, GE Healthcare).

2.2. SDS-Polyacrylamide Gel Electrophoresis

1. *Gel.* A precast polyacrylamide gel (20 × 80 × 1 mm 4% stacking gel, and 80 × 80 × 1 mm 12.5% separating gel) with ten wells. Commercially available from BIO CRAFT Co. Ltd., Tokyo, Japan.
2. *Running buffer.* 25 mM Tris–HCl, 192 mM glycine and 0.1% (w/v) SDS.
3. *Sample buffer.* 50 mM Tris–HCl (pH6.8), 10% (v/v) Glycerol, 2% (w/v) SDS, 5% (w/v) 2-mercaptoethanol, and 0.0025% (w/v) bromophenol blue.
4. *Dye reagents.* Coomassie® G-250 (Simple Blue SafeStain). Commercially available from Invitrogen (Carlsbad, CA, USA).

5. *Prestained molecular weight markers.* Precision Plus Protein™ All Blue Standards (Bio-Rad Laboratories, Inc.).
6. *Instrument.* Vertical Slab gel electrophoresis unit (BIO CRAFT). Power supply.

2.3. Extraction of Glycoproteins and Proteolytic Digestion

1. *Extraction buffer.* 20 mM Tris–HCl (pH 8.0) containing 1% (w/v) SDS. [1% (w/v) octylglucoside can be used instead.]
2. Solvent for glycoprotein precipitation: Cold acetone.
3. *Filter for removal of gel pieces.* Ultrafree-MC (0.22 μm, Millipore Corporation, Bedford, USA).
4. *Protease.* Trypsin-Gold (Promega, Madison, WI, USA) (*see* **Note 1**).
5. *Proteolytic digestion buffer.* 0.1 M Tris–HCl (pH 8.0).
6. *Vacuum centrifuge.* SpeedVac concentrator (Thermo Fisher Scientific, Waltham, MA, USA).

2.4. LC/MS^n and Database Search Analysis

1. *LC equipment.* HPLC system capable of binary gradient formation at a flow rate of μl/min or nl/min (e.g., Paradigm MS4, Michrom BioResources Inc., Auburn, CA, USA).
2. *Column.* A reversed-phase (RP) column (0.2 mm i. d. × 50 mm) (*see* **Note 2**). A C18 or C30 column is commonly used for the peptide/glycopeptide mapping. Graphitized carbon column (Thermo Fisher Scientific) can be used for the analysis of small hydrophilic glycopeptides.
3. *Tip.* Nanospray tip and *x–y–z* translational stage. Commercially available from AMR (Tokyo, Japan).
4. An ion trap type mass spectrometer (e. g., LTQ, Thermo Fisher Scientific) equipped with ESI.
5. *LC mobile phase.* HPLC grade acetonitrile and formic acid. Solvent A: 0.1% formic acid in 2% acetonitrile. Solvent B: 0.1% formic acid in 90% acetonitrile.
6. *Search engine.* TurboSEQUEST (Thermo Fisher Scientific) or Mascot (Matrix Sciences, London UK).
7. *Database.* Swiss-Prot or NCBI-nr.

3. Methods

3.1. Reduction and Alkylation of Glycoproteins (See Note 3)

1. Dissolve the crude fraction (360 μg) in 360 μl of reduction and carboxymethylation buffer.
2. Incubate the glycoprotein sample with 2-mercaptethanol (2.6 μl) at room temperature for 2 h.

3. To the reaction mixture, add 7.56 mg of sodium iodoacetate and incubate the mixture at room temperature for 2 h in the dark.
4. Apply the mixture (2.5 ml) to the PD-10 column equilibrated with water. Pass through the column to remove the excess reagent.
5. Wash the column with 3.5 ml of water and collect the eluant.
6. Lyophilize the eluant.

3.2. SDS-Polyacrylamide Gel Electrophoresis

1. Set a polyacrylamide gel in the slab gel electrophoresis unit.
2. Pour the running buffer into the upper and lower chambers of the gel unit.
3. Load the alkylated proteins (~25 μg/~25 μl) in a well. Include one well for a prestained molecular weight marker.
4. Connect the assembly of the gel tank to a power supply. Run at a 20 mA constant current until the dye front reaches the end of the gel.
5. Stain the proteins in the gel with Coomassie® G-250. The typical electrophoretic separation is shown in **Fig. 1**.

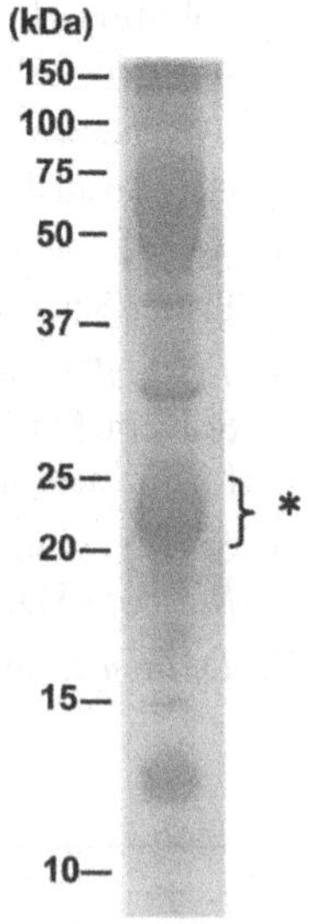

Fig. 1. SDS-PAGE of crude fraction containg the target glycoprotein (Thy-1). GPI-anchored proteins were prepared by phosphatidylinositol-specific phospholipase C treatment of a rat brain membrane fraction solubilized with Triton X-114. The crude fraction containing lipid-free soluble GPI-anchored proteins were separated by SDS-PAGE. The protein at 20–25 kDa indicated by an asterisk is a target glycoprotein, Thy-1. This band was excised, and the proteins extracted from the gel were digested with trypsin.

3.3. Extraction of the Glycoprotein from the Gel and Proteolytic Digestion (See Note 4)

1. Excise the gel bands of Coomassie-stained proteins and crush the gel pieces.
2. Extract the glycoproteins by shaking the crushed gel vigorously in 500 μl of extraction buffer overnight.
3. Filter the extracted solution with Ultrafree-MC to remove the gel pieces
4. To the filtrated solution, add four volumes of cold acetone (2 ml) to precipitate the proteins.
5. Centrifuge the sample and remove the supernatant.
6. Incubate the precipitated glycoproteins with trypsin-gold (1 μg) in 20 μl of proteolytic digestion buffer at 37°C overnight.

3.4. LC/MS^n

1. Equilibrate the RP-column by running through 95% of solvent A at a flow rate of 3 μl/min for 20–30 min.
2. Attach the nanospray tip to the outlet of the column, and set it in the *x–y–z* translational stage. Align the spray tip with the capillary of the mass spectrometer for accurate spray direction by moving the *x–y–z* translational stage.
3. Set up the mass spectrometer (e. g. LTQ: ESI voltage, 2.0 kV; capillary temperature, 200°C, etc.) and check the stable spray with elution buffer.
4. Inject samples (5–10 μl) onto the column and simultaneously start both the chromatography gradient and the mass spectrometer data collection. The column is eluted typically by a linear gradient from 2% of B to 90% of solvent B in 60 min. A full MS scan (*m/z* 300–2,000) followed by MS/MS and MS/MS/MS scans are acquired in a data-dependent manner (*see* **Note 5**). Simultaneous in-source collision-induced dissociation (CID) (*m/z* 100–500) is recommended for locating the glycopeptides in the mass spectrometric peptide map (*see* **Note 6**) *(3)*. As a typical separation of peptides, the total ion chromatogram (TIC) obtained by a full MS scan of the tryptic digested Thy-1 is illustrated in **Fig. 2A**.

3.5. Assignment of Glycopeptides

3.5.1. Database Search Analysis

1. Begin the database search analysis using a search engine such as TurboSEQUEST or Mascot to match the acquired spectra with predicted spectra for the known sequence in a database for protein identification. For the identification of glycopeptides which have a single remaining GlcNAc attached in the MS/MS spectrum, use the following parameters: carboxymethylation (58 Da) at Cys, and a possible modification of HexNAc (203 Da) at Asn (*see* **Note 7**).
2. After the glycopeptides have been identified by the database searching, pick out the MS/MS spectra of identified

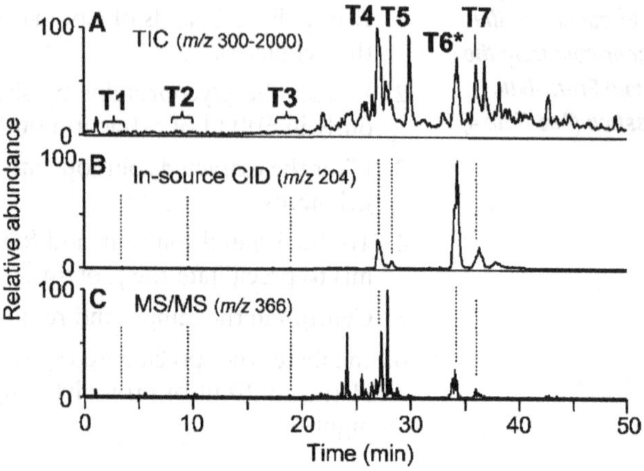

Fig. 2. Separation of tryptic digest of extracted Thy-1. (**A**) TIC obtained by a full MS scan (*m/z* 300–2,000) in the positive ion mode. (**B**) EIC of *m/z* 204 obtained by in-source CID. (**C**) EIC of *m/z* 366 obtained by data-dependent MS/MS. LC: instrument, Paradigm MS4; column, C18 (Magic C18, 0.2 × 50 mm, Michrom Bioresources); MS: instrument, LTQ (Thermo Fisher Scientific); ion mode, positive. T1 – T7 and the *dotted lines* depict the elution position of glycopeptides. Tryptic digestion of Thy-1 provides three major glycopeptides and some minor glycopeptides arising from Thy-1 by failed digestion.

glycopeptides from all product ion spectra. The carbohydrate sequence can sometimes be deduced from the fragment ions, such as the B and Y ions in the MS/MS spectra depending on the complexity of the glycan structure *(7, 8)*. The peptide sequence can be confirmed by the b- and y-ions produced by MS/MS/MS *(5, 9)*.

As the typical product ion spectra, MS/MS and MS/MS/MS spectra of glycopeptide from Thy-1 are indicated in **Fig. 3**. **Figure 3A** shows the MS/MS spectrum acquired from $[M + 2H]^{2+}$ (*m/z* 1,512.2) as a precursor ion. The most intense ion $[\text{peptide} + \text{GlcNAc} + H]^{+}$ (*m/z* 1,310.7) was subjected to MS/MS/MS data-dependently and identified as VLTLA(GlcNAc-)N FTTK based on the b- and y-ions by the database search analysis (**Fig. 3B**). From the calculated glycopeptide mass (3,022.4 Da) and the peptide theoretical mass (1,106.6), the mass of *N*-glycan moiety was calculated as 1,933.8 Da ($dHex_2Hex_5HexNAc_4$). One of two dHex (Fuc) is suggested to be attached to Gal-GlcNAc at the non- reducing end as the Lewis a/x antigen (Gal-(Fuc-)GlcANc-) or the blood group H-determinant (Fuc-Gal-GlcNAc-) by the presence of distinctive ions, such as $dHex_1Hex_1HexNAc_1^+$ ($B_{2\alpha}$, *m/z* 512) and $dHex_1Hex_2HexNAc_1^+$ ($B_{3\alpha}$, *m/z* 674). The presence of $Y_{1\alpha}$ (*m/z* 1,456.7) reveals that the other Fuc is attached to the inner trimannosyl core GlcNAc. The

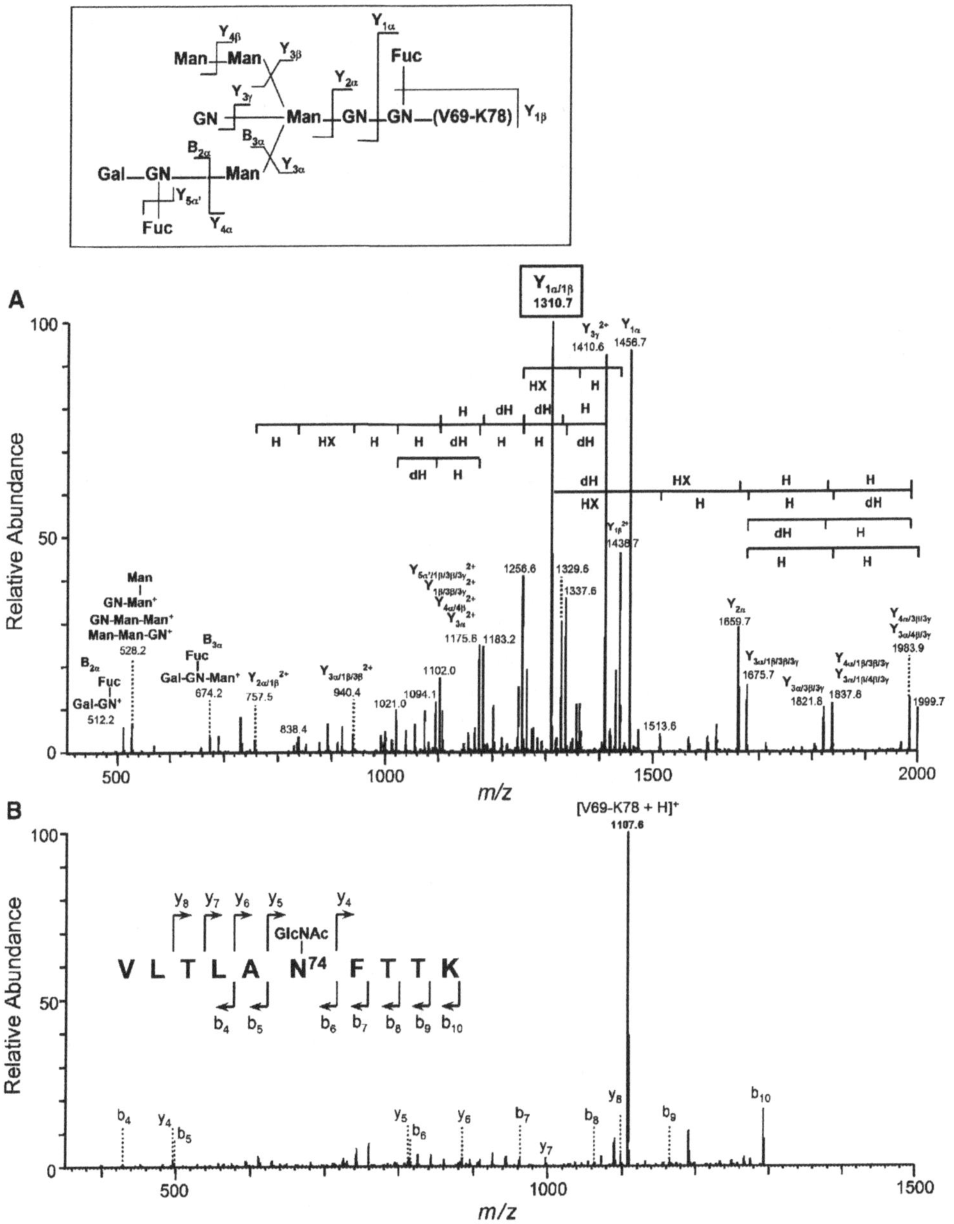

Fig. 3. Product ion spectra of Thy-1 glycopeptide eluted at 34.52 min, (**A**) MS/MS spectrum of the glycopeptide precursor ion: *m/z* 1,512.2 (2+), (**B**) MS/MS/MS spectrum of the glycopeptide precursor ion: *m/z* 1,310.7 (1+). A fucosylated hybrid type oligosaccharide was deduced from the MS/MS/MS spectrum, and Val69-Lys78 in Thy-1 was identified by database search analysis using the MS/MS/MS spectrum. *H* Hexose; *HX N*-acetylhexosamine; *dH* Deoxyhexose.

product ion (*m/z* 1,410.6) that arose from the precursor ion (*m/z* 1,512.2) by loss of 101.6 u (HexNAc) and [GlcNAc-Man-GlcNAc-GlcNAc-peptide (V69-K78) + H]$^+$ ($Y_{3\alpha/1\beta/3\beta}$, *m/z* 940.4) suggests a linkage of a bisecting GlcNAc. From these product ions, it was deduced that the structure of the N-glycan moiety is a hybrid type oligosaccharide as indicated in **Fig. 3**, inset.

3.5.2. Assignment of Glycopeptides Not Identified by Database Search Analysis

1. Most of the glycopeptides can be located in the chromatogram using oxonium ions generated by CID. To estimate the elution positions of glycopeptides, including unidentified glycopeptides by database searching, draw the extracted ion chromatogram (EIC) of *m/z* 204 (HexNAc$^+$) or *m/z* 366 (HexHexNAc$^+$) produced by in-source CID or CID-MS/MS. The peaks appearing in the EIC can be assumed as glycopeptide peaks. **Fig. 2B, C** show the EICs of HexNAc$^+$ (*m/z* 204) produced by in-source CID of the tryptic digest of Thy-1 and HexHexNAc$^+$ (*m/z* 366) that were obtained by data-dependent CID-MS/MS of the digest, respectively *(10–12)*. After the elution times of glycopeptides have been assumed, pick out the MS/MS spectra acquired when the glycopeptides were eluted. Ascertain whether the MS/MS spectra were really acquired from glycopeptides or not, by making a check on the presence of carbohydrate distinctive ions, such as HexHexNAc$^+$ (*m/z* 366) and NeuAcHexHexNAc$^+$ (*m/z* 657) (*see* **Note 8**).
2. Peptides which have a single remaining GlcNAc are often detected as intense ions in the picked MS/MS spectra. The peptide moiety can be assigned by matching the *m/z* value of the intense ion with the theoretical value of the peptide bearing GlcNAc. The peptide sequence could be ascertained if an additional LC/MS/MS/MS would be performed by setting the *m/z* values of peptide-related ions as precursors. Once the peptide can be assigned, the carbohydrate structure may be deduced from the MS/MS spectra.

4. Notes

1. *Enzyme specificity is as follows.* Trypsin, Lys/Arg-↓-X; Lys-C, Lys-↓-X; Glu-C, Glu (Asp)-↓-X; AspN, X-↓-Asp (Glu). Glu-C digests without cleavage of Asp-X at a 1:50 enzyme-to-substrate ratio at pH 8.0, and Asp-X can be hydrolyzed by Glu-C at a 1:4 enzyme-to-substrate ratio.

2. *Typical flow rate against the column diameter is as follows.* 0.3 mm, 5–10 μl/min; 0.2 mm, 3 μl/min; 0.1 mm, 0.5 μl/min; 0.075 mm, 0.2 μl/min.
3. In-gel reduction and carboxymethylation could be employed instead of solution reduction and carboxymethylation as follows *(13)*:
   (a) Incubate destained and dried gel pieces with 10 mM dithiothreitol in 25 mM NH_4HCO_3 at 56°C for 1 h.
   (b) After washing gel pieces with 25 mM NH_4HCO_3, incubate gel pieces with 55 mM sodium monoiodoacetic acid in 25 mM NH_4HCO_3 at room temperature in the dark for 45 min.
4. SDS/acetone is used for the extraction and precipitation of most glycoproteins. For soluble glycoproteins, extraction with 1% octylglucoside and gel filtration with Sephadex G-25 for the removal of octylglucoside can be used instead.
5. A data-dependent scan is performed for the most intense ion at each scan. The positive ion mode is used for neutral and mono/disialylated glycopeptides, and the additional negative ion mode is recommended for tri/tetrasialylated and sulfated glycopeptides.
6. Monitoring of the oxonium ions at *m/z* 204 ($HexNAc^+$) and 366 ($HexHexNAc^+$) is effective to locate both *N*- and many but not all *O*-glycosylated peptides *(14)*. In most cases, an ion trap type mass spectrometer cannot detect small fragment ions, such as $HexNAc^+$, due to its low mass cut-off system. In-source CID is recommended for glycopeptide mapping when an ion trap type mass spectrometer is used.
7. For the identification of peptides, care should be taken that an *N*-terminal glutamine residue is converted to pyroglutamic acid, which leads to a 17 Da mass loss from the predicted peptide. A similar mass loss is observed on an *N*-terminal carboxyamidomethylated cysteine residue (alkylated with iodoacetamide) *(15)*.
8. Addition to the oxonium ions at *m/z* 204 and 366, monitoring of *m/z* 292 ($NeuAc^+$) and *m/z* 512 (Hex-(dHex) $HexNAc^+$, Lewis a/x motif) is also useful to locate the glycopeptides. For sulfated, sialylated, and fucosylated oligosaccharides, a neutral loss of 80, 291, and 146, respectively, are often used to assume their elution position in the chromatogram.

References

1. Wuhrer, M., A.M. Deelder, and C.H. Hokke (2005) Protein glycosylation analysis by liquid chromatography-mass spectrometry: *J Chromatogr B Analyt Technol Biomed Life Sci* 825, 124–33.
2. Zaia, J. (2004) Mass spectrometry of oligosaccharides: *Mass Spectrom Rev* 23, 161–227.
3. Guzzetta, A.W., et al. (1993) Identification of carbohydrate structures in glycoprotein peptide maps by the use of LC/MS with selected ion extraction with special reference to tissue plasminogen activator and a glycosylation variant produced by site directed mutagenesis: *Anal Chem* 65, 2953–62.
4. Ling, V., et al. (1991) Characterization of the tryptic map of recombinant DNA derived tissue plasminogen activator by high-performance liquid chromatography-electrospray ionization mass spectrometry: *Anal Chem* 63, 2909–15.
5. Itoh, S., et al. (2005) Characterization of a gel-separated unknown glycoprotein by liquid chromatography/multistage tandem mass spectrometry: analysis of rat brain Thy-1 separated by sodium dodecyl sulfate-polyacrylamide gel electrophoresis: *J Chromatogr A* 1094, 105–17.
6. Parekh, R.B., et al. (1987) Tissue-specific N-glycosylation, site-specific oligosaccharide patterns and lentil lectin recognition of rat Thy-1: *Embo J* 6, 1233–44.
7. Domon, B. and C.E. Costello (1988) A systematic nomenclature for carbohydrate fragmentation in FAB-MS/MS spectra of glycoconjugates: *Glycoconjugate J* 5, 397–409.
8. Demelbauer, U.M., et al. (2004) Determination of glycopeptide structures by multistage mass spectrometry with low-energy collision-induced dissociation: comparison of electrospray ionization quadrupole ion trap and matrix-assisted laser desorption/ionization quadrupole ion trap reflectron time-of-flight approaches: *Rapid Commun Mass Spectrom* 18, 1575–82.
9. Zhang, S. and D. Chelius (2004) Characterization of protein glycosylation using chip-based infusion nanoelectrospray linear ion trap tandem mass spectrometry: *J Biomol Tech* 15, 120–33.
10. Harazono, A., et al. (2005) Site-specific glycosylation analysis of human apolipoprotein B100 using LC/ESI MS/MS: *Glycobiology* 15, 447–62.
11. Sullivan, B., T.A. Addona, and S.A. Carr (2004) Selective detection of glycopeptides on ion trap mass spectrometers: *Anal Chem* 76, 3112–8.
12. Wuhrer, M., et al. (2005) Protein glycosylation analyzed by normal-phase nano-liquid chromatography – mass spectrometry of glycopeptides: *Anal Chem* 77, 886–94.
13. Kikuchi, M., et al. (2004) Proteomic analysis of rat liver peroxisome: presence of peroxisome-specific isozyme of Lon protease: *J Biol Chem* 279, 421–8.
14. Huddleston, M.J., M.F. Bean, and S.A. Carr (1993) Collisional fragmentation of glycopeptides by electrospray ionization LC/MS and LC/MS/MS: methods for selective detection of glycopeptides in protein digests: *Anal Chem* 65, 877–84.
15. Krokhin, O.V., W. Ens, and K.G. Standing (2003) Characterizing degradation products of peptides containing N-terminal Cys residues by (off-line high-performance liquid chromatography)/matrix-assisted laser desorption/ionization quadrupole time-of-flight measurements: *Rapid Commun Mass Spectrom* 17, 2528–34.

Subsection C

O-GlcNAcomics

Chapter 19

Detecting the "O-GlcNAc-ome"; Detection, Purification, and Analysis of O-GlcNAc Modified Proteins

Natasha E. Zachara

Summary

The modification of Ser and Thr residues of cytoplasmic and nuclear proteins with a monosaccharide of O-linked β-*N*-acetylglucosamine is an essential and dynamic post-translational modification of metazoans. Deletion of the O-GlcNAc transferase (OGT), the enzyme that adds O-GlcNAc, is lethal in mammalian cells highlighting the importance of this post-translational modification in regulating cellular function. O-GlcNAc is believed to modulate protein function in a manner analogous to protein phosphorylation. Notably, on some proteins O-GlcNAc and O-phosphate modify the same Ser/Thr residue, suggesting that a reciprocal relationship exists between these two post-translational modifications. In this chapter we describe the most robust techniques for the detection and purification of O-GlcNAc modified proteins, and discuss some more specialized techniques for site-mapping and detection of O-GlcNAc during mass spectrometry.

Key words: Metabolic sensor, Signal transduction, Cellular stress, Post-translational modification, Affinity purification, Site-mapping, Mass spectrometry.

1. Introduction

More than 400 proteins of the cytosol and nucleus are modified by monosaccharides of O-linked β-*N*-acetylglucosamine (O-GlcNAc) (*1*). This simple sugar modification is common in metazoans and, with the possible exception of plant nuclear pore proteins, is not elongated into more complex structures (*1*). Unlike other carbohydrates, the O-GlcNAc protein modification is dynamic, that is, it turns over faster than the protein backbone. Moreover, the levels of O-GlcNAc respond dynamically to numerous stimuli, including (1) lymphocyte activation

Nicolle H. Packer and Niclas G. Karlsson (eds.), *Methods in Molecular Biology, Glycomics: Methods and Protocols, vol. 534*

DOI: 10.1007/978-1-59745-022-5_19

with phorbol esters, Con A, ionomycin (*2*); (2) cell cycle (*3*) and development (*4, 5*); (3) changing extracellular glucose/glucosamine concentrations (*1*); and (4) the induction of cellular stress with diverse agents (*6, 7*). These data, and those showing that O-GlcNAc modifies the same Ser/Thr residues utilized by kinases, suggests that O-GlcNAc is a regulatory post-translational modification analogous to protein phosphorylation (*8*). Notably, O-GlcNAc is required for life in mammals in single cells, whole tissues, and animals (*9, 10*). Recent studies have shown that the dynamic cycling of O-GlcNAc is important in sensing both nutrients and cellular stress (*8*); and, that the mis-regulation of O-GlcNAc may be involved in the etiology of type II diabetes, cardiovascular disease, and neurodegenerative disease (*1, 8*).

O-GlcNAc was initially detected in lymphocytes using bovine milk β(1–4)-galactosytransferase, which will transfer a galactose residue from uridine diphosphate-galactose (UDP-Gal) to any terminal β-*N*-acetylglucosamine (βGlcNAc) residue, resulting in βGal1-4βGlcNAc-X (*11*). Interestingly, while Torres and Hart expected to find the label restricted to cell surface proteins, they actually detected the majority of the label on cytosolic and nuclear proteins. Using tritiated UDP-galactose (^{3}H-Gal) they subsequently showed that the galactose was incorporated onto a monosaccharide of GlcNAc O-linked to nuclear and cytoplasmic proteins (*11*). This labeling process, known as galactosyltransferase labeling, remains the gold standard for the detection and characterization of O-GlcNAc modified proteins. Galactosyl-transferase labeling is the only strategy that subsequently characterizes the released product, showing that the label was on a monosaccharide of O-linked βGlcNAc. Recently, a mutant of galactosyl-transferase (Y289L) has been used to incorporate a ketone-biotin tag, rather than a galactose residue (*12–14*). The resulting tag proves very useful for enriching proteins, and for the rapid and sensitive detection of O-GlcNAc proteins by immunoblot (*12–14*). Unlike the galactosyl-transferase labeling methodology, product analysis is more challenging.

Numerous other techniques have been reported for the detection and analysis of O-GlcNAc modified proteins, and these are summarized in **Table 1**(*12–33*). The techniques range from simple immunoblotting procedures that require no specialized equipment (*23, 24, 28*), to those that facilitate high throughput detection of O-GlcNAc modified proteins by mass spectrometry (MS) (*12–15, 17–22, 25–27, 29–33*). The methodology of choice depends on the focus of the study, and the equipment available. To answer the simple question "is my protein of interest glycosylated with O-GlcNAc", immunoblotting procedures are the most straightforward. Immunoblotting is also the most compatible with 2-dimensional gels, as the other techniques (except for

Table 1
Strategies for the detection and analysis of O-GlcNAc modified proteins

Method	Common	Derivatives	Site-mapping	Enrichment	High-throughput	Definitive	Indicative	Suggestive	Comments	Reference
sWGA	√			√				√	False positives possible, best controls +/–PUGNAc & +/–hexosaminidase	(*16,18,23,24,28*)
									Can be used to enrich proteins, best done with nuclear and cytoplasmic extracts to reduce plasma membrane contamination	
									Not all O-GlcNAc modified proteins are recognized	
CTD110.6	√			√			√		False positives possible, best controls +/–PUGNAc & +/–hexosaminidase	(*16,18,24,28*)
									Not all O-GlcNAc modified proteins are recognized	
									Compatible with IF	
RL2	√			√				√	False positives possible, best controls +/–PUGNAc & +/–hexosaminidase	(*41,42*)
									Not all O-GlcNAc modified proteins are recognized	
									Compatible with IF	

(continued)

Table 1 (continued)

Method	Common	Derivatives	Site-mapping	Enrichment	High-throughput	Definitive	Indicative	Suggestive	Comments	Reference
ITT	√						√		Useful for low abundance proteins	(*16,24,28*)
									False positives possible, best controls +/ –hexosaminidase	
									Best in combination with CTD110.6 affinity chromatography	
Galactosyl transferase	√[a]	√	√	√		√			One step enzymatic derivatization. ^{3}H label not sensitive	(*16,24,28*)
									Can be used to enrich proteins with RCA1 affinity chromatography	
									Label is useful for following proteins through purification steps	
									Best done with nuclear and cytoplasmic extracts to reduce plasma membrane protein contamination, can be performed on immunoprecipitates	

BEMAD	√	√	√	√		√	False positives possible, appropriate control peptides required	(*17,25,28,30*)
							Proline rich peptides/proteins are problematic	
							Density labeling for quantification possible	
							Independent method should be used to confirm proteins of interest	
GlcNAz/ Staudinger ligation	√		√	√	√		Low risk of incorporation of label into cell surface proteins	(*12–14*)
							Detection in cell extract requires western blot with anti-flag	
							Useful for pulse chase studies	
							GlcNAz is commercially available (Invitrogen, Carlsbad, CA)	
							Excellent for high-throughput studies	
							Independent method should be used to confirm proteins of interest	
Galtransferase (Y289L)	√		√	√	√		False positives possible, best controls +/–PUGNAc & +/–hexosaminidase	(*27,29,31,32*)

(continued)

Table 1 (continued)

Method	Common	Derivatives	Site-mapping	Enrichment	High-throughput	Definitive	Indicative	Suggestive	Comments	Reference
									Can be used to enrich proteins, best done with nuclear and cytoplasmic extracts to reduce plasma membrane protein contamination	
									Reagents commercially available (Invitrogen, Carlsbad, CA)	
									Excellent for high-throughput studies	
									Independent method should be used to confirm proteins of interest	
DIONEX								√	Quantitative	
									Both monosaccharide analysis and alditol analysis is required	
									Most confidence when combined with galactosyltransferase labeling	
									Independent method should be used to confirm proteins of interest	

MS	√	√	√	O-GlcNAc highly labile in MS	(*15,19–22,26,33*)
				MS*n* is most compatible with site-mapping	
				Neutral loss can be used	
				Mass spectrometer does not differentiate between sugars with isobaric masses	
				Independent method should be used to confirm proteins of interest	

BEMAD Beta-elimination followed by Michael addition with dithiothreitol; *GlcNAz N*-azidoacetylglucosamine; *IF* Immunofluorescence; *ITT In vitro* transcription translation; *MS* Mass spectrometry; *PUGNAc O*-(2-acetamido-2-deoxy-d-glucopyranosylidene)amino-*N*-phenylcarbamate.
Common. Denotes techniques for which the procedure is quick, the reagents are commercially available, and no specialized equipment is required.
Derivatization. Technique requires one or more chemical/enzymatic derivatizations.
High throughput. Denotes techniques that are especially designed for proteomic type studies.
Site-mapping. This technique has been used to successfully to map O-GlcNAc attachment sites.
Enrichment. Techniques that can be used to enrich proteins/peptides prior to subsequent analysis.
Definitive. When performed correctly this technique definitively identifies O-GlcNAc modified proteins.
Indicative. A positive reaction with this technique suggests that proteins are modified by O-GlcNAc, but additional experiments should be performed to confirm this data.
Highly suggestive. When performed with the appropriate controls a positive results suggests that proteins are likely to be modified
[a]Final product analysis by DIONEX HPLC required specialized equipment.

classical galactosyl-transferase labeling) derivatize GlcNAc residues with compounds that can alter the pI of proteins.

For high throughput proteomics, several techniques facilitate the enrichment of O-GlcNAc modified proteins, including both galactosyl-transferase methods described above. Perhaps the most specific is that utilizing the Staudinger ligation (*27, 29, 31, 32*). Here a *N*-azidoacetylglucosamine (GlcNAz; either peracetylated or not) is introduced into proteins by metabolic labeling (*27*). Interestingly, GlcNAz is converted to UDP-GlcNAz and incorporated into proteins by the O-GlcNAc transferase. Importantly, GlcNAz does not appear to be incorporated into carbohydrates attached to proteins in the endoplasmic reticulum or the Golgi apparatus (*27*). The GlcNAz tag can then be modified and either detected using standard western blotting techniques, or used to enrich O-GlcNAz modified proteins (*31, 32*). All of these techniques (galactosyl-transferase, Y289L, galactosyl-transferase, and GlcNAz) retain the GlcNAc(z)-protein bond, which is potentially useful for site-mapping. However, the detection of O-GlcNAc modified peptides in most mass spectrometers is challenging for two reasons (1) the β-O-glycosidic bond is highly labile; (2) the signal of O-GlcNAc modified peptides is suppressed compared to that of unmodified peptides. Of course, techniques that enrich for the O-GlcNAc glycopeptides render the latter point moot.

To overcome challenges associated with the MS labile linkage, researchers have beta-eliminated O-GlcNAc from the protein backbone, converting Thr and Ser to α-aminobutyric acid or 2-aminopropenoic acid respectively (*34*). These modified residues are more readily identified in the mass spectrometer (*34*). Wells and co-workers expanded on this technique, adding dithiothreitol (DTT) back to the protein via a Michael addition (BEMAD) (*25, 28, 30*). The resulting chemical bond is MUCH more stable in the mass spectrometer, and can also be used to enrich peptides/proteins. The main disadvantage of this technique is the possible identification of peptides modified by other forms of glycosylation; and/or the identification of phosphorylation sites. Other techniques that have been used for site-mapping include galactosyl-transferase labeling in combination with: manual Edman degradation (*16, 24, 28*); ion trap mass spectrometry (*19, 22*); quadrupole time of flight mass spectrometry (*21, 22, 25, 26*); and, fast atom bombardment mass spectrometry (*15*). As discussed above these techniques are challenged by the loss of O-GlcNAc during analysis. Recently, Electron-capture dissociation tandem mass spectrometry has been introduced and reportedly this form of mass spectrometry maintains labile protein modifications such as phosphorylation and potentially O-GlcNAc.

In this paper we review many of the more general techniques for the detection and analysis of O-GlcNAc modified proteins.

2. Materials

2.1. Altering the Levels of O-GlcNAc in Cell Culture

1. *O*-(2-acetamido-2-deoxy-D-glucopyranosylidene)amino-*N*-phenylcarbamate (PUGNAc: Catalog #A157250; Toronto Research Chemicals Inc, Ontario, Canada), made up at 20 mM in water. Filter sterilize, aliquot, and store at –20°C for 3–6 months.
2. Cells of interest grown in media of choice; 1 g/L glucose is preferable (*see* **Note 1**).

2.2. Detecting O-GlcNAc Modified Proteins by Immunoblot with CTD110.6

1. Control proteins Ovalbumin (100 ng per lane; SIGMA-Aldrich, MO, USA) and Bovine Serum Albumin (BSA)-GlcNAc (5–10 ng per lane; *see* **Note 2**).
2. Purified or crude proteins (*see* **Note 3**) separated by SDS-PAGE and transferred to either nitrocellulose or polyvinylidene fluoride (PVDF). Note that duplicate blots are required (*see* **Note 4**).
3. *Blocking agent.* 2% w/v skim milk in Tris Buffered Saline with Tween-20 (TBST): 10 mM Tris-HCl, 150 mM NaCl, 0.05% v/v Tween-20, pH 7.5.
4. *Wash buffer.* TBST.
5. *Antibody incubation buffer.* 3% w/v BSA, in TBST.
6. *GlcNAc competition buffer.* 100 mM GlcNAc, in antibody incubation buffer (*see* **Notes 4** and **5**, **Fig.1**).
7. *Glucose or galactose competition buffer.* 100 mM glucose or galactose, in antibody incubation buffer (*optional;* **Fig.1**).
8. *Primary antibody.* CTD 110.6 (antigen purified; *see* **Subheadings 2.3** and **3.3**) at 1 mg/mL in 1:1 PBS:Glycerol (*see* **Note 6**). CTD110.6 ascites is available from Pierce (Catalog #24565, Pierce Biotechnology Inc., IL, USA) and Covance (Catalog #MMS-248R, Covance Research Products, CA, USA) (*see* **Note 7**).
9. *Secondary antibody.* Anti-Mouse IgM HRP (Catalog #A8786, SIGMA-Aldrich).
10. ECL Reagent, Autoradiography/ECL film or other detection system.

2.3. Antigen Purification of CTD110.6

1. GlcNAc-agarose or GlcNAc-sepharose (Catalog #CG-003-5, EY labs, CA, USA; Catalog #01143, SIGMA-Aldrich; *see* **Note 8**).
2. Column housings, such as BIORAD (CA, USA) ECONO-columns (Catalog #7370506) or BIORAD disposable ECONO-PAC columns (Catalog #7321010).

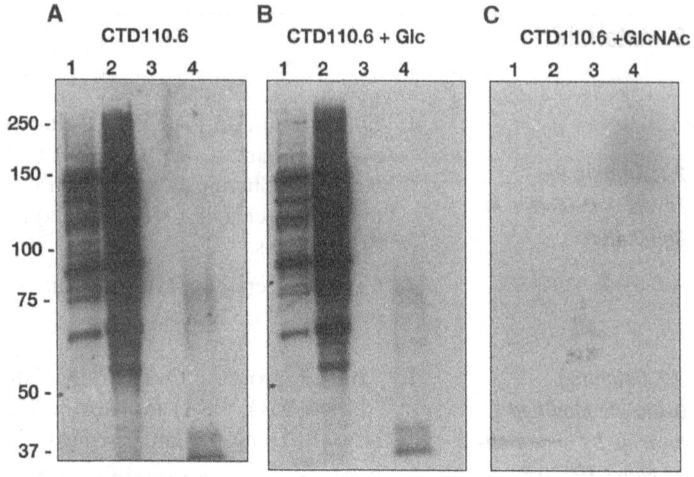

Fig. 1. The O-GlcNAc protein modification of Cos-7 cells (±PUGNAc) detected by immunoblotting with CTD 110.6. Proteins were separated by SDS-PAGE (7.5%) and blotted to nitrocellulose. Protein samples were loaded as follows *(1)* 40 μg whole Cos-7 cell lysate; *(2)* 40 μg PUGNAc (100 μM, 18 h) treated whole Cos-7 cell lysate; *(3)* 5ng BSA-aminophenol; and *(4)* 5 ng BSA-aminophenyl-GlcNAc. Three identical immunoblots are shown, they were probed as follows: (**A**) 1 μg/mL CTD110.6 in 3% w/v BSA in TBST; (**B**) 1 μg/mL CTD110.6 in 3% w/v BSA in TBST, 100 mM glucose; and (**C**) 1 μg/mL CTD110.6 in 3% w/v BSA in TBST, 100 mM GlcNAc.

3. 1 M Tris-HCl pH 8.0.
4. 1 μM filters (non-sterile).
5. Phosphate buffered saline (PBS; 10 mM phosphate buffer, 136 mM sodium chloride, 2.6 mM potassium chloride).
6. 100 mM galactose or glucose, in PBS (*see* **Note 9**).
7. 1M GlcNAc, in PBS.
8. 0.05% w/v sodium azide, in PBS.
9. Dialysis equipment of choice.
10. 80% v/v glycerol, autoclaved.
11. Protease inhibitors of choice (*see* **Note 10**).

2.4. Immunoblotting with Succinylated Wheatggerm Agglutinin (sWGA)

1. Purified or crude protein separated by SDS-PAGE and transferred to nitrocellulose or PVDF. Duplicate blots required (*see* **Note 4**).
2. TBST.
3. *Blocking agent.* 3% w/v BSA, in TBST (*see* **Note 11**).
4. 1 μg/mL sWGA-HRP (EY Labs; San Mateo, CA), in 3% BSA in TBST (*see* **Notes 12** and **13**)

5. 100 mM GlcNAc, in TBST.
6. *High salt TBST (HS-TBST).* 10 mM Tris-HCl, 1 M NaCl, 0.05% (v/v) Tween-20, pH 7.5.
7. TBS.
8. ECL reagent (or equivalent), autoradiography/ECL film or other detection system.

2.5. Detecting O-GlcNAc Modified Proteins by Galactosyltransferase Labeling

1. Protein sample(s), ovalbumin positive control (2 μg).
2. UDP-[^{3}H]Gal, 1.0 mCi/mL (specific activity 17.6 Ci/mmol) in 70% v/v ethanol.
3. 25 mM 5 -adenosine monophosphate (5 -AMP) in Milli-Q water pH 7.0 (*see* **Note 14**).
4. *Buffer H.* 50 mM HEPES pH 6.8, 50 mM NaCl, 2% v/v Triton X-100.
5. *10 × Labeling buffer.* 100 mM HEPES pH 7.5, 100 mM galactose, 50 mM $MnCl_2$.
6. UDP-Gal (not tritiated).
7. *Stop solution.* 10% w/v SDS, 100 mM EDTA.
8. *Desalting column.* Sephadex G-50 (20–80 μm mesh; SIGMA-Aldrich) column (30 × 1 cm) equilibrated in 50 mM ammonium formate, 0.1% (w/v) SDS (*see* **Note 15**).
9. Acetone.

2.6. Autogalactosylated Galactosyltransferase

1. *10× galactosyltransferase buffer.* 100 mM HEPES pH 7.4, 100 mM galactose and 50 mM $MnCl_2$.
2. 30–50 mL centrifuge tubes.
3. Aprotinin.
4. β-Mercaptoethanol.
5. UDP-Gal, not tritiated.
6. *Saturated ammonium sulfate.* At least 17.4 g $(NH_4)_2SO_4$ in 25 mL Milli-Q water.
7. *85% ammonium sulfate.* 14 g $(NH_4)_2SO_4$ in 25 mL Milli-Q water.
8. *Galactosyltransferase storage buffer.* 2.5 mM HEPES pH 7.4, 2.5 mM $MnCl_2$, 50% v/v glycerol.

2.7. Detecting O-GlcNAc Modified Proteins by Immunofluorescence

1. Cells grown on cover slips or in chamber slides.
2. PBS (ice cold).
3. 4% w/v formaldehyde, in PBS (*see* **Note 16**).
4. 2 mg/mL BSA, in PBS.
5. 0.075% v/v Triton X-100, in 2 mg/mL BSA, in PBS.

6. Antigen purified CTD110.6 (1 mg/mL) in PBS:Glycerol 1:1 (*see* **Subheadings 2.3** and **3.3**).
7. Non-specific IgM (*see* **Note 17**).
8. 100 mM galactose or glucose in 2 mg/mL BSA, in PBS (*see* **Note 9**).
9. 100 mM GlcNAc in 2 mg/mL BSA, in PBS.
10. Anti-Mouse IgM with the fluorophore of choice.
11. DAPI at 0.1 μg/mL in 2 mg/mL, BSA in PBS (optional).
12. Antifade of choice.
13. Slides of choice.

2.8. Purifying O-GlcNAc Modified Proteins

1. Protein extracts (*see* **Note 18**).
2. CTD-110.6 covalently coupled to agarose or sepharose (*see* **Note 19**).
3. Anti-IgM agarose or sepharose (Catalog #A4540 SIGMA-Aldrich; Catalog #046841 Invitrogen).
4. Sepharose or agarose preferably coupled to a non-specific IgM (*see* **Note 17**).
5. 15 and 50 mL tubes.
6. *Column housings.* Such as BIORAD ECONO-PAC columns (Catalog #7321010), BIO-SPIN columns (Catalog #7326204 or 7326008), or Poly-Prep columns (Catalog #7311550).
7. 0.1% v/v NP-40 or Triton X-100 in PBS (*see* **Note 20**).
8. 100 mM galactose or glucose, in PBS.
9. 1 M GlcNAc, in PBS.
10. 1 M GlcNAc, in 100 mM sodium acetate buffer, 500 mM NaCl, pH 4.5 (*see* **Note 21**).
11. 0.05% w/v sodium azide, in PBS.
12. Methanol.

2.9. In Vitro Transcription Translation

1. cDNA of nuclear pore protein p62, and your protein(s) of interest, subcloned into an expression vector with an SP6 or T7 promoter (~0.5–1 μg/μL; *see* **Note 22**).
2. Rabbit reticulocyte lysate (RRL) *in vitro* transcription translation kit (Promega; WI, USA).
3. Label, ^{35}S-Met, or ^{35}S-Cys, or ^{14}C-Leu.
4. sWGA-agarose (Catalog #AL-1023S, Vector Labs; CA, USA; Catalog #A-2102-5, EY labs, CA, USA).
5. 1 mL tuberculin syringe with a glass wool frit (100 μL) or a Bio-Spin disposable chromatography column (BIORAD, CA, USA).

6. Sephadex G-50 (20–80 μm mesh; SIGMA-Aldrich).
7. PBS.
8. 0.1% v/v Triton X-100 or NP-40, in PBS.
9. 1 M galactose, in PBS 0.1% v/v triton X-100 or NP-40.
10. 1 M GlcNAc, in PBS 0.1% v/v Triton X-100 or NP-40.
11. Liquid scintillation counter.
12. SDS-PAGE equipment and buffers.
13. Gel dryer.

2.10. Removing O-GlcNAc with Hexosaminidase

1. *N*-acetyl-β-D-glucosaminidase (Sigma-Aldrich; V-Labs Inc, LA, USA; New England Biolabs, MA, USA).
2. 2% w/v SDS.
3. *2× reaction mixture.* 80 mM citrate-phosphate buffer pH 4.0, 1–10U *N*-acetyl-β-D-glucosaminidase, 8% v/v triton X-100 or v/v NP-40, 0.01 U aprotinin, 1 μg of leupeptin, and 1 μg α_2-macroglobulin.

2.11. Removing N-Linked Glycosylation with PNGase F

1. Purified or crude protein samples, and a control protein such as ovalbumin (2–5 μg).
2. *10× PNGase F denaturing buffer.* 5% w/v SDS, 10% v/v β-mercaptoethanol, in 50 mM sodium phosphate buffer pH 7.5.
3. 10% v/v NP-40.
4. *10× PNGase F reaction buffer.* 500 mM sodium phosphate buffer, pH 7.5.
5. *Peptide N.* Glycosidase F (PNGase F, EC 3.2.2.18; New England Biolabs).

3. Methods

3.1. Altering the Levels of O-GlcNAc in Cell Culture

The endogenous enzyme that removes O-GlcNAc, the O-GlcNAcase, can be inhibited in tissue culture resulting in increased levels of the O-GlcNAc protein modification. This is a useful way of increasing the stoichiometry of O-GlcNAc on your protein of interest before site-mapping, and also provides a useful control for O-GlcNAc specific antibodies and lectins.

Numerous inhibitors have been reported in the literature (*35–37*); of these two are commercially available (*35, 36*). We prefer PUGNAc to streptozotocin (STZ) for two reasons (1) PUGNAc is a more potent inhibitor of O-GlcNAcase *in vitro*, with a Ki of 52 nM, compared to 2.5 mM for STZ (*38*) and (2) STZ appears to initiate apoptosis via an O-GlcNAc independent mechanism in min6 cells (*38*). Studies have shown that both stationary and dividing cells take up PUGNAc with no apparent toxicity over short periods of time (*39*). However, prolonged treatment reduces cells growth, and this has been attributed to alterations in the cell cycle and defects in cytokinesis (*3*).

1. Take cells grown in dishes.
2. Add PUGNAc to 40–100 μM (*see* **Note 23**).
3. Incubate the cells for 4–18 h (*see* **Note 24**).
4. Harvest the cells, and extract as desired (*see* **Note 10**).

3.2. Detecting O-GlcNAc Modified Proteins by Immunoblotting

Numerous O-GlcNAc specific antibodies have been characterized, they include HGAC85 *(40)*, RL2 *(41, 42)*, MY95 *(43)*, and CTD110.6 *(23)*. Of these, CTD110.6 appears to have the least peptide-dependence and recognizes the widest range of O-GlcNAc modified proteins. Importantly, it was shown that CTD110.6 does not react with O-αGlcNAc, a common modification of protozoan proteins, or GalNAc modified peptides *(23)*. Recently, site/protein specific glycosylation antibodies have been reported for the c-Myc proto-oncogene *(44)* and neurofilament proteins *(45)*.

A protocol for the use of CTD110.6 is given, but the controls described are appropriate for any of the anti-O-GlcNAc antibodies. To confirm that the antibody signal is specific the following controls are ideal (1) pretreating immunoprecipitates or cell extract with hexosaminidase (*see* **Subheadings 2.10** and **3.10**); (2) increasing the levels of O-GlcNAc in cell culture with PUGNAc (**Fig.1**; *see* **Subheadings 2.1** and **3.1**); (3) running appropriate positive and negative control proteins such as ovalbumin (negative) and BSA-GlcNAc (positive; **Fig. 1**); and (4) competing away the antibody signal with a specific sugar (GlcNAc) and a non-specific sugar (glucose or galactose; **Fig. 1**).

1. Block the membranes in blocking agent for 1 h at room temperature with shaking/rocking.
2. Wash the membranes 2 × 10 min in TBST at room temperature with shaking/rocking.
3. Incubate the membranes with antibody (*below*) overnight at 4°C with shaking/rocking:
   (a). *Blot 1.* CTD110.6 1 μg/mL in antibody incubation buffer;
   (b). *Blot 2.* CTD110.6 1 μg/mL in GlcNAc competition buffer.
4. Wash the membranes 3 × 10 min in TBST at room temperature with shaking/rocking.

5. Incubate the membranes with secondary antibody (1/5,000) in antibody incubation buffer for 1 h at room temperature with shaking/rocking.
6. Wash the membranes 4 × 10 min in TBST at room temperature with shaking/rocking.
7. Wash the membranes 1 × 10 min in TBS at room temperature with shaking/rocking (*see* **Note 25**).
8. Develop the horse radish peroxidase reaction with ECL reaction and detect (*see* **Note 26**).

3.3. Antigen Purification of CTD110.6 (See Note 6)

1. Defrost CTD110.6 ascites on ice.
2. Dilute the antibody tenfold with PBS, add protease inhibitors, centrifuge at 10,000 × *g* for 20 min at 4°C.
3. Recover the supernatant and filter it through a 1 μM filter. Check that the pH is between pH 7.0 and 8.0.
4. Pack 4 mL of GlcNAc-agarose or sepharose slurry (50% v/v) in a column housing (*see* **Note 27**), resulting in a column with a 2 mL bed volume (*see* **Note 28**).
5. Wash the column with at least 10 column volumes (20 mL) of PBS, with a flow rate of 0.5–1 mL/min (*see* **Note 29**).
6. Apply the antibody to the column with a flow rate of 0.2–0.5 mL/min (*see* **Note 30**). Collect the material that flows through the column, designate this fraction as the "flowthrough".
7. Reapply the "flowthrough" to the column, designate this fraction as "flowthrough 2".
8. Wash the column with 10 column volumes of PBS with (~20 mL).
9. Wash the column with 5 column volumes (10 mL) of PBS with 100 mM, galactose.
10. Wash the column with 5 column volumes (10 mL) of PBS.
11. Elute the bound antibody from the column by adding 10 × 2 mL of PBS, with 1 M GlcNAc. Collect 2 mL fractions, designated "eluant 1–10".
12. Wash the column with 5 column volumes (10 mL) of PBS.
13. The column should be stored in PBS with 0.05% w/v sodium azide.
14. Assay the fractions for protein to determine which of the elution fractions contain the CTD110.6.
15. Combine the fractions containing antibody, and dialyze against PBS at 4°C (*see* **Note 31**).
16. Measure the protein concentration of the dialyzed antibody. Concentrate if necessary. Dilute to 2 mg/mL and then add

an equal volume of glycerol, giving a final antibody concentration of 1 mg/mL and a final glycerol concentration of 50% (v/v). The antibody can be stored at –20°C for at least 1 year.

17. Test the "flowthrough" fractions to determine if all the CTD110.6 was recovered (*see* **Note 32**).

3.4. Detecting Proteins Modified by O-GlcNAc with sWGA

1. Wash duplicate blots for 10 min in TBST (*see* **Note 4**).
2. Incubate the blots in 3% w/v BSA in TBST for 60 min at room temperature (*see* **Note 11**).
3. Wash the blots 3 × 10 min in TBST.
4. Incubate the blots in 1 µg/mL sWGA-HRP 3% BSA in TBST with and without 1 M GlcNAc overnight at 4°C.
5. Wash the blots in HS-TBST 6 × 10 min.
6. Wash the blots in TBS 1 × 10 min.
7. Develop the HRP-reaction.

3.5. Detecting O-GlcNAc Modified Proteins by Galactosyltransferase Labeling

Galactosyltransferase labeling is definitive for the detection of O-GlcNAc modified proteins if the following experiments are performed (1) treatment of samples with PNGase F to remove N-linked carbohydrates; (2) confirmation that the label is incorporated in a disaccharide, by size exclusion chromatography; and (3) confirmation that the disaccharide is βGal1-4βGlcNAc, which is typically performed by DIONEX chromatography on a CarboPac PA100 column.

Like many glycosyltransferases, bovine milk galactosyltransferase requires the divalent cation Mn^{2+}. However high concentrations of Mn^{2+} (>5 mM) and Mg^{2+} are inhibitory. Galactosyltransferase works best on denatured proteins, and this can be achieved by boiling samples in 10 mM DTT and 0.5% w/v SDS for 5 min. While SDS inhibits galactosyltransferase, it can be titrated out with ten times more v/v NP-40. Thus, the reaction mixture is brought to 5% v/v NP-40, and then diluted to reduce the total NP-40 concentration to 2%. Digitonin should be avoided, as this can react with galactosyltransferase.

1. Remove the solvent from label in a Speed-Vac or under a stream of nitrogen, dry ~1–2µCi/reaction (*see* **Note 33**).
2. Resuspend the label in 25 mM 5 -AMP, 50 µL per reaction (*see* **Note 34**).
3. Set the reactions up as follows (*see* **Note 35**):

[a]*see* **Note 35**

Reagent	Volume
Sample (0.5–5 mg)	Up to 50 μL
Buffer H	350 μL
10× labeling buffer	50 μL
Calf intestinal alkaline phosphatase[a]	1–4 U
UDP-[^{3}H]Gal/5 -AMP	50 μL
Galactosyltransferase 30–50 U/mL	2–5 μL
Milli Q water, to a final volume of	500 μL

[a] See Note 36

4. Incubate the samples for 2 h at 37°C or overnight at 4°C.
5. of 0.5–1.0 mM and 2–5 μL of fresh galactosyltransferase (*see* **Note 37**). Incubate for a further 2–4 h.
6. Add 50 μL of stop solution to each sample.
7. Heat to 100°C for 5 min.
8. Resolve the protein(s) from the unincorporated label using a Sephadex G-50 column (30 × 1 cm), equilibrated in 50 mM ammonium formate, 0.1% w/v SDS. Collect 1 mL fractions (*see* **Note 38**).
9. Count an aliquot (50 μL) of each fraction using a liquid scintillation counter (*see* **Note 39**)
10. Combine the Vo and lyophilize the sample to dryness.
11. Resuspend the sample(s) in 100–1,000 μL of Milli-Q water and acetone precipitate as follows:
    (a) Add at least three volumes of ice-cold acetone to the sample
    (b) Incubate for 2–18 h at -20°C
    (c) Pellet protein at 4°C for 10 min at a minimum of 3,000 × *g* (*see* **Note 40**)
    (d) Tip of the acetone, and air-dry pellets
12. Resuspend the sample(s) in your buffer of choice.

3.6. Autogalactosylation of Galactosyltransferase

Bovine milk galactosyltransferase is modified by N-linked carbohydrates, and it is necessary to block these with cold galactose before using this enzyme.

1. Resuspend 25 U of galactosyltransferase (Catalog #G5507; SIGMA-Aldrich) in 1 mL of 1× galactosyltransferase buffer.

2. Transfer the sample to a 30–50 mL centrifuge tube.
3. Remove a 5 μL aliquot for an activity assay.
4. Add 10 μL of aprotinin, 3.5 μL of β-Mercaptoethanol and 1.5–3.0 mg of UDP-Gal.
5. Incubate the sample on ice for 30–60 min.
6. Add 5.66 mL of pre-chilled saturated ammonium sulfate in a dropwise manner, mix well, and incubate on ice for 30 min.
7. Centrifuge at > 10,000 × *g* for 15 min at 4°C.
8. Resuspend the pellet in 5 mL of cold 85% ammonium sulfate and incubate on ice for 30 min.
9. Centrifuge at > 10,000 × *g* for 15 min at 4°C.
10. Resuspend the pellet in 1 mL of galactosyltransferase storage buffer.
11. Aliquot the enzyme (50 μL), store at –20°C for up to 1 year.
12. Assay 5 μL of the autogalactosylated and non-galactosylated enzyme to determine the activity (*above*).

3.7. Detecting O-GlcNAc Modified Proteins by Immunofluorescence

1. Wash the cells (3×) with cold PBS.
2. Fix the cells in 4% w/v formaldehyde, in PBS, for 20 min.
3. Wash the cells with cold PBS, 3 × 3 min.
4. Permeabilize the cells with 0.075% v/v triton X-100 in 2 mg/mL BSA, in PBS, for 30 min.
5. Wash the cells with cold PBS, 3 × 3 min.
6. Block the cells with 2 mg/mL BSA, in PBS, for 1 h at room temperature.
7. Incubate the cells with antibodies diluted in 2 mg/mL BSA, in PBS, for 2 h as follows:
   (a) CTD110.6 (1 mg/mL) 1/200
   (b) CTD110.6 (1 mg/mL) 1/200 CTD110.6 (1 mg/mL) 1/200 plus 100 mM glucose (*see* **Note 9**)
   (c) CTD110.6 (1 mg/mL) 1/200 plus 100 mM GlcNAc
8. Wash the cells with cold PBS, 3 × 3 min.
9. Keep as dark as possible from this point on.
10. Incubate the cells with secondary antibody (DARK), diluted in 2 mg/mL BSA, in PBS.
11. Wash the cells with cold PBS, 1 × 3 min.
12. Incubate the cells with DAPI 0.1 μg/mL for 10 min.
13. Wash the cells with cold PBS, 3 × 3 min.
14. Mount the coverslip/slide in anti-fade.

15. Use standard techniques to collect immunofluorescence data.

3.8. Purifying O-GlcNAc Modified Proteins

Proteins modified by O-GlcNAc can be affinity purified using sWGA-agarose/sepharose or CTD110.6-agarose/sepharose. While this can be done using typical immunoaffinity/immunoprecipitation procedures, the column format gives cleaner data and is more appropriate for large-scale purifications. The protocol described here is for large-scale purification/enrichment of O-GlcNAc modified proteins (~10 mg of cell extract), and the procedure is compatible with sWGA, RCA1 or CTD110.6 affinity chromatography (*see* **Note 41**). This technique is also compatible with stable density labeling of amino acids in cell culture (SILAC) (*46*).

1. Equilibrate sepharose/agarose or IgM-sepharose/agarose in PBS 0.1% v/v triton X-100 (or NP-40; at least 10 column volumes, 20 mL; *see* **Note 42**).
2. Preclear lysate by adding 1 mL of IgM-sepharose/agarose or unmodified sepharose/agarose to 10 mg of cell extract. Incubate with rocking for 2 h at 4°C.
3. Apply the sample to a column and collect the flowthrough. This is the precleared supernatant.
4. Equilibrate 0.5 mL CTD110.6-agarose/sepharose (2 mg/mL) and IgM-agarose/sepharose (2 mg/mL) in independent column housings in PBS 0.1% v/v triton X-100 (at least 10 column volumes, 5 mL).
5. Add 5 mg of cell extract to the CTD110.6-affinity column and 5 mg to the IgM-affinity column. Incubate with mixing overnight at 4°C.
6. Set up anti-IgM post columns (200 μL) with frits (*see* **Fig. 2**. Equilibrate in PBS 0.1% v/v triton X-100 (*see* **Note 43**).
7. Add the extract/affinity-matrix from overnight incubation to anti-IgM columns.
8. Wash the columns with 10 column volumes (7 mL) of PBS (*see* **Note 44**), with a flow rate of ~0.2–0.5 mL/min.
9. Wash the columns with 10 column volumes (7 mL) of 100 mM Gal, in PBS (*see* **Note 9**).
10. Wash the columns with 10 column volumes (7 mL) of PBS.
11. Elute O-GlcNAc modified proteins with 2 column volumes (1.4 mL) of 1 M GlcNAc, in PBS.
12. Repeat **step 11** four more times. Collect eluant fractions individually.

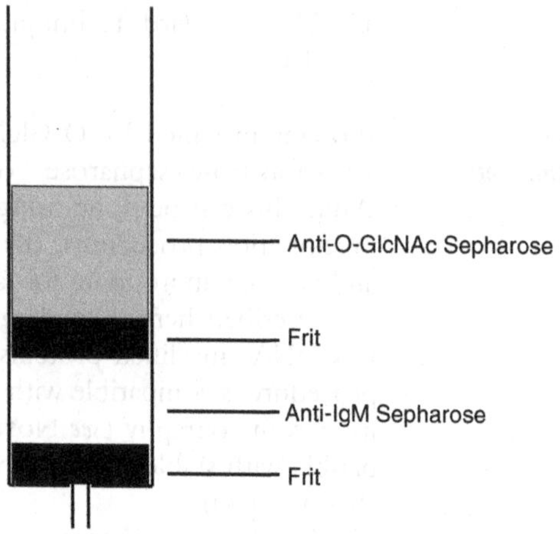

Fig. 2. CTD110.6 affinity chromatography. A schematic of the CTD110.6 affinity column, with anti-IgM trap column is shown.

13. Wash the columns with 10 column volumes (7 mL) of 100 mM sodium acetate pH4, 0.5 M NaCl, 0.5 M GlcNAc (*see* **Note 21**).
14. Wash the columns with 10 column volumes (7 mL) of PBS.
15. Store the columns in PBS, 0.05% sodium azide.
16. Determine the elution pattern of the O-GlcNAc modified proteins by SDS-PAGE immunoblot.
17. Typically the 1 M GlcNAc is removed from the samples at this point. Alternatively, the samples can be acidified and injected directly onto a reversed phase column for a subsequent round of separation. Desalting methods include:
    (a) Precipitation with methanol (*see* **Note 45**)
    (b) Spin filtration;
    (c) Dialysis

3.9. In Vitro Transcription Translation

It is possible to screen for the O-GlcNAc protein modification on low abundance proteins, by making these proteins in a rabbit reticulocyte lysate (RRL) *in vitro transcription translation* system and determining if they bind either sWGA or CTD110.6. Typically ^{13}C-leucine, ^{35}S-cysteine or ^{35}S-methionine is incorporated during translation, and elution of the proteins from either sWGA or CTD110.6 can be followed by liquid scintillation counting. Due to the high concentration of unincorporated label, this experiment is best done in column format, with a no DNA control, a negative control (luciferase, in the kit) and a positive control (plasmid encoding a known O-GlcNAc modified protein, for example p62).

1. Synthesize proteins of choice using a RRL ITT system, according to the manufacturer's instructions. Include a negative control (for example, luciferase supplied with kit), a positive control (for example, nuclear pore protein p62), a no DNA control, and your protein(s) on interest.
2. Treat half of each sample with hexosaminidase (*see* **Note 46**)
3. Desalt samples using a 1mL G50 column equilibrated in PBS, 0.1% v/v Triton X-100, as follows:
   (a) Pour exactly 1mL Sephadex G-50 in a tuberculin syringe that has been packed with a small glass wool frit (~100μL).
   (b) Wash the column with 5mL PBS, 0.1% v/v Triton X-100.
   (c) Load the RRL reaction onto the column. The volume of sample can be up to 200μL.
   (d) Wash the column with PBS, 0.1% v/v Triton X-100 so that the total volume of this wash and the protein sample is 350μL (for example: if sample volume is 150μL, add 200μL PBS, 0.1% v/v Triton X-100).
   (e) Transfer the column to a clean pre-chilled tube. Elute protein with 200μL PBS, 0.1% v/v Triton X-100. This is the desalted sample.
4. Perform the following steps at 4°C or with ice cold buffers.
5. Equilibrate ~150μL of sWGA-agarose in PBS, 0.1% v/v triton X-100.
6. Cap the column.
7. Apply the desalted sample to the column and stand at 4°C for 30min.
8. Uncap the column and run the unbound material through the column.
9. Wash the column with 15mL of PBS, 0.1% v/v Triton X-100 at ~0.5mL/min, collecting 0.5mL fractions.
10. Cap the column. Load the column with 500μL of 100mM galactose, in PBS, 0.1% v/v Triton X-100 and stand at 4°C for 20min.
11. Wash the column with 5mL of 100mM galactose, in PBS, 0.1% v/v Triton X-100, collecting 0.5mL fractions.
12. Repeat **steps 5** and **6** using the 1M GlcNAc, in PBS, 0.1% v/v Triton X-100.
13. Count 25μL of each fraction using a liquid scintillation counter.

14. Pool the positive fractions that elute in the presence of GlcNAc and precipitate using TCA or methanol (*see* **Note 45**).
15. Analyze the pellet by SDS-PAGE and the autoradiography to confirm that the label has been incorporated into a protein of an appropriate size.

3.10 Removing O-GlcNAc with Hexosaminidase

Terminal GlcNAc residues can be removed using commercial hexosaminidases. While this will show that the reactivity is GlcNAc specific, it will not confirm that the signal was O-GlcNAc dependent unless used in conjunction with an O-GlcNAc specific antibody.

1. Protein samples (include a positive control such as ovalbumin).
2. Mix sample 1:1 with 2% w/v SDS and boil for 5min.
3. Mix sample 1:1 with reaction mixture and incubate at 37°C for 4–24h.

3.11. Removing N-Linked Glycosylation with PNGase F

1. Add 1/10 sample volume of 10× PNGase F denaturing buffer to each sample and heat to 100°C for 10min.
2. Add 1/10 sample volume of 10× PNGase F reaction buffer and 10% v/v NP-40, mix (*see* **Note 47**).
3. Add 1μL of PNGase F and incubate samples at 37°C for 1h to overnight.
4. Analyze deglycosylation by SDS-PAGE, sWGA immunoblot (*see* **Note 48**).

4. Notes

1. Cells appear to be more responsive to PUGNAc in physiological levels of glucose (1g/L), this may be for two reasons *(1)* the levels of O-GlcNAc are very responsive to extracellular glucose concentrations, and thus when cells are grown in media at 1g/L rather than the typical 4.5g/L, the baseline levels of O-GlcNAc are lower; *(2)* PUGNAc may be transported into the cells through the glucose transporter, and at 1g/L it is more efficiently transported as there is less competition.
2. Ovalbumin contains terminal GlcNAc residues. It should react with sWGA and WGA, but NOT any of the O-GlcNAc specific antibodies (RL2, CTD110.6, MY95). BSA-GlcNAc can easily be made using aminophenyl-GlcNAc (Catalog #CG-029-5, EY Labs; Catalog #B6893; SIGMA-Aldrich).

3. It is possible to detect O-GlcNAc on as little at 10–15μg of cell extract, although this can be dependent on cell type, the cell cycle, and the nutritional state. Typically, the best data is obtained when at least 30μg of protein is loaded onto mini- and midi-sized SDS-PAGE gels. Note, O-GlcNAc is concentrated on nuclear proteins and 10–15μg of nuclear extract is more than sufficient
4. To adequately control for lectin/antibody specificity, the signal from the antibody should be competed away by free GlcNAc. An additional blot with a non-specific sugar, such as galactose or glucose (**Fig.1**), will provide a higher level of confidence.
5. GlcNAc is stored at –20°C. Solutions should be prepared fresh. As GlcNAc is hydroscopic, the bottle should be removed from the –20°C and allowed to warm to room temperature before opening.
6. Serum can be up to 50mg/mL albumin. Purification of CTD110.6 removes albumin and other proteins. While antigen purification is preferable before covalent coupling, it also appears to result in immunoblots of a higher quality.
7. Pierce Biotechnology sells CTD110.6 as part of an O-GlcNAc detection kit, while Covance Research Products sells ascites. Neither company purifies the antibody before sale.
8. We have used both products to successfully purify CTD110.6.
9. Glucose or galactose provides a specificity control. This control is MOST essential with immunofluorescence.
10. We typically include protease inhibitors, phosphatase inhibitors, and the O-GlcNAcase inhibitor, PUGNAc, in all cell extract, incubation, and elution buffers. Any cocktail of protease inhibitors known to inhibit mammalian proteases is sufficient.
11. Milk proteins contain terminal GlcNAc residues that will react with sWGA, thus membranes are blocked in BSA, which is NOT glycosylated.
12. Before succinnylation WGA will recognize both terminal sialic acid and GlcNAc residues.
13. sWGA can be stored in PBS:Glycerol (1:1) at –20°C for at least 1 year.
14. Aliquot and store at –20°C for up to 6 months.
15. Commercial desalting columns, such as PD-10 columns (GE Healthcare), can be used. However, the long thin column described here results in the best resolution between the labeled proteins and the unincorporated label.

16. Toxic gases are released when this reagent is made, thus it should be prepared in a fume hood. Bring PBS to the boil in a microwave and transfer it to the fume hood. Add the appropriate amount of formaldehyde, and wait for it to dissolve.
17. A non specific IgM provides a specificity control.
18. This procedure is compatible with 1% NP-40 extracts and RIPA extracts. Typically we dilute these out 1:1 with 50mM Tris-HCl before loading extracts onto the affinity columns. For sWGA and RCA1, the extracts should be diluted out fivefold with 50mM Tris-HCl.
19. Typically we couple CTD110.6 to cyanogen bromide activated Sepharose (GE Healthcare) at 2–3mg/mL. The coupled antibody is reusable and less antibody is released into protein extracts, which is preferable upstream of MS.
20. Both Triton X-100 and NP-40 are compatible with this method.
21. A low pH wash with GlcNAc is used to clean the column so it can be reused.
22. The Novagen pCITE vector is designed to work at high efficiency in *in vitro* transcription translation reactions.
23. Different cells lines have altered responsiveness to PUGNAc.
24. Most cell lines see elevations in O-GlcNAc by 2–4h, with robust increases by 8h. Cells should not be treated for longer than 18h, as defects in cell growth and division have been observed (*39*).
25. Tween-20 can cause increased background and spottiness in the ECL reaction.
26. Some cross reactivity with prestained markers may be observed, this is normal.
27. To avoid developing air bubbles in the column bed, add the resin to water already in the housing.
28. The volumes described in this protocol are appropriate for 1mL of ascites. The yield is usually 2–3mg of purified antibody.
29. This protocol is either performed in the cold room, or with buffers that have/are kept on ice. All fractions are maintained in the cold room or on ice to avoid degradation.
30. For an open system the flow rate can be altered by attaching a needle with a narrow gauge to the end of the column. In a closed system running under gravity, the height of the buffer reservoir can be altered.

31. For some applications PBS is not the best dialysis buffer of choice. For instance, for covalent coupling to CNBR-activated Sepharose the antibody should be dialyzed against 200mM sodium bicarbonate, 500mM NaCl, pH 8.3–8.5.
32. To determine if all of the CTD110.6 has been removed from the flowthrough, run a western blot or ELISA using BSA-GlcNAc as a positive control and determine if all of the reactivity has been removed from the flowthrough fraction.
33. Ethanol can inhibit the galactosyltransferase, no more than 4μL should be added to a 500μL reaction.
34. AMP is included to inhibit any phosphodiesterase activity, which would consume the label.
35. In addition to the protein samples, there should be a positive control (ovalbumin 2μg) and a blank.
36. Free UDP is a potent inhibitor of galactosyltransferase and alkaline phosphatase degrades UDP. UDP is both a by-product of the reaction, and a common contaminant of UDP-Gal.
37. For site-mapping studies, where complete labeling of terminal GlcNAc residues is required, reactions are chased with cold UDP-Gal.
38. Size exclusion chromatography is used to separate the unincorporated label from the labeled protein. The method given provides the best results, although TCA precipitation, spin filtration/buffer exchange and other forms of size exclusion chromatography can be used.
39. ~2×10^6 DPM should be incorporated into ovalbumin (2 μg).
40. Most 15 mL polypropylene tubes can be used for this, although centrifuging faster than 3,000 × *g* is NOT recommended.
41. For RCA1 affinity chromatography, the order of GlcNAc and Gal washes/elutions should be reversed.
42. For RCA1 or sWGA affinity chromatography the extract should be pre-cleared with resin alone, or another non-specific lectin. The use of a ConA affinity chromatography step would remove many N-linked modified carbohydrates, before sWGA chromatography.
43. The post columns are used to trap any CTD110.6 leaching off the affinity column. This is especially important for samples being prepared for mass spectrometry.
44. Triton X-100 and NP-40 can be removed from subsequent steps for samples being prepared for mass spectrometry.

45. We typically precipitate proteins bound for SDS-PAGE or 2-dimensional gels using methanol. At least 10 volumes of ice-cold methanol are added to samples, and incubated overnight at -20°C. Proteins are precipitated at > 3,000 × *g* at 4°C for 30 min. Note, acetone precipitates free GlcNAc and should not be used.
46. This is optional, but provides an additional level of confidence.
47. PNGase F is inhibited by SDS. It is essential to add NP-40.
48. Ovalbumin contains one N-linked glycosylation site, and treatment with PNGase F should increase its mobility by several kilodaltons on a 10–12% SDS-PAGE gel. This shift is difficult to detect on a 7.5% SDS-PAGE gel. Ideally, samples would be transferred to a membrane and detected with sWGA.

Acknowledgements

The author would like to acknowledge Prof. Gerald W. Hart and Dr Chad Slawson (Johns Hopkins University School of Medicine) for comments on this manuscript; and the technical help of Katie Zoey Ho (Johns Hopkins Singapore). Some of the data presented in this paper were collected in the laboratory of Prof. Gerald W. Hart (Johns Hopkins University School of Medicine) and were supported by NIH grants HD R37-13563, CA R01-42486, DK-R01-61671, DK-R21/33-71280 (GWH), the National Heart, Lung, and Blood Institute, National Institutes of Health, contract No. N01-HV-28180 (GWH). Under a licensing agreement between Covance Research Products and The Johns Hopkins University, Dr. Hart receives a share of royalties received by the university on sales of the CTD 110.6 antibody. The terms of this arrangement are being managed by The Johns Hopkins University in accordance with its conflict of interest policies. NEZ is supported by an A*Star Research grant to Johns Hopkins Singapore.

References

1. Love, D. C., and Hanover, J. A. (2005) The hexosamine signaling pathway: deciphering the "O-GlcNAc code. " *Sci STKE* 2005, re13.
2. Kearse, K. P., and Hart, G. W. (1991) Lymphocyte activation induces rapid changes in nuclear and cytoplasmic glycoproteins. *Proc Natl Acad Sci U S A* 88, 1701–5.
3. Slawson, C., Zachara, N. E., Vosseller, K., Cheung, W. D., Lane, M. D., and Hart, G. W. (2005) Perturbations in O-linked beta-*N*-acetylglucosamine protein modification cause severe defects in mitotic progression and cytokinesis. *J Biol Chem* 280, 32944–56.
4. Slawson, C., Shafii, S., Amburgey, J., and Potter, R. (2002) Characterization of the

O-GlcNAc protein modification in *Xenopus laevis* oocyte during oogenesis and progesterone-stimulated maturation. *Biochim Biophys Acta* 1573, 121–9.

5. Lefebvre, T., Baert, F., Bodart, J. F., Flament, S., Michalski, J. C., and Vilain, J. P. (2004) Modulation of O-GlcNAc glycosylation during Xenopus oocyte maturation. *J Cell Biochem* 93, 999–1010.
6. Zachara, N. E., O'Donnell, N., Cheung, W. D., Mercer, J. J., Marth, J. D., and Hart, G. W. (2004) Dynamic O-GlcNAc modification of nucleocytoplasmic proteins in response to stress. A survival response of mammalian cells. *J Biol Chem* 279, 30133–42.
7. Liu, J., Pang, Y., Chang, T., Bounelis, P., Chatham, J. C., and Marchase, R. B. (2006) Increased hexosamine biosynthesis and protein O-GlcNAc levels associated with myocardial protection against calcium paradox and ischemia. *J Mol Cell Cardiol* 40, 303–12.
8. Zachara, N. E., and Hart, G. W. (2004) O-GlcNAc a sensor of cellular state: the role of nucleocytoplasmic glycosylation in modulating cellular function in response to nutrition and stress. *Biochim Biophys Acta* 1673, 13–28.
9. Shafi, R., Iyer, S. P., Ellies, L. G., O'Donnell, N., Marek, K. W., Chui, D., Hart, G. W., and Marth, J. D. (2000) The O-GlcNAc transferase gene resides on the X chromosome and is essential for embryonic stem cell viability and mouse ontogeny. *Proc Natl Acad Sci USA* 97, 5735–9.
10. O'Donnell, N., Zachara, N. E., Hart, G. W., and Marth, J. D. (2004) Ogt-dependent X-chromosome-linked protein glycosylation is a requisite modification in somatic cell function and embryo viability. *Mol Cell Biol* 24, 1680–90.
11. Torres, C. R., and Hart, G. W. (1984) Topography and polypeptide distribution of terminal *N*-acetylglucosamine residues on the surfaces of intact lymphocytes. Evidence for O-linked GlcNAc. *J Biol Chem* 259, 3308–17.
12. Khidekel, N., Arndt, S., Lamarre-Vincent, N., Lippert, A., Poulin-Kerstien, K. G., Ramakrishnan, B., Qasba, P. K., and Hsieh-Wilson, L. C. (2003) A chemoenzymatic approach toward the rapid and sensitive detection of O-GlcNAc posttranslational modifications. *J Am Chem Soc* 125, 16162–3.
13. Khidekel, N., Ficarro, S. B., Peters, E. C., and Hsieh-Wilson, L. C. (2004) Exploring the O-GlcNAc proteome: direct identification of O-GlcNAc-modified proteins from the brain. *Proc Natl Acad Sci USA* 101, 13132–7.
14. Tai, H. C., Khidekel, N., Ficarro, S. B., Peters, E. C., and Hsieh-Wilson, L. C. (2004) Parallel identification of O-GlcNAc-modified proteins from cell lysates. *J Am Chem Soc* 126, 10500–1.
15. Reason, A. J., Morris, H. R., Panico, M., Marais, R., Treisman, R. H., Haltiwanger, R. S., Hart, G. W., Kelly, W. G., and Dell, A. (1992) Localization of O-GlcNAc modification on the serum response transcription factor. *J Biol Chem* 267, 16911–21.
16. Roquemore, E. P., Chou, T. Y., and Hart, G. W. (1994) Detection of O-linked *N*-acetylglucosamine (O-GlcNAc) on cytoplasmic and nuclear proteins. *Methods Enzymol* 230, 443–60.
17. Greis, K. D., Hayes, B. K., Comer, F. I., Kirk, M., Barnes, S., Lowary, T. L., and Hart, G. W. (1996) Selective detection and site-analysis of O-GlcNAc-modified glycopeptides by beta-elimination and tandem electrospray mass spectrometry. *Anal Biochem* 234, 38–49.
18. Greis, K. D., and Hart, G. W. (1997) *in* "Methods in Molecular Biology, Vol. XX: Glycoanalysis Protocols" (Hounsell, E. F., Ed.), Humana Press, Totowa, NJ.
19. Cole, R. N., and Hart, G. W. (1999) Glycosylation sites flank phosphorylation sites on synapsin I: O-linked *N*-acetylglucosamine residues are localized within domains mediating synapsin I interactions. *J Neurochem* 73, 418–28.
20. Haynes, P. A., and Aebersold, R. (2000) Simultaneous detection and identification of O-GlcNAc-modified glycoproteins using liquid chromatography-tandem mass spectrometry. *Anal Chem* 72, 5402–10.
21. Chalkley, R. J., and Burlingame, A. L. (2001) Identification of GlcNAcylation sites of peptides and alpha-crystallin using Q-TOF mass spectrometry. *J Am Soc Mass Spectrom* 12, 1106–13.
22. Cole, R. N., and Hart, G. W. (2001) Cytosolic O-glycosylation is abundant in nerve terminals. *J Neurochem* 79, 1080–9.
23. Comer, F. I., Vosseller, K., Wells, L., Accavitti, M. A., and Hart, G. W. (2001) Characterization of a mouse monoclonal antibody specific for O-linked *N*-acetylglucosamine. *Anal Biochem* 293, 169–77.
24. Zachara, N. E., Gao, Y., Cole, R. N., and Hart, G. W. (2001) Detection and analysis of proteins modified by O-linked *N*-acetylglucosamine. *Curr Protocols Prot Sci* 2, 12. 8. 1–12. 8. 25.
25. Wells, L., Vosseller, K., Cole, R. N., Cronshaw, J. M., Matunis, M. J., and Hart, G. W.

(2002) Mapping sites of O-GlcNAc modification using affinity tags for serine and threonine post-translational modifications. *Mol Cell Proteomics* 1, 791–804.

26. Chalkley, R. J., and Burlingame, A. L. (2003) Identification of novel sites of *O-N*-acetylglucosamine modification of serum response factor using quadrupole time-of-flight mass spectrometry. *Mol Cell Proteomics* 2, 182–90.
27. Vocadlo, D. J., Hang, H. C., Kim, E. J., Hanover, J. A., and Bertozzi, C. R. (2003) A chemical approach for identifying O-GlcNAc-modified proteins in cells. *Proc Natl Acad Sci USA* 100, 9116–21.
28. Zachara, N. E., Cheung, W. D., and Hart, G. W. (2003) *in* "Methods in Molecular Biology, Vol. 284: Signal Transduction Protocols" (Dickson, R. C., Ed.), Humana Press, Totowa, NJ.
29. Sprung, R., Nandi, A., Chen, Y., Kim, S. C., Barma, D., Falck, J. R., and Zhao, Y. (2005) Tagging-via-substrate strategy for probing O-GlcNAc modified proteins. *J Proteome Res* 4, 950–7.
30. Vosseller, K., Hansen, K. C., Chalkley, R. J., Trinidad, J. C., Wells, L., Hart, G. W., and Burlingame, A. L. (2005) Quantitative analysis of both protein expression and serine/threonine post-translational modifications through stable isotope labeling with dithiothreitol. *Proteomics* 5, 388–98.
31. Liu, H. D., Li, Y. M., Du, J. T., Hu, J., and Zhao, Y. F. (2006) Novel acetylation-aided migrating rearrangement of uridine-diphosphate-N-acetylglucosamine in electrospray ionization multistage tandem mass spectrometry. *J Mass Spectrom* 41, 208–15.
32. Nandi, A., Sprung, R., Barma, D. K., Zhao, Y., Kim, S. C., Falck, J. R., and Zhao, Y. (2006) Global identification of O-GlcNAc-modified proteins. *Anal Chem* 78, 452–8.
33. Vosseller, K., Trinidad, J. C., Chalkley, R. J., Specht, C. G., Thalhammer, A., Lynn, A. J., Snedecor, J. O., Guan, S., Medzihradszky, K. F., Maltby, D. A., Schoepfer, R., and Burlingame, A. L. (2006) O-linked *N*-acetylglucosamine proteomics of postsynaptic density preparations using lectin weak affinity chromatography and mass spectrometry. *Mol Cell Proteomics* 5, 923–34.
34. Greis, K. D., and Hart, G. W. (1998) Analytical methods for the study of O-GlcNAc glycoproteins and glycopeptides. *Methods Mol Biol* 76, 19–33.
35. Dong, D. L., and Hart, G. W. (1994) Purification and characterization of an O-GlcNAc selective *N*-acetyl-beta-d-glucosaminidase from rat spleen cytosol. *J Biol Chem* 269, 19321–30.
36. Roos, M. D., Xie, W., Su, K., Clark, J. A., Yang, X., Chin, E., Paterson, A. J., and Kudlow, J. E. (1998) Streptozotocin, an analog of *N*-acetylglucosamine, blocks the removal of O-GlcNAc from intracellular proteins. *Proc Assoc Am Physicians* 110, 422–32.
37. Stubbs, K. A., Zhang, N., and Vocadlo, D. J. (2006) A divergent synthesis of 2-acyl derivatives of PUGNAc yields selective inhibitors of O-GlcNAcase. *Org Biomol Chem* 4, 839–45.
38. Gao, Y., Parker, G. J., and Hart, G. W. (2000) Streptozotocin-induced beta-cell death is independent of its inhibition of O-GlcNAcase in pancreatic Min6 cells. *Arch Biochem Biophys* 383, 296–302.
39. Haltiwanger, R. S., Grove, K., and Philipsberg, G. A. (1998) Modulation of O-linked *N*-acetylglucosamine levels on nuclear and cytoplasmic proteins in vivo using the peptide O-GlcNAc-beta-*N*-acetylglucosaminidase inhibitor *O*-(2-acetamido-2-deoxy-d-glucopyranosylidene)amino-*N*-phenylcarbamate. *J Biol Chem* 273, 3611–7.
40. Turner, J. R., Tartakoff, A. M., and Greenspan, N. S. (1990) Cytologic assessment of nuclear and cytoplasmic O-linked *N*-acetylglucosamine distribution by using anti-streptococcal monoclonal antibodies. *Proc Natl Acad Sci USA* 87, 5608–12.
41. Holt, G. D., Snow, C. M., Senior, A., Haltiwanger, R. S., Gerace, L., and Hart, G. W. (1987) Nuclear pore complex glycoproteins contain cytoplasmically disposed O-linked *N*-acetylglucosamine. *J Cell Biol* 104, 1157–64.
42. Snow, C. M., Senior, A., and Gerace, L. (1987) Monoclonal antibodies identify a group of nuclear pore complex glycoproteins. *J Cell Biol* 104, 1143–56.
43. Matsuoka, Y., Shibata, S., Yasuhara, N., and Yoneda, Y. (2002) Identification of Ewing's sarcoma gene product as a glycoprotein using a monoclonal antibody that recognizes an immunodeterminant containing O-linked *N*-acetylglucosamine moiety. *Hybrid Hybridomics* 21, 233–6.
44. Kamemura, K., Hayes, B. K., Comer, F. I., and Hart, G. W. (2002) Dynamic interplay between O-glycosylation and O-phosphorylation of nucleocytoplasmic proteins: alternative glycosylation/phosphorylation of THR-58, a known mutational hot spot of c-Myc in lymphomas, is regulated by mitogens. *J Biol Chem* 277, 19229–35.
45. Ludemann, N., Clement, A., Hans, V. H., Leschik, J., Behl, C., and Brandt, R. (2005)

O-glycosylation of the tail domain of neurofilament protein M in human neurons and in spinal cord tissue of a rat model of amyotrophic lateral sclerosis (ALS). *J Biol Chem* 280, 31648–58.

46. Ibarrola, N., Kalume, D. E., Gronborg, M., Iwahori, A., and Pandey, A. (2003) A proteomic approach for quantitation of phosphorylation using stable isotope labeling in cell culture. *Anal Chem* 75, 6043–9.

Subsection D

Glycobioinformatic Tools

Chapter 20

Preparation of a Glycan Library Using a Variety of Glycosyltrasferases

Hiromi Ito, Akihiko Kameyama, Takashi Sato, and Hisashi Narimatsu

Summary

We have comprehensively cloned genes associated with the synthesis of glycans in the human body, then translated those glycogenes into recombinant glycosyltransferases using a variety of expression systems. The library of glycans or glycopeptides having various structures can be synthesized in a test tube using various enzymes. Using the glycan library, we have recently constructed a multistage tandem mass (MS^n) spectral library containing observed spectra. This library may be utilized as a novel tool for glycomics research, as it enables users to identify oligosaccharides very easily and quickly by spectral matching.

Key words: Glycosyltransferase, Glycopeptide, *O*-linked glycan, *N*-linked glycan, MW-tagged library, Mass spectrometry, MS^n spectral library.

1. Introduction

Currently, the identification of many proteins can be accomplished easily using mass spectrometry (MS). The reason for this is not only technical advances in MS, but also the presence of genomes as templates. We have comprehensively cloned genes associated with the synthesis of glycans in the human body (*1, 2*). They are summarized in a database, named GGDB, http://riodb.ibase.aist.go.jp/rcmg/ggdb. These glycogenes were translated into recombinant glycosyltransferases using a variety of expression systems. Glycans or glycopeptides having various structures can be synthesized in a test tube utilizing various enzymes. The goal of obtaining structural information for most of the oligosac-

Nicolle H. Packer and Niclas G. Karlsson (eds.), *Methods in Molecular Biology, Glycomics: Methods and Protocols, vol. 534*

DOI: 10.1007/978-1-59745-022-5_20

charides present in the human body appears to be within reach through combining oligosaccharide structural analysis with the corresponding genomic information.

To produce diverse sugar chains simultaneously in a short time, a synthetic approach utilizing libraries was developed, which was named the enzyme-cue Synthesis (*3*). According to this method, each reaction is stopped midway before starting the following reaction. Therefore, when the reaction is repeated n times, it is possible to obtain theoretically 2^n number of mixtures. This method has been applied mainly to the construction of an *O*-glycan library, because in order to avoid obtaining isomeric mixtures of *N*-glycans, the resulting mixture of enzyme reactions using the *N*-glycans must be isolated at each step.

Glycomics has lagged far behind proteomics because of difficulties arising from their structural complexity, such as variations in branching, linkage, and stereo-chemistry. Recently, tandem mass spectrometry has revealed that glycans might have characteristic fragment patterns in their collision-induced dissociation (CID) spectra. Using a glycan library, we constructed a multistage tandem mass (MS^n) spectral library containing observed spectra. This library may be utilized as a novel tool for glycomics research, as it enables users to identify oligosaccharides very easily and quickly by spectral matching (*4*).

2. Materials

2.1. Preparation of Recombinant Enzymes

1. Expression system: human embryonic kidney (HEK) 293T cells.
2. Transfected plasmids: pFLAG-CMV-1-ST3GalIV, pFLAG-CMV-1-FUT6, pFLAG-CMV-1-β4GalNAc-T3, pFLAG-CMV-3-DEST-β3Gn-T2, pFLAG-CMV-3-DEST-β4GalT-I, and pFLAG-CMV-3-DEST-FUT4.
3. 50 mM Tris-buffered saline (TBS): 50 mM Tris-HCl, pH7.4 and 150 mM NaCl prepared in MilliQ water.

2.2. Construction of an O-Linked Glycopeptide Library

1. Reaction Buffers: 500 mM HEPES (pH 7.0) or 200 mM $MnCl_2$ in MilliQ water.
2. Each sugar nucleotide (CMP-Neu5Ac, UDP-GlcNAc, GDP-Fuc, and UDP-Gal; Sigma-Aldrich) was prepared at an appropriate concentration in MilliQ water and stored frozen.
3. A carboxyfluorescein (FAM)-labeled Muc1a tandem repeat peptide (FAM-AHGVTSAPDTR) was prepared by custom peptide synthesis.

2.3. Conversion of a Glycopeptide Library to an Oligosaccharide Library

1. A solution of 500 mM $NaBH_4$ in 50 mM NaOH should be freshly prepared and used immediately.

2.4. Preparation of an N-Linked Glycan Library

1. Each sugar nucleotide (UDP-GalNAc, GDP-Fuc, and UDP-Gal; Sigma-Aldrich) is prepared in MilliQ water at an appropriate concentration and stored frozen.
2. The starting material (**1**; **Fig. 2** for an acceptor substrate is commercially available (TaKaRa or Seikagaku Corporation).
3. The basic reaction mixture: 50 mM HEPES buffer (pH 7.0), 10 mM $MnCl_2$, and an appropriate concentration of sugar nucleotide (*see* **Note 1**).

2.5. HPLC-Based Separation and Purification of Products

1. Reversed-phase column: PALPAK Type R column (4.6 × 250 mm, 5 μm; TaKaRa).
2. Solvent A: 10 mM ammonium acetate (pH 4.0) in MilliQ water.
3. Solvent B: 10 mM ammonium acetate (pH 4.0) in 10% acetonitrile (v/v).

2.6. Construction of a Multistage Tandem Mass Spectral (MS^n) Library

1. Sample solution: Approximate 1 μM glycan solution in double distilled water.
2. Calibrant solution: Peptide calibration standard II (Bruker Daltonics, Billerica, MA) in 200 μL of 0.1% TFA in 50% acetonitrile.
3. Matrix solution: 10 mg/mL of 2,5-dihydroxybenzoic acid (2,5-DHB, proteomics grade, Wako, Japan) in 30% ethanol.
4. Recrystallization solution: 99.5% Ethanol (HPLC grade, Wako, Japan).
5. Collision gas: Argon.

3. Methods

Recombinant glycosyltransferases are useful synthetic tools that we have used to produce a diverse set of oligosaccharides easily. Our general strategy for the construction of a glycan library using these glycosyltransferses is as follows. An *O*-linked glycopeptide library, which contains a variety of glycan structures, is prepared by carrying out sequential, incomplete glycosyltransferse reactions, i.e. by intentionally terminating each reaction at the halfway point, and carrying out all reactions in a single tube. An illustration of our strategy is shown in **Fig. 1**. The structures of the resulting

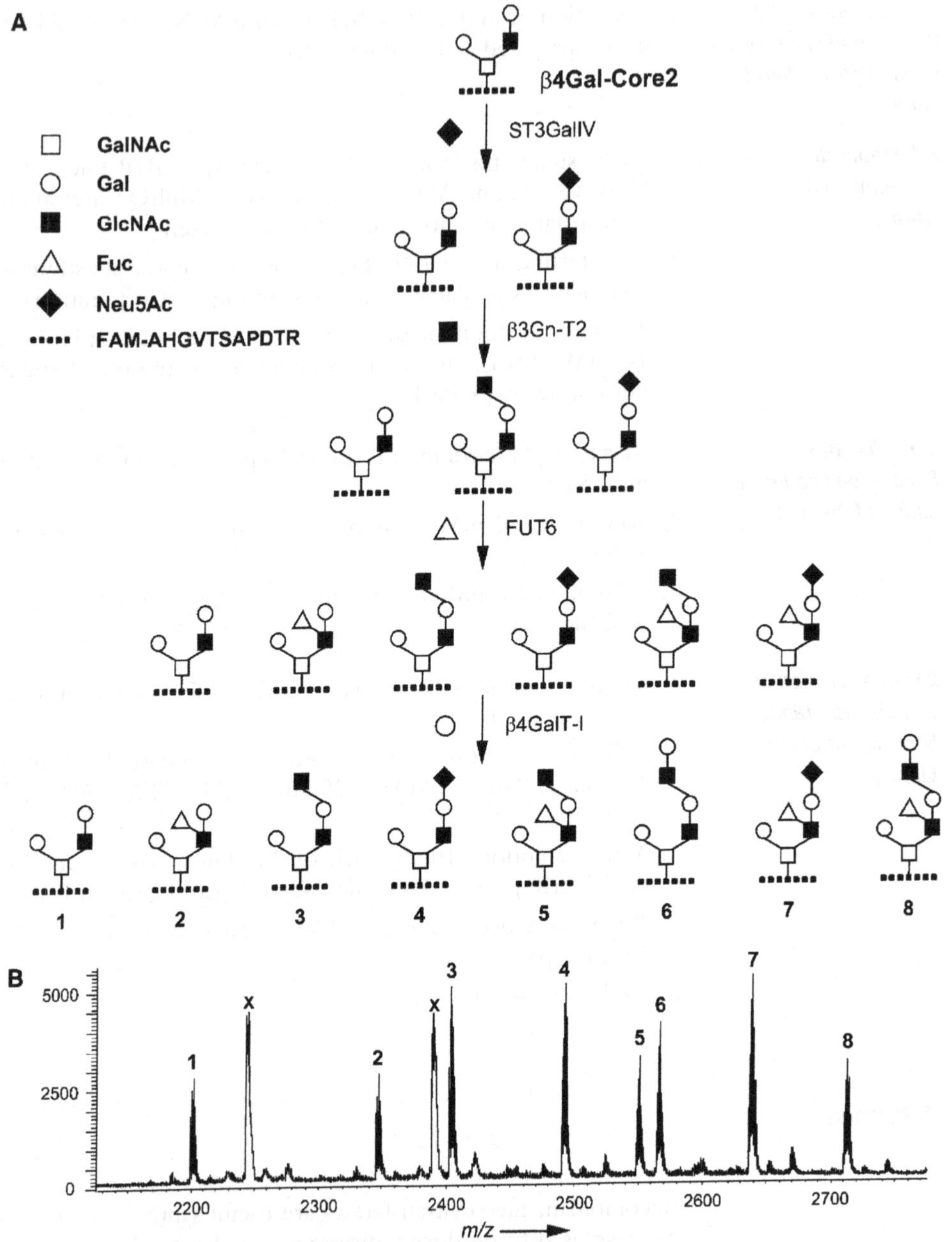

Fig. 1. Construction of an *O*-linked glycopeptide library. **(A)** Synthesis of an *O*-linked glycopeptide library including Le^x and sialyl-Le^x structures. **(B)** MALDI-TOF mass spectrum from the prepared library. The X indicates metastable ions from the sialylated compounds.

products can be instantly assigned from the *m/z* values, because the structures and molecular weights (MW) of the components in the library are designed to correspond one to one with MW-tagged libraries **(Fig. 1B)**. Additionally, an *O*-linked glycopeptide

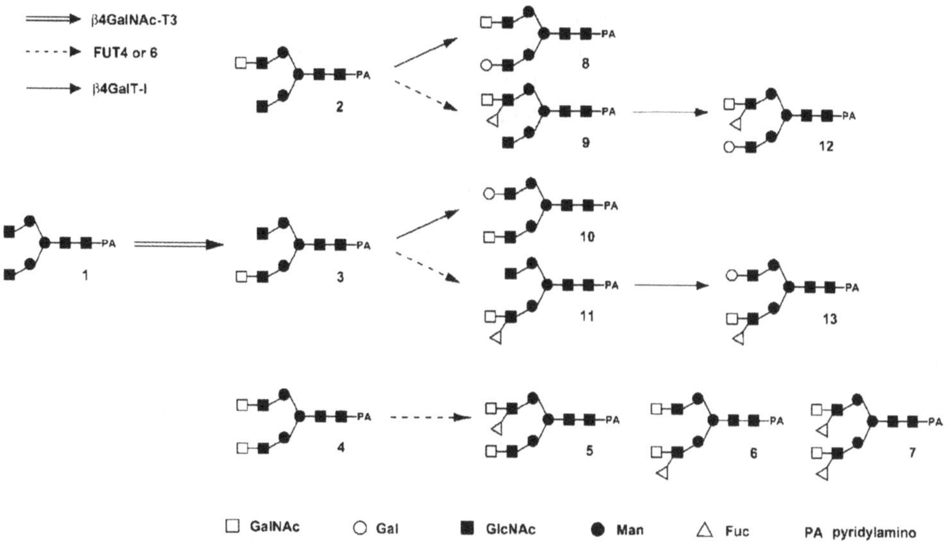

Fig. 2. Preparation of an *N*-linked glycan library.

library can be easily converted to an oligosaccharide library by reductive β-elimination, using standard samples to construct an MS^n spectral library containing observed spectra. On the other hand, since *N*-linked glycans have multi-antennary structures with the same sugar sequence at the nonreducing terminal ends (i.e. GlcNAc, LacNAc), glycosylation of *N*-linked glycans yields positional isomers. Therefore, an *N*-linked glycan library containing isomers may be synthesized by sequential transfers of sugars to a starting substrate. A flow chart showing the preparation of an *N*-linked glycan library containing the LacdiNAc structure (5) is shown in **Fig. 2**. By following this procedure, it is possible to synthesize non-commercially available *O*- and *N*-glycans, using a recombinant glycosyltransferase library.

The MS^n spectral library contains structurally defined *O*- and *N*-glycan spectra, and it should prove very useful for the rapid and accurate identification of oligosaccharide structures. This library stores triplicate MS^n spectra of each standard oligosaccharide, sample information, and experimental conditions in a relational database. The strategy of rapid identification is based on a comparison of the signal intensity profiles of MS^n spectra between the analyte and a library of stored mass spectra.

3.1. Preparation of Recombinant Enzymes

1. Glycosyltransferase genes are engineered for heterologous expression in mammalian cells as fusion proteins containing a FLAG Tag at the *N*-terminus.
2. The putative catalytic domain of each glycosyltransferase gene, without the transmembrane domain, is subcloned into a mammalian expression vector.

3. 30 μg of each resulting plasmid is transfected into HEK293T cells (2 × 10^6) using LIPOFECTAMINE 2000 (Invitrogen) according to the manufacturer's instructions.
4. After incubation at 37°C for 72 h, the culture medium is collected.
5. 10–50 mL of each culture medium from HEK293T cells is mixed with anti-FLAG M1 antibody resin (Sigma-Aldrich) and incubated with rotating at 4°C overnight.
6. The resin is washed 3–5 times with 50 mM TBS buffer containing 1 mM $CaCl_2$ and suspended in 100 μL of 50 mM TBS buffer with 1 mM $CaCl_2$.

3.2. Construction of O-Linked Glycopeptide Library

1. The reaction mixture (40 μL) contains 50 mM HEPES buffer (pH 7.0), 500 μM CMP-Neu5Ac, 10 mM $MnCl_2$, 25 μM of starting glycopeptide having Galβ1-4GlcNAcβ1-6(Galβ1-3) GalNAc on the FAM-labeled MUC1a tandem repeat peptide, and 5 μL of the purified ST3GalIV (*see* **Note 1**). After incubation at 37°C for 20 h, enzyme is inactivated at 100°C for 5 min (*see* **Notes 2–4**).
2. 1 μL of 1 mM UDP-GlcNAc and 2 μL of the purified β3Gn-T2 are added to the reaction solution (*see* **Note 1**). The reaction solution is incubated at 37°C for 2 h and then heated at 100°C for 5 min (*see* **Notes 2–4**).
3. 1 μL of 1 mM GDP-Fuc and 1 μL of the purified FUT6 are added to the reaction mixture (*see* **Note 1**). The reaction mixture is incubated at 37°C for 30 min and the reaction is then stopped by heating (*see* **Notes 2–4**).
4. 1 μL of 500 μM UDP-Gal and 1 μL of the purified β4GalT-I are added to the mixture (*see* **Note 1**). The reaction mixture is incubated at 25°C for 2 h and heated (*see***Notes 2** and **3**).
5. The resulting mixture (**1–8**); **Fig. 1**, *see* **Note 5**) is filtered through an Ultrafree-MC column (Millipore), and then the glycopeptides are purified with a reverse-phased micro-column, ZipTip C18 (Millipore) or Oasis HLB (10 mg/1 cc; Waters).

3.3. Conversion of the Glycopeptide Library to an O-linked Oligosaccharide Library

1. 20 μL of 500 mM $NaBH_4$ in 50 mM NaOH is added to the desalted glycopeptide library and the mixture is incubated at 50°C overnight.
2. The reaction solution is neutralized by adding 1 μL of acetic acid (LC/MS grade, Wako).
3. The released, and reduced *O*-linked oligosaccharides are applied to the cation exchange column, Oasis MCX (30 mg/1 cc; Waters).
4. Alditols are eluted with 300 μL of MilliQ water and dried with a Speed-Vac.

5. The remaining borate is removed by adding 100 μL of 1% acetic acid in methanol and drying in a Speed-Vac several times.
6. A homemade micro-column is packed with 20 μL of porous graphitized carbon (PGC; Alltech Associates) onto the top of a SPE C-TIP (200 μL, Nikkyo Technos).
7. The micro-column is washed and equilibrated with 2 × 100 μL of 50% aqueous acetonitrile, 80% aqueous acetonitrile, and MilliQ water prior to loading the samples.
8. The samples are redissolved in 20 μL of MilliQ water and applied to the pretreated micro-column.
9. Salts are removed by washing with 2 × 100 μL of MilliQ water, followed by elution of alditols with 50% aqueous acetonitrile (3 × 100 μL) and dried with a Speed-Vac.

3.4. Preparation of an N-Linked Glycan Library

1. The reaction mixture (100 μL) contains 50 mM HEPES buffer (pH 7.0), 1 mM UDP-GalNAc, 10 mM $MnCl_2$, 5 μM of PA-labeled oligosaccharides (**1**; **Fig. 2**) as an acceptor substrate, and 20 μL of the purified β4GalNAc-T3. After incubation at 37°C for 24 h and inactivation at 100°C for 5 min (*see* **Note 6**), the GalNAc-transferred products (**2–4**; two mono- and one di-glycosylated products) are separated using HPLC (*see* **Subheading 3.5**).
2. The purified di-GalNAc-transferred product (**4**; **Fig. 2**) as an acceptor substrate and 10 μL of FUT4 are added to the basic reaction mixture for fucosylation. After incubation at 37°C for 20 h and inactivation at 100°C for 5 min (*see* **Note 6**), the fucosylated products (**5–7**) are separated using HPLC (*see* **Subheading 3.5**).
3. Each mono-GalNAc-transferred product (**2** and **3**; **Fig. 2**) is subjected to fucosylation or galactosylation using FUT6 or β4GalT-I, respectively. After each glycosylation is complete, the product (**8–11**) is purified using HPLC (*see* **Subheading 3.5**).
4. Each fucosylated product (**9** and **11**; **Fig. 2**) and β4GalT-I are added to the basic reaction mixture. After incubation at 37°C for 3 h, each Gal-transferred product is purified using HPLC (*see* **Subheading 3.5**).

3.5. HPLC-Based Separation and Purification of the Products

1. The reaction products are filtered through an Ultrafree-MC column (Millipore).
2. The mixture is subjected to reversed-phase HPLC on a PALPAK Type R column (4.6 × 250 mm; TaKaRa).
3. The products are eluted with a linear gradient ranging from 5 to 55% of solvent B in solvent A at a flow rate of 1 mL/min for 60 min at 40°C.

4. The elution is detected by the fluorescence intensity at 400 nm (excitation, 320 nm) with a fluorescence detector, RF-10AXL (Shimadzu).
5. The separated samples are dried with a Speed-Vac and redissolved in MilliQ water for MALDI-TOF MS analysis (Reflex IV, Bruker-Daltonik).

3.6. Construction of a MS^n Library Containing Observed Spectra

1. These instructions assume the use of a MALDI-QIT-TOF MS equipped with an Ultra Cooling Kit (AXIMA-QIT; Shimadzu, Kyoto, Japan) plus operation software Kompact Version 2.5.1 (Kratos Analytical Ltd., Manchester, UK).
2. Deposit 0.5 μL of a sample solution on the target plate (special surface stainless steel plate, 2 mm diameter) and allow to dry. Next, cover the dried analyte on the target plate with a 0.5 μL of the matrix solution and allow to dry. Finally, recrystallize the dried material by depositing a 0.15 μL of 99.5% ethanol on the matrix-analyte mixture on the target plate.
3. Acquire MS spectra of the glycans in the positive mode.
4. Acquire MS^2 spectra of sodium adduct ions derived from the glycans. The MS^2 spectra of the same ion should be acquired three times from different wells on the target plate. For the acquisition of CID spectra, adjust a CID energy so that the intensity of parent ions almost disappears; that is, when the intensities of parent ions in the spectra are more than 15% of the base peak, the spectra must be disc arded and reacquired with a larger CID energy. Use an automatic acquisition function with the regular raster that governs the laser shot patterns.
5. Acquire MS^3 spectra of all the major fragment ions in the MS^2 spectra in the same manner described above for MS^2 acquisition.
6. Output the data including m/z values and their intensities in these CID spectra. Match with the corresponding glycan ID (*see* **Note 7**), and input these data to the relational database, RDBMS Oracle9i (Oracle Corporation Japan, Tokyo, Japan).

4. Notes

1. If possible, the use of detergents in enzyme reactions should be avoided when preparing samples for MS analysis.
2. The time course of each reaction is monitored using a MALDI-TOF MS (ReflexIV; Bruker-Daltonik) by directly applying the reaction solution to the target plate, and the enzyme reaction is stopped at around a 50% yield by heating.

3. If purification is necessary for MS analysis, an aliquot of the reaction solution may be desalted with a reversed-phase micro-column such as a ZipTip C18 (Millipore).
4. After cooling the incubation temperature, donor and enzyme are added to the reaction solution.
5. The structure of each component in the mixture can be identified immediately using MS alone, because each product has a different molecular weight (*see* **Fig. 1b**).
6. This enzyme reaction is stopped by heating before the sample is perglycosylated.
7. Glycan IDs are linked to the corresponding data in the glycan structure DB, which was constructed independently (*6*).

Acknowledgements

The authors would like to thank Professor Yoshiaki Nakahara for providing the glycopeptides as starting materials. This work was performed as a part of the R&D Project of the Industrial Science and Technology Frontier Program supported by the New Energy and Industrial Technology Development Organization (NEDO).

References

1. Narimatsu, H. (2004) Construction of a human glycogene library and comprehensive functional analysis. *Glycoconj. J.* 21, 17–24
2. Taniguchi, N., Honke, K., and Fukuda, M. (eds.) (2001) *Handbook of Glycosyltransferases and Related Genes* (Springer-Verlag Tokyo Berlin Heidelberg New York)
3. Ito, H., Kameyama, A., Sato, T., Kiyohara, K., Nakahara, Y., and Narimatsu, H. (2005) Molecular-weight-tagged glycopeptide library: efficient construction and applications. *Angew. Chem. Int. Ed. Engl.* 44, 4547–4549
4. Kameyama, A., Kikuchi, N., Nakaya, S., Ito, H., Sato, T., Shikanai, T., Takahashi, Y., Takahashi, K., and Narimatsu, H. (2005) A strategy for identification of oligosaccharide structures using observational multi-stage mass spectral library. *Anal. Chem.* 77, 4719–4725
5. Sato, T., Gotoh, M., Kiyohara, K., Kameyama, A., Kubota, T., Kikuchi, N., Ishizuka, Y., Iwasaki, H., Togayachi, A., Kudo, T., Ohkura, T., Nakanishi, H., and Narimatsu, H. (2003) Molecular cloning and characterization of a novel human beta 1,4-*N*-acetylgalactosaminyl-transferase, beta 4GalNAc-T3, responsible for the synthesis of *N,N*-diacetyllactosediamine, galNAc beta 1-4GlcNAc. *J. Biol. Chem.* 278, 47534–47544
6. Kikuchi, N., Kameyama, A., Nakaya, S., Ito, H., Sato, T., Shikanai, T., Takahashi, Y., Narimatsu, H. (2005) The carbohydrate sequence markup language (CabosML): an XML description of carbohydrate structures. *Bioinformatics* 21, 1717–1718

Chapter 21

Data Mining the PDB for Glyco-Related Data

Thomas Lütteke and Claus-W. von der Lieth

Summary

The 3D structural data of glycoprotein or protein-carbohydrate complexes that are found in the Protein Data Bank (PDB) are an interesting data source for glycobiologists. Unfortunately, carbohydrate components are difficult to find with the means provided by the PDB. The GLYCOSCIENCES.de internet portal offers a variety of tools and databases to locate and analyze these structures. This chapter describes how to find PDB entries that feature a specific carbohydrate structure and how to locate carbohydrate residues in a 3D structure file and to check their consistency. In addition to this, methods to statistically analyze torsion angles and the abundance of amino acids both in the neighborhood of glycosylation sites and in the spatial vicinity of non-covalently bound carbohydrate chains are summarized.

Key words: Bioinformatics, 3D structure, Glycoprotein, Protein-carbohydrate interaction, Torsion angle, Glycosylation.

1. Introduction

For a complete understanding of the molecular basis of protein-carbohydrate interactions as well as the impact of glycosylation on protein properties, knowledge of the 3D structure of the protein–carbohydrate complexes or the glycosylated protein, respectively, is often indispensable. Such structures can be found in the Protein Data Bank (PDB) *(1)*, which is the largest freely available resource of biomolecular 3D structure data. Although the main focus of the PDB is to provide access to protein structures, it also constitutes a large publicly available resource of glycan structures, which are either found covalently attached to proteins or as non-covalently bound ligands. Unfortunately, there is no standard notation for carbohydrate residues in the PDB, and information on how the residues are linked is often missing completely in

Nicolle H. Packer and Niclas G. Karlsson (eds.), *Methods in Molecular Biology, Glycomics: Methods and Protocols, vol. 534*

DOI: 10.1007/978-1-59745-022-5_21

the notation. Therefore, with the means provided by the PDB, it is difficult to locate entries that feature a certain carbohydrate structure (*2*).

The GLYCOSCIENCES.de web portal (*3*) offers a variety of tools and databases which bridge this gap and facilitate the search for PDB entries containing specific carbohydrate structures, as well as an analysis of 3D-structural properties like torsion angles and the distribution of amino acids in the spatial vicinity of carbohydrate residues. In this chapter, the use of these resources and some examples of results that can be obtained from them are described.

2. Materials

2.1. Source and Updates of Data

The analyses described in this chapter are based on 3D structural data, which is taken from the protein databank (PDB) (*1*). In total, about 8–10% of all PDB entries contain carbohydrates. Although it is estimated that about 50% of all proteins are glycosylated (*4*), less than 5% of all PDB entries contain covalently bound glycan chains. There are several reasons for this low ratio of glycosylated proteins in the PDB. First, many of the protein structures in that database originate from bacteria or are recombinantly expressed in bacterial expression systems. In these cases, the protein is not glycosylated, even if the original protein is e.g., a mammalian protein that is known to be glycosylated. Second, glycan chains often hamper crystal growth and therefore are cleaved off before starting to grow the crystals (*5*). Third, even if glycan chains are present in the crystal, they are located on the protein surface and are quite flexible. Therefore, in many cases it is not possible to obtain sufficient electron density to resolve the glycan chain (*2*). This tendency increases with increasing the distance from the protein attachment site. As a result, often only the first residues close to the protein surface are reported in the PDB. Furthermore, the experimentalists who resolve the protein structures are often mainly interested in the protein part and do not spend much effort in resolving the glycan chains as well.

The PDB is updated weekly with new entries. Consequently, the related resources within the GLYCOSCIENCES.de web portal are also updated every week. With the existence of a check tool, which finds potentially erroneous entries by checking the consistency of residue notation and residues actually present in the 3D structure (*see* **Subheading 3.2.2**), the update process is mainly automatic and requires only minimal human intervention (*6*).

For the analysis of protein sequences in the neighborhood of glycosylation sites (*see* **Subheading 3.5**), in addition to the PDB data, sequence information from SwissProt (*7*) is also available.

2.2. Required Software

The methods illustrated in this chapter are based on bioinformatics, not on experimental biology. Therefore, only a computer with internet access is necessary. Standard software, i.e., a web browser, is sufficient.

3D structures can be displayed in two different ways. The Java applet JMol (jmol.sourceforge.net) does not require any software installation on the client computer; only Java has to be enabled in the web browser's security settings. Alternatively, the Chime plugin can be used. This plugin is only available for Windows and Mac. It can be downloaded freely at www.mdlchime.com/products/framework/chime/and has to be installed locally.

Graphics that are generated by the analysis tools (diagrams, plots) are offered in two different formats. The Scalable Vector Graphics (SVG) format provides figures in a quality suitable for publication. However, to display this format in web browsers, an additional plugin is required. If figures are not needed for printing purposes, the GIF format can be chosen, which can be displayed by virtually all current web browsers without any further software (*see* **Note 1**).

3. Methods

The databases and tools described in this chapter are all available online at the GLYCOSCIENCES.de web portal (www.glycosciences.de). Instead of complete URLs, which are rather lengthy in some instances, the menu items to be clicked to reach the interfaces from the portal's start page are given in the following sections in the format →X → Y → Z.

3.1. Searching for PDB Entries that Contain Specific Carbohydrate Structures

The PDB does not offer any option to perform a comprehensive search for carbohydrate structures. There are two possible ways to find entries containing carbohydrate residues on the PDB website. One is to do a text search in the structure titles/descriptions, ligand names, or the titles of the primary references related to the entries. This option, however, only works with carbohydrate structures for which a common name is regularly used, or which are small enough to be mentioned in IUPAC notation in these fields. Besides, such a search mostly yields only a few, but not all available entries featuring the structure to be searched, because not all these entries will have the carbohydrate structure mentioned in the texts which are to be referred to. The second alternative is to search for entries which contain the separate residues which make up the carbohydrate structure of interest. However, residue names used for carbohydrate residues in PDB entries are often ambiguous or bear little resemblance to the ones used by glycoscientists, and linkages cannot be included in this search.

To enable scientists to specifically search for PDB entries containing carbohydrate structures, data found using the pdb2linucs software (*see* **Subheading 3.2.1**) were integrated into the GLYCOSCIENCES.de databases. This information can be searched in several ways, which will be briefly outlined in the following sections. To find only those database entries which are associated with PDB data, the checkbox "*with PDB entries*" has to be ticked in the search interfaces (*see* **Note 2**).

3.1.1. Search by Substructure

The substructure search (→Databases → Structure → Substructure Search) is available in a beginner and an advanced mode. These modes mainly differ in the input of residue names. In the beginner mode, a limited number of frequent residues and linkage types can be selected from pull-down menus, while the advanced mode allows free text input of residues. This allows searching for residues that are not listed in the pull-down menus. However, the correct spelling of the residue names (*see* **Note 3**) has to be known when using the free text input (as shown in **Fig1A**). In addition, the advanced mode permits the user to enter two substructures, which can be combined by "AND" or "OR" operators. Since the handling of both modes is the same except for the residue input, here only the advanced mode is described.

To search for N-glycan structures, which consist of at least the trimannosyl core, the core structure has to be entered into the search fields **(Fig.1A)**. This substructure is too large to be entered completely, but it is sufficient to enter the mannose residues and one b-D-Glc*p*NAc residue and change the first pull-down menu from "*all chains*" to "*glycans only.*"

In the search results, the structures are displayed in CarbBank (*8*) notation (*see* **Note 4**) with the substructures matching the ones entered in the search fields highlighted **(Fig. 1B)**. The "*Explore*" buttons lead to the related database entry pages. If PDB entries which contain the carbohydrate structure described in that entry are available, a button "*pdb-entries*" is displayed, which directly leads to the respective part of the entry page **(Fig. 1C)**. Information on up to ten PDB entries is listed here, ordered by the resolution of the structure. If more than ten entries are available that contain the carbohydrate structure, the additional entries can be reached by the "*more*" link.

3.1.2. Search by Structure Motifs

A simple way to search for carbohydrate structures that contain frequent substructures like Lewis antigens, *O*-glycan core structures, LacDiNAc motifs or glycosphingolipid core structures is to

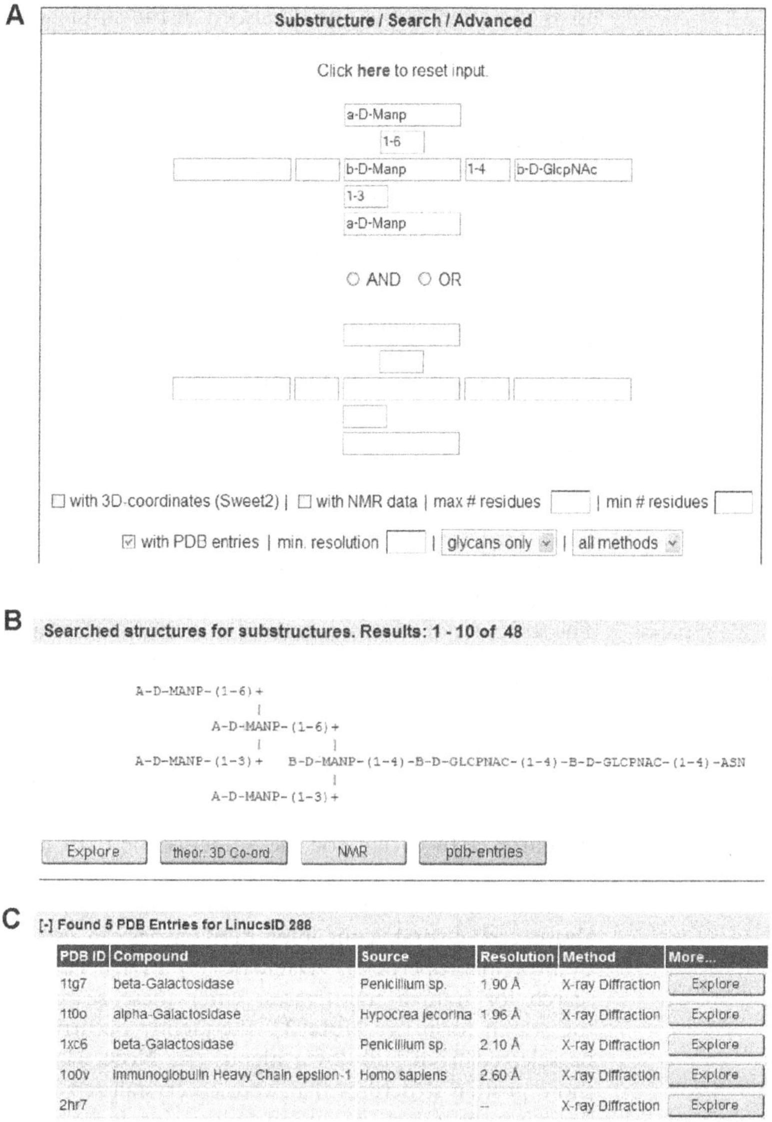

Fig. 1. GLYCOSCIENCES.de substructure search. The substructure search interface allows the input of up to two pentasaccharides. To include only those structures which are associated with PDB entries, the checkbox *with PDB entries* has to be ticked (**A**). In the result list, the structures matching the query are displayed in IUPAC notation (**B**). Links to the entry start page and – if present – to specific properties like PDB entries (**C**) are also included.

search by structure motifs (→ Databases → Structure → Motifs) (found under "*other*"). The desired motif can be selected from a pull-down menu.

The motif search mode enables the user to find e.g., all structures which contain the Lewis^X substructure, without the need to type in the residues in the substructure search mode. The result list is the same as the one described in the substructure search mode (*see* **Subheading 3.1.1**).

3.1.3. Search by N-Glycan Classification

An alternative way to search for *N*-glycan structures in PDB entries is provided by the *N*-glycan classification search (→Databases → Structure → *N*-Glycan Classification). In this mode, *N*-glycan structures can be searched by properties like subfamily or class (complex, high mannose, hybrid), number of antenna or of terminal residues (Neu5Ac, Gal, GlcNAc, Fuc, Man, other), and core-fucosylation. A motif search as described in the previous section can also be incorporated.

In addition to the information described in the results of substructure searches (*see* **Subheading 3.1.1**), the results list in this mode contains some information about the *N*-glycan classification such as subfamily and number of antennae.

3.1.4. Exact Structure Search

The exact structure search mode (→Databases → Structure → Exact Structure Search) differs from the previously described modes insofar as it does not reveal a list of structures matching the selected properties but directly leads to a database entry page. The structure of interest is entered into the interface in CarbBank (*8*) notation (examples are given below the search field, also *see* **Note 4**). If a database entry for the requested structure is available, the entry page is directly displayed. If PDB entries featuring that carbohydrate structure exist, these are listed in a separate section of the entry page. This is expanded by clicking on the "*PDB entries*" link in the shortcuts listed at the top and the bottom of the entry, or by clicking on the header of the PDB section.

3.1.5. Search by PDB Entry Properties

In addition to the structure-related search modes described above, carbohydrate containing PDB entries can also be searched by protein properties (→Databases → PDB). For example, to search for human transferases, enter these search terms into the search form (*see* **Note 3**). Frequently occurring terms can be selected from pull-down menus (*see* **Note 5**). To find only those protein structures that were co-crystallized with carbohydrate ligands, select the option "*must contain ligands*" from the "*Chain type*" menu.

In the results list, the PDB entries matching the query are listed together with some basic information on the entry. Using the "*Explore*" button, an entry page is displayed showing more detailed information about the PDB entry, including a list of carbohydrate structures present in the 3D coordinates (*see* **Note 6**). These are linked to the respective GLYCOSCIENCES.de carbohydrate entries. Links to external sources such as the PDB and the Jena Image Library are also provided together with a link to the pdb2linucs tool (*see* **Subheading 3.2.1**).

3.2. Analysis of Single PDB Entries

3.2.1. Detection of Carbohydrate Residues

Because the PDB file format is lacking a proper notation for carbohydrate residues, an algorithm was developed that detects carbohydrates in PDB structures based on atom types and coordinates only (*9*). This algorithm is implemented in the pdb2linucs software (→Tools → pdb2linucs). To check a 3D structure for carbohydrates, the PDB ID of the structure is entered or alternatively a file in PDB format is uploaded from the user's local computer.

Standard PDB files distinguish between protein/nucleic acid atoms (stored in ATOM records) and other atoms (stored in HETATM records). Therefore, in these files carbohydrate residues are only found within the HETATM section. If a file is being uploaded from the user's local computer, e.g., a structure generated with the Sweet2 software (*10*), carbohydrates might also be present in the ATOM section. In such a case, "*ATOM and HETATM*" is selected from the "*Find carbohydrates in*" pull-down menu.

In standard PDB files, connectivities for atoms stored in HETATM records should be given explicitly in the CONECT records. However, in some PDB files the information in the CONECT records is missing or incomplete. In such cases pdb2linucs can be forced to determine atom bonds based on the distances between atoms. To do so, the record type (*HETATM/ATOM/both*) is selected from the pull-down menu "*Assign connections by atom distances*" (*see* **Note 7**).

The next options relate to the display of the 3D structure in the result page. It can be displayed using the JMol Java applet or the Chime plugin (*see* **Subheading 2.2**). The size of the viewer window is adjusted to the user's screen resolution by typing in appropriate width and height values. To highlight the detected carbohydrate residues in the displayed structure, two color modes are available. In the *chain* mode carbohydrate residues are colored by chain type (*N*-glycan, *O*-glycan, ligand), while in the *residue* mode the residue type (GlcNAc, Gal, Man, Fuc, ...) determines the highlighting color.

By default, phosphate groups are handled as separate residues in pdb2linucs. This makes it possible to correctly detect and name carbohydrate chains in which two monosaccharide residues are linked via a phosphate group. But pdb2linucs also offers an option to treat phosphates as substituents. To activate this option, set "*Phosphate handling*" to "*as Substituents.*"

The last option regards the notation in which the detected carbohydrate structures are displayed. Since the LINUCS notation (*11*) is optimized for computer readability, it is recommended to select the IUPAC extended code here, which is easier to survey for humans.

On the results page, the 3D structure is displayed according to the options selected. Below this, the detected carbohydrate structures are listed in the notation selected on the search page. If a corresponding entry is available in GLYCOSCIENCES.de, the "*Explore*" button directly links to that entry page (*see* **Note 6**).

If carbohydrates covalently attached to a protein are detected in the PDB entry, the protein's primary sequence is also displayed, with the attachment sites highlighted.

3.2.2. Check for Consistency of Carbohydrate Structures and Residue Notation

As the PDB file format allows only three characters for the residue name, it is difficult to properly name carbohydrate residues. The carbohydrate residue names used in PDB files are often cryptic, redundant or ambiguous. Also, the scientists who resolve the protein 3D structures are not often well-trained glycobiologists, which makes it even more difficult for them to assign the correct residue names, or to find errors in the carbohydrate parts of glycoprotein or protein–carbohydrate complex 3D structures. Thus, according to a recent study, about 30% of all PDB entries containing carbohydrates have errors in the carbohydrate parts of the 3D structure (*9*).

To check a 3D structure file before submitting it to the PDB, or to check an existing PDB entry for errors, the pdb-care (PDB CArbohydrate REsidue check) software (*12*) (→Tools → pdb-care) can be used. First the PDB ID of an existing entry is entered or a file to be uploaded from the local computer is chosen. Afterwards, the checks to be performed are selected.

The "*Bond length/valence check*" analyzes the distances between atoms, which are listed in the CONECT records of the PDB file. If a bond length differs by more than the user-selected tolerance from the default value for the elements involved in the bond, a warning will be displayed in the results. The same applies to atoms which are connected to more other atoms than is maximally allowed for that element type. Connections to ions usually are not covalent but complex bonds. Therefore, it is recommended to exclude these bonds from the check by ticking "*Ignore connections to ions.*"

The *Nomenclature* check compares the residue type defined by the residue name used in the PDB file with the residue type actually present in the 3D structure. In case of mismatches, an error message is displayed in the output. To display a PDB residue name matching the residue present in the coordinates, "*Find correct nomenclature*" has to be ticked in the "*Additional nomenclature options.*" The "*Suggest non-ambiguous residue names*" option in this section is only recommended if the user is checking a file to be submitted to the PDB and wants to avoid ambiguously defined residue names (*see* **Note 8**).

3.3. Carbohydrate Torsion Angles

From the carbohydrate residues found in the PDB, a series of torsion angles can be determined. These include linkage torsions, ring torsions, omega torsions and the torsions of NAc and

Asn sidechains. The tools to statistically analyze these angles are described in the following sections.

3.3.1. Statistical Distribution of Torsions in Carbohydrate 3D Structures

The GlyTorsion web interface (*13*) (→Tools → glytorsion) was built to analyze the statistical distribution of carbohydrate torsion angles. The use of this interface will be exemplified for linkage torsions; the other torsion types are analyzed analogously.

In the general settings, the preferred structure viewer (JMol or Chime, *see* **Subheading 2.2**) can be selected and the width and height of the viewer adjusted to the user's screen size. Setting a resolution cut-off excludes low resolution structures from the analysis. Furthermore, the analysis can be limited to structures resolved by a specific method (X-ray or NMR) and to *N*-glycans, *O*-glycans, or non-covalently bound ligands.

As an example, to investigate the linkage torsions between β-D-Man*p* 1–4 linked to β-D-Glc*p*NAc residues in *N*-glycan structures, which are resolved by X-ray crystallography, "*x-ray*" is selected from the "*Experimental Method*" menu and "*N-glycan*" from the "*Chain type*" menu. In the "*Linkage Torsion Analysis*" section, the residues "*b-D-Manp*" and "*b-D-GlcpNAc*" and the linkage "*1–4*" are selected from the pull-down menus (*see* **Note 5**) or the data are directly typed into the fields below the menu. "*Phi*" is selected as angle 1 and "*Psi*" as angle 2. In this example, the *scattered plot* is chosen from the "*Output Type*" menu. If an SVG plugin is installed (*see* **Subheading 2.2**), the "*SVG scattered plot*" can be selected, otherwise the "*GIF scattered plot*" should be selected.

For linkage torsions, different definitions of the atoms forming phi, psi and omega angles exist. In the NMR-like definition, hydrogen atoms are involved (phi: H_1–C_1–O_1–C'_X, psi: C_1–O_1–C'_X–H'_X). Structures resolved by X-ray usually do not contain any hydrogens. Therefore, for such structures, often the crystallographic definition is used, in which the atoms of the carbohydrate backbone are used instead of hydrogens (phi: O_5–C_1–O_1–C'_X, psi: C_1–O_1–C'_X–C'_{X+1}). The NMR-like definition can be calculated from the crystallographic definition by adding or subtracting 120°, so GlyTorsion offers the use of either definition for all PDB entries (under "*Definition of linkage angles*"). In the current example, the crystallographic definition is selected.

After submitting the request, a plot depicting the torsions that match the given restraints is shown. If the majority of torsions is located at the border of the plot **(Fig. 2A)**, the angle ranges can be changed within the general settings at the GlyTorsion start page ("*Angle Range*" settings) and the request resubmitted, so that the preferred conformation is located more in the center of the plot **(Fig. 2B)**.

To get more information on specific torsions (e.g., those that differ from the preferred conformation), the user can enter

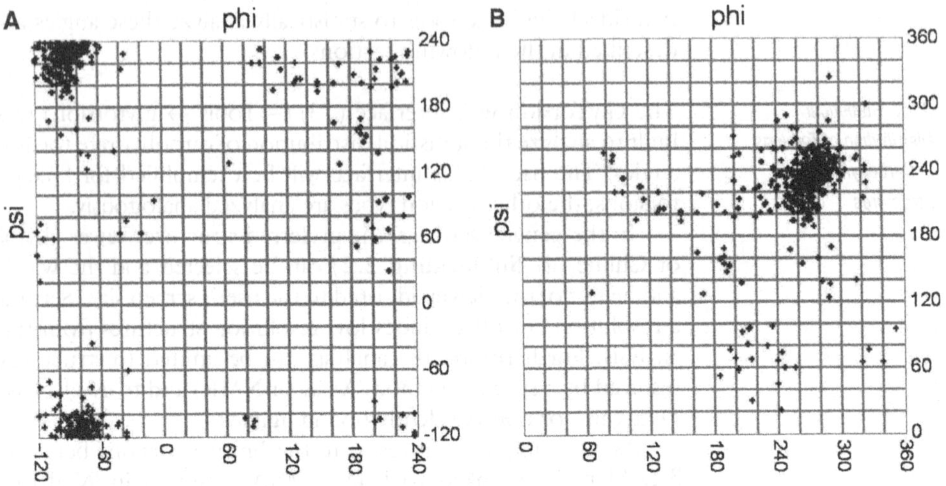

Fig. 2. Linkage torsion analysis. The conformation of the β-D-Man*p*-(1-4)-β-D-Glc*p*NAc disaccharide fragment mainly shows phi angles between −120 and −60° and psi angles of about 240°. (**A**) *Settings*: Angle1 Range −120 to +240; Angle2 Range −120 to +240. If the majority of torsions are located at the edges of the plot, the display range can be adjusted to shift them to the center of the displayed area (**B**) *Settings*: Angle1 Range 0–360; Angle2 Range 0–360).

a range of torsions below the plot. After submitting that query, a list of the torsions that fall into the selected area is displayed. Within this list, the PDB IDs are linked to the pdb2linucs interface (*see* **Subheading 3.2.1**). A click on a torsion value leads to a 3D structure view of the structure, with the selected torsion highlighted. By this means it is possible to pick unusual torsions and directly visualize the 3D structure to investigate possible reasons for the unusual values.

3.3.2. Comparing Experimental and Theoretical Data

Apart from experimentally resolved structures, linkage torsions can also be retrieved from theoretical data, i.e., molecular dynamics (MD) simulations. A collection of such data is the GlycoMapsDB (→Modeling → GlycoMapsDB). To find phi–psi maps for a specific linkage, the user has to click on "*search database*" and enter a disaccharide fragment by selecting residues and linkage type from the provided pull-down menus (*see* **Note 5**) or by directly entering the data into the search fields below the menus.

To analyze the same linkage as in the section above, "*b-D-Manp*", "*1-4*" and "*b-D-GlcpNAc*" is entered. In the results list, a preview of the map together with some information on the structure and the simulation the map was derived from is displayed. The entry page for a map is accessed by clicking on the map id or on the preview image. If the disaccharide fragment that is analyzed in the map is present in the PDB, the entry page contains a link "*compare to PDB data.*" Again, the user can choose between SVG and GIF format. The link leads to a page where the map and the PDB data provided by GlyTorsion (*see* **Subheading 3.3.1**) are overlaid. In addition, some statistics about the conformance of theoretical and experimental data are provided. Two kinds of hits to the PDB data are distinguished here: torsions found in structures that contain the same disaccharide fragment as the one analyzed in the map, and torsions found in carbohydrate chains that match not only the disaccharide fragment but also the entire chain of the structure that the map was derived from. In the example of map id 8243, which represents the β-D-Man*p*-(1-4)-β-D-Glc*p*NAc linkage of the α-D-Man *p*-(1-3)-β-D-Man*p*-(1-4)-β-D-Glc*p*NAc trisaccharide, at the time of writing only two matches of the entire chain were found in the PDB, both of which were located within low energy regions of 0–1.5 kcal/mol. Of the 737 occurrences (at the time of writing) of this linkage in other carbohydrate chains in PDB structures, about 70% were located in these regions as well, and more than 80% were found in regions of 0–3.0 kcal/mol.

The *show details* link on this page leads to a list of the torsion values that are plotted onto the map and the respective PDB entries. It is the same list as provided by the torsion selection in GlyTorsion (*see* **Subheading 3.3.1**).

3.3.3. Carbohydrate Ramachandran Plot

The Ramachandran Plot (*14*) is a method frequently used to evaluate the quality of protein 3D structures (*15*). In this plot, the phi and psi torsion angles of the protein backbone are plotted against each other. A similar approach is possible to evaluate carbohydrate conformation. In contrast to proteins, however, it is not possible to put all the torsions into one plot, as the preferred conformations depend on the linkage type and the residues involved in the glycosidic linkages (*16*). Therefore, for each kind of disaccharide fragment a separate plot has to be generated.

CARP, the CArbohydrate Ramachandran Plot (*13*) (→Tools → carp), is dedicated to the analysis of the conformations of glycosidic linkages present in a PDB entry or a 3D structure in PDB file format. To perform such an analysis, the ID of a PDB entry is entered or a file in PDB format are uploaded from the local computer. To decide whether a linkage is in a preferred conformation or not, some data for comparison are necessary. Such data can be derived from GlyTorsion (*see* **Subheading 3.3.1**) or the GlycoMapsDB (*see* **Subheading 3.3.2**). The user has to select one of these sources ("*Select data source for plot background*"), the graphics output format (SVG or GIF) and the 3D structure viewer (JMol or Chime), and submit the request.

On the results page, one plot for each type of disaccharide fragment found in the 3D structure is displayed. The torsions present in the structure are indicated by red crosses, which are referenced by numbers. Below each plot, the carbohydrate chains in which the linkages are present are displayed in IUPAC notation, with the residues forming the linkage highlighted. The linkage path from the reducing end to the analyzed linkage, the PDB residue names of the monosaccharides involved in the linkage and the actual torsion values are also displayed. By clicking on the PDB residue names, a 3D structure view is obtained in which the atoms forming the glycosidic bond are highlighted. This feature enables the user to easily access the 3D structure, e.g., to check for possible reasons for unusual torsion values.

3.4. Amino Acids in the Spatial Vicinity of Carbohydrate Residues

Carbohydrate binding proteins such as lectins and glycoenzymes exhibit a high specificity for certain carbohydrate residues. The molecular basis for this specificity can be investigated by an analysis of the amino acids in the spatial vicinity of specific carbohydrate residues which are present in protein–carbohydrate complexes in the PDB. Such an analysis can be done using GlyVicinity (*13*) (→Tools → glyvicinity).

In the "*General Settings*" section, the user can select the preferred 3D structure viewer (JMol or Chime, *see* **Subheading 2.2**) and adjust the sizes of the structure viewer window and of the graphical results to their screen size. In the "*Analysis Parameters*," a resolution cut-off can be set to avoid including low-resolution structures into the analysis. In addition, the user can define the radius around the carbohydrate atoms in which the amino acids are taken into account, set a minimum carbohydrate chain length, and specifically select carbohydrate residues within a chain by setting the depth, i.e., the distance of the analyzed residues from the reducing end of a carbohydrate chain. Using the latter option, it is possible to e.g., distinguish between the β-D-Glc*p*NAc residue at the reducing end of an *N*-glycan chain (depth = 1) and the one attached to that (depth = 2).

Below this, information about the carbohydrate residue type that is to be examined is entered. Again, the residue name can be entered directly or chosen from a list of frequent residues (*see* **Note 5**). To analyze only residues that are present in non-covalently bound carbohydrate chains, "*ligands only*" is selected from the "*Chain type*" menu.

The results can be given in three different forms ("*Output*"): number of amino acids of each type found in the selected vicinity of the given carbohydrate residue type ("*absolute numbers*"), relative portions of amino acids compared to the total number of residues counted ("*percentage*"), and the deviations of the relative portions from the average values for the amino acids ("*deviation from standard percentage values*"). Combinations of these values are also possible. The deviation values indicate which amino acids are over- or under-represented in the vicinity of the selected carbohydrate residue type. In addition to a textual display of the values, the results can also be graphically represented by diagrams in SVG or GIF format. The amino acids in the output can be ordered by polarity, by hydrophobicity or alphabetically.

For example, if the distribution of amino acids within 0.0–4.0 Å of β-D-Gal*p* residues of non-covalently bound carbohydrate chains is examined, it becomes obvious that, in general, polar amino acids are over-represented, while non-polar residues are under-represented. The positively charged amino acids Lys and Arg as well as the aromatic residues Tyr and Trp form an exception (**Fig. 3**).

To refine the analysis to single atoms instead of residues, the user can click on the name of an amino acid within the results table. From the interactions between β-D-Gal*p* and Trp it is apparent that the main interactions are formed between the C3-C4-C5-C6 atoms of β-D-Gal*p* and the CE2-CD2-CZ2 atoms of the Trp sidechain. From this information, it is possible to draw conclusions about the type of interaction preferably formed between these residues. The C3-C4-C5-C6 axis is part of the hydrophobic B-face of β-D-Gal*p*, while CE2, CD2 and CZ2 are located in the center of the Trp indole ring. The high frequency of interactions between these atoms is possible because β-D-Gal*p* forms stacking interactions to Trp. Polar residues, in contrast, mainly interact with carbohydrates via the atoms located at the tips of the sidechains rather that those in the sidechain center.

More details about the PDB entries, the carbohydrate chains and the residues that the results were obtained from, and the

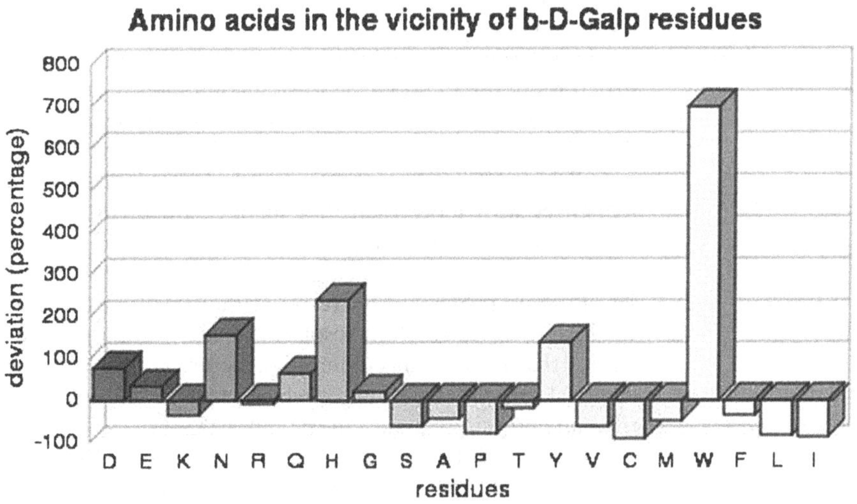

Fig. 3. Amino acids in the spatial vicinity of non-covalently bound β-D-Gal*p* residues. With the exception of positively charged amino acids and the aromatic residues Trp and Tyr, polar amino acids (*left, gray*) are over-represented, while non-polar residues (*right, white*) occur less frequently than their natural abundance.

atoms involved can be gained by clicking on an atom name or a count number in the results table. When the user clicks on an atom name, details are listed for all entries in which that atom is participating in the selected kind of interaction. In contrast, clicking on a count value lists only those entries where both the amino acid atom and the carbohydrate atom associated with that count interact with each other. To obtain a complete detailed list for all kinds of atoms a link is offered below the table that lists the atom interaction counts.

Apart from receiving information about the counted interactions, 3D structure views can be obtained from the detailed list by clicking on a residue or atom name. Furthermore, the analysis can be refined to a subset of PDB entries by selecting the entries to be included in the analysis and resubmitting the query using the button at the bottom of the list.

3.5. Amino Acids in the Neighborhood of Glycosylation Sites

Glycosylation is by far the most common and most complex co- and posttranslational modification of proteins. It is assumed that about 50% of all proteins are glycosylated. While *O*-glycosylation often is a bulk property (*17*), *N*-glycosylation is very site-specific. A prerequisite for the presence of *N*-glycan chains is the so-called sequon, a sequence motif of the amino acids Asn-Xxx-Ser/Thr, where Xxx can be any amino acid but proline (*18*). However, not all sequons are actually glycosylated (*19*). The presence of a glycan chain in the 3D structure clearly indicates that the respective amino acid indeed is glycosylated (*20*). However, since the absence of a glycan chain can have several reasons (*see* **Subheading 2.1**), the opposite conclusion is not possible. To analyze the amino acids that are present in the neighborhood of glycosylation sites, the GlySeq tool (*13*) (→Tools → glyseq) was established.

In addition to protein sequences derived from glycoprotein structures in the PDB, GlySeq also incorporates data from SwissProt (*7*). Thus, the user can choose between different data sets: PDB data, SwissProt data and complete data. Each of these sets is available as full data and as non-redundant data. In the latter, sequences with more than 95% identity are excluded to avoid biased results due to multiple entries with the same or nearly the same sequence. This is of particular importance for the PDB data set, as some proteins were resolved multiple times with different ligands or at different resolutions, and there are some PDB entries that feature an oligomeric structure consisting of several identical protein chains, all of which leads to multiple copies of the same protein sequence in the full data set.

The data set to be used in the analysis is chosen from the pull-down menu on the GlySeq start page. On this page, the user can also set the type of glycosylation sites (Asn, Ser, Thr, Ser/Thr) and the range of residues preceding and following the glycosylation sites. GlySeq determines the frequency of the different amino acids at the positions relative to the glycosylation site within the given range. The analysis can be limited to amino acids with certain properties only, like aromatic, acidic or small residues. A residue set can also be defined by entering the one letter codes of the amino acids. The output options (absolute numbers, relative numbers, deviation from natural abundance) are defined in the same way as in GlyVicinity (*see* **Subheading 3.4**). To make results of different residue sets (e.g., aromatic and acidic amino acids) better comparable, the values are divided by the number of amino acids in the set.

As an example, to investigate the distribution of aromatic residues in the neighborhood of *N*-glycosylation sites in the non-redundant data set of PDB and SwissProt data, that dataset ("*all non-redundant sequences*"), the glycan type "*Asn*" and the residue group "*predefined: aromatic*" are selected from the GlySeq menus. Instead of using the predefined residue group, the user can also select "*user defined*" and type in the one letter codes of aromatic amino acids (FYW). To investigate the possible influence of aromatic amino acids on the occupancy of potential glycosylation sites, "*deviation from standard percentage values*" is chosen as output, and the request is submitted. To do the same for *O*-glycosylation sites, the glycan type is changed to *Ser/Thr* and the request is resubmitted. Graphical results for these requests and for proline are summarized in **Fig. 4**. It becomes obvious that aromatic amino acids are over-represented around occupied

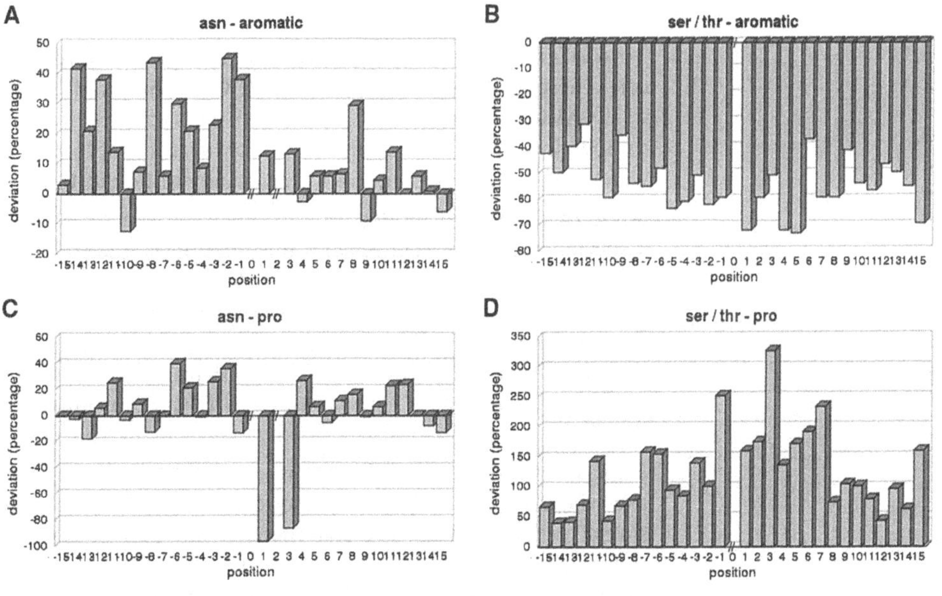

Fig. 4. Occurrence of aromatic residues and proline in the neighborhood of occupied N- and O-glycosylation sites. Around N-glycosylation sites, aromatic amino acids are over-represented (**A**), while proline occurs less frequently than in the average of protein sequences (**C**). With O-glycosylation sites, the opposite effects are observed (**B**, **D**).

N-glycosylation sites, while they are under-represented in the neighborhood of *O*-glycans. This indicates that aromatic residues may favor *N*-glycosylation, while they hamper *O*-glycosylation. Proline shows the opposite effect. This residue is hardly found in the close neighborhood of *N*-glycans, especially at the positions +1 and +3, while it is over-represented around *O*-glycosylation sites. Proline at position +3 does not, in contrast to position +1, exclude *N*-glycosylation, but strongly hampers it (*19, 21*).

Instead of investigating the occurrences of a group of amino acids at the different positions around glycosylation sites, it is also possible to analyze the abundances of all amino acids at one particular position. To check which residues are preferably present at position +1 of occupied *N*-glycosylation sites (the Xxx of the Asn-Xxx-Ser/Thr sequon), the user has to select "*position*" as "*Residue group*" and type in "1".

For all kinds of analyses in GlySeq, it is also possible to set some filters, i.e., to examine only those glycosylation sites where a certain residue is or is not present at a specific position. To see the difference between the amino acids at position +1 of Asn-Xxx-Ser and Asn-Xxx-Thr sequons, for example, the user can set a "*Filter*" to "*pos. 2 must be S*", submit the request, and then resubmit it after changing the filter to "*pos. 2 must be T*".

4. Notes

1. Most of the tools described which provide graphical results, offer these in SVG and GIF format. If GIF is selected, a link to the SVG file is additionally provided. This enables the user to download that file and import it into a vector graphics software, e.g., to generate figures in the quality of publication.
2. Some of the search modes for carbohydrate structures offer a specific species search. Species information is related to all data associated with a carbohydrate entry, not necessarily with the PDB structure only. Therefore, this option is not suited for searching for PDB entries related to a specific species. At the time of writing this can only be done using the search by PDB entry properties (→Databases → PDB, *see* **Subheading 3.1.5**).
3. Most web-interfaces which require a free text input support the wildcards "?" and "*". The question mark matches any single character, while the star matches any run of text. In the exact structure search mode of the GLYCOSCIENCES.de portal (*see* **Subheading 3.1.4**), however, no wildcards are supported as these may define a set of structures rather than an exact structure.

4. For more information on the CarbBank notation see http://www.boc.chem.uu.nl/sugabase/help_nomenclature.html.
5. In those web-interfaces where a value can be directly typed in or alternatively selected from a list of common values listed in a pull-down menu, Java Script has to be activated in the web browser's security settings. If Java Script is disabled, the values chosen from the pull-down menus will not be copied into the text fields below.
6. In some cases, analyzing a PDB entry with pdb2linucs reveals more carbohydrate residues than are displayed in the GLYCOSCIENCES.de entry page for that PDB ID. This can be caused by mismatches between the PDB residue name and the residue actually present in the 3D structure or by non-carbohydrate residues attached to the reducing end of the carbohydrate chain, for which no unique residue name could be assigned by the detection software.
7. If residues are overlapping in the 3D structure file (e.g., if a protein was co-crystallized with both the α- and the β-anomer of a carbohydrate residue), a distance-based determination of atom bonds in the carbohydrate detection process might lead to errors. This also applies to structures in which (due to unusual conformations and/or errors in the coordinates) atoms, which are actually not linked to each other, are located so close to each other that they are detected to be connected.
8. When pdb-care finds a residue name mismatch, the decision about whether this is due to a wrong residue name or resulting from erroneous 3D structure data is up to the user. Especially in the case of non-covalently bound ligands, further knowledge about the structure, i.e., knowledge of the type of residue the protein was co-crystallized with, is necessary. In case of existing PDB entries, this information can often be retrieved from the primary publication of that entry. The citation of this publication can be looked up at the PDB entry page for the structure (go to www.pdb.org and enter the PDB ID into the search field at the top of the page).

Acknowledgment

The authors thank Dr Robin Thomson, Institute for Glycomics, Griffith University, Australia for carefully reading the manuscript, testing the applications and many useful suggestions to improve its readability. The development of GLYCOSCIENCES.de at the German Cancer Research Center was supported by a Research Grant of the German Research Foundation (DFG BIB 46 HDdkz 01-01) within the digital library program.

References

1. Berman, H.M., Westbrook, J., Feng, Z., Gilliland, G., Bhat, T.N., Weissig, H., Shindyalov, I.N., and Bourne, P.E. (2000) The Protein Data Bank. *Nucleic Acids Res.* 28, 235–242
2. Petrescu, A.J., Petrescu, S.M., Dwek, R.A., and Wormald, M.R. (1999) A statistical analysis of N- and O-glycan linkage conformations from crystallographic data. *Glycobiology* 9, 343–352
3. Lütteke,T.,Bohne-Lang,A.,Loss,A.,Goetz,T., Frank, M., and von der Lieth, C.W. (2006) GLYCOSCIENCES.de: an Internet portal to support glycomics and glycobiology research. *Glycobiology* 16, 71R–81R
4. Apweiler, R., Hermjakob, H., and Sharon, N. (1999) On the frequency of protein glycosylation, as deduced from analysis of the SWISS-PROT database. *Biochim. Biophys. Acta* 1473, 4–8
5. Imberty, A. and Pérez, S. (1995) Stereochemistry of the *N*-glycosylation sites in glycoproteins. *Protein Eng.* 8, 699–709
6. Lütteke, T. and von der Lieth, C.W. (2006) The protein data bank (PDB) as a versatile resource for glycobiology and glycomics. *Biocatal Biotransformation* 24, 147–155
7. Boeckmann, B., Bairoch, A., Apweiler, R., Blatter, M.C., Estreicher, A., Gasteiger, E., Martin, M.J., Michoud, K., O'Donovan, C., Phan, I., Pilbout, S., and Schneider, M. (2003) The SWISS-PROT protein knowledgebase and its supplement TrEMBL in 2003. *Nucleic Acids Res.* 31, 365–370
8. Doubet, S., Bock, K., Smith, D., Darvill, A., and Albersheim, P. (1989) The complex carbohydrate structure database. *Trends Biochem. Sci.* 14, 475–477
9. Lütteke, T., Frank, M., and von der Lieth, C.W. (2004) Data mining the protein data bank: automatic detection and assignment of carbohydrate structures. *Carbohydr. Res.* 339, 1015–1020
10. Bohne, A., Lang, E., and von der Lieth, C.W. (1999) SWEET – WWW-based rapid 3D construction of oligo- and polysaccharides. *Bioinformatics* 15, 767–768
11. Bohne-Lang, A., Lang, E., Forster, T., and von der Lieth, C.W. (2001) LINUCS: linear notation for unique description of carbohydrate sequences. *Carbohydr. Res.* 336, 1–11
12. Lütteke, T. and von der Lieth, C.W. (2004) pdb-care (PDB CArbohydrate REsidue check): a program to support annotation of complex carbohydrate structures in PDB files. *BMC Bioinformatics* 5, 69
13. Lütteke, T., Frank, M., and von der Lieth, C.W. (2005) Carbohydrate Structure Suite (CSS): analysis of carbohydrate 3D structures derived from the Protein Data Bank. *Nucleic Acids Res.* 33, D242–D246
14. Ramachandran, G.N., Ramakrishnan, C., and Sasisekharan, V. (1963) Stereochemistry of polypeptide chain configurations. *J. Mol. Biol.* 7, 95–99
15. Hooft, R.W., Sander, C., and Vriend, G. (1997) Objectively judging the quality of a protein structure from a Ramachandran plot. *Comput. Appl. Biosci.* 13, 425–430
16. Wormald, M.R., Petrescu, A.J., Pao, Y.L., Glithero, A., Elliott, T., and Dwek, R.A. (2002) Conformational studies of oligosaccharides and glycopeptides: complementarity of NMR, X-ray crystallography, and molecular modeling. *Chem. Rev.* 102, 371–386
17. Julenius, K., Molgaard, A., Gupta, R., and Brunak, S. (2005) Prediction, conservation analysis and structural characterization of mammalian mucin-type O-glycosylation sites. *Glycobiology* 15, 153–164
18. Marshall, R. (1972) Glycoproteins. *Annu. Rev. Biochem.* 41, 673–702
19. Ben-Dor, S., Esterman, N., Rubin, E., and Sharon, N. (2004) Biases and complex patterns in the residues flanking protein N-glycosylation sites. *Glycobiology* 14, 95–101
20. Petrescu, A.J., Milac, A.L., Petrescu, S.M., Dwek, R.A., and Wormald, M.R. (2004) Statistical analysis of the protein environment of N-glycosylation sites: implications for occupancy, structure and folding. *Glycobiology* 14, 103–114
21. Gavel, Y. and von Heijne, G. (1990) Sequence differences between glycosylated and nonglycosylated Asn-X-Thr/Ser acceptor sites: implications for protein engineering. *Protein Eng.* 3, 433–442

Section II

Carbohydrate-Protein Interactions

Chapter 22

Saccharide Microarrays for High-Throughput Interrogation of Glycan–Protein Binding Interactions

Andrew K. Powell, Zheng-liang Zhi, and Jeremy E. Turnbull

Summary

This chapter describes two methods for fabricating microarrays of saccharides for display and interrogation with binding proteins, using fluorescence detection. The first approach is based on the rapid immobilization of heparan sulphate glycans upon commercially available aminosilane slides via their reducing ends. The second approach is based on the use of a hydrazide-derivatized self-assembled monolayer (SAM) on a gold-coated slide surface. Both provide for efficient and chemoselective attachment and anchoring of oligosaccharide probes via their reducing ends, enabling the large-scale arraying of natural saccharides without cumbersome pre-derivatization. The latter platform, in particular, also has the potential for use with other biophysical readout methods including matrix-assisted laser desorption/ionization time-of-flight mass spectroscopy, surface plasmon resonance, and quartz crystal microbalances. These microarray platforms provide a facile approach for interrogating multiple carbohydrate-protein interactions in a high-throughput manner using minimal quantities of reagents. They provide an essential new experimental strategy in the growing armoury of the glycomics toolkit.

Key words: Microarray, Biochip, Glycan, Specificity, Saccharide, High-throughput, Interaction, Heparan sulphate, Heparin, Self-assembled monolayer.

1. Introduction

Carbohydrate microarrays, which display many structurally diverse oligosaccharide probes on a single substrate, are a very efficient means to map out interactions of carbohydrates and proteins in a high-throughput manner (*1*). A key step in the establishment of carbohydrate microarrays is the development of reliable and reproducible chemistries for the immobilization of chemically and structurally diverse carbohydrate probes onto

Nicolle H. Packer and Niclas G. Karlsson (eds.), *Methods in Molecular Biology, Glycomics: Methods and Protocols, vol. 534*

DOI: 10.1007/978-1-59745-022-5_22

a substrate with retention of their functionality. The previously reported carbohydrate microarrays have mainly used derivatized oligosaccharide probes to link covalently to the appropriately modified surfaces (*2*). However, all these strategies, while apparently functional, require multiple chemical steps of derivatization and clean-up of each individual carbohydrate probe that makes microarray fabrication a laborious and complex procedure. Meanwhile, the potential for structural decomposition or reorganization and low reactivity of certain types of saccharides such as heparin-type saccharides makes the application of pre-derivatizing approaches for microarray generation difficult, and direct immobilization of non-derivatized oligosaccharide probes is obviously preferable for handling large numbers of structures which can be prepared from natural tissue sources.

Our goal has been to develop simple generic microarray platforms for display of non-derivatized oligosaccharides, potentially on a large scale. We describe here two different array approaches. Firstly, the fabrication and interrogation of heparan sulphate saccharide microarrays based upon commercially available aminosilane slides, permitting qualitative and semi-quantitative interaction analysis using fluorescence detection. Secondly we describe an additional strategy based on a hydrazide-derivatized self-assembled monolayer (SAM) on a gold chip surface, and its detection using fluorescence (*3*). The gold surface-based saccharide microarray platforms also have the potential for exploitation using other biophysical detection techniques including matrix-assisted laser desorption/ionization time-of-flight mass spectroscopy, surface plasmon resonance, and quartz crystal microbalances. They establish flexible platforms for direct chemoselective covalent attachment of non-derivatized oligosaccharides to glass-slide surfaces, facilitating carbohydrate microarray construction for interrogation with multiple target proteins using multiple detection strategies for the large-scale study of carbohydrate-protein recognition events. They provide an essential new experimental strategy in the growing armoury of the glycomics toolkit.

2. Materials

2.1. Aminosilane Glass Slide Microarrays

General solvents used for aminosilane slide microarray generation and utilization are tabulated below.

1. Sodium phosphate buffered saline (sPBS) pH 7.2 prepared by dissolving 1 tablet (Oxoid, Basingstoke, UK) in 100 mL deionised water, autoclaved and used for resuspension of lyophilized protein stocks.

2. Potassium phosphate buffered saline (pPBS): 10 mM KH_2PO_4/K_2HPO_4, 137 mM NaCl, pH 7.2 (HiPerSolv reagents: BDH, Poole, UK) dissolved in deionised water.
3. Potassium phosphate buffered saline (pPBS)/ 1 or 2% (w/v) bovine serum albumin (BSA): BSA (Fraction V, Sigma-Aldrich, St Louis, USA) dissolved on ice with gentle mixing in pPBS and filtered using a 0.2 μm Minisart® single-use syringe filter (Sartorius AG, Goettingen, Germany).
4. pPBST/1% BSA or pPBS2T/ 2% BSA solutions: 0.05 or 0.1% (v/v) Tween-20 (Sigma-Aldrich, St Louis, USA) dissolved in filtered pPBS/ 1 or 2% BSA by gentle inversion. pPBST/ 1% BSA is used for high dilution of protein samples, whilst pPBS2T/ 2% BSA is used for 1:1 dilution of protein samples.

2.1.1. Oligosaccharide Probe Preparation

1. 250 mM nitrous acid (HNO_2) solution made from mixing equal volumes of 500 mM sodium nitrite (Sigma-Aldrich, St Louis, USA) and 500 mM HCl (Analar grade; BDH, Poole, UK).
2. 1 M sodium hydrogen carbonate (Analar Grade: BDH, Poole, UK).
3. Heparinase *I* (IBEX Technologies Inc, Montreal, Canada).
4. Porcine mucosal heparin (PMH: Celsus Ltd, Cincinnati, USA).
5. Bovine lung heparin (BLH: Merck Chemicals Ltd, Beeston, UK).
6. ™Superdex 30 prep grade (GE Healthcare, Little Chalfont, UK) 3 × 200 cm column poured in-house using two XK16/100 columns (GE Healthcare, Little Chalfont, UK).
7. Running buffer: 0.5 M ammonium hydrogen carbonate (Sigma Aldrich, Steinheim, Germany).
8. AKTA purifier 10 FPLC system (GE Healthcare, Little Chalfont, UK) (at least 0.5 mL/min flow rate) with UV-detection (232 nm).
9. Fraction collector (1 mL fractions).
10. ™HiPrep 26/10 desalting columns (GE Healthcare, Little Chalfont, UK).
11. Heto Power Dry PL3000 lyophilizer with VLP200 Valu-Pump (Thermo, Basingstoke, UK).
12. Propac PA1 (0.4 × 25 cm) (Dionex UK Ltd, Camberley, UK).
13. HPLC (Shimadzu, Milton Keynes, UK) with UV-detection (232 nm).
14. 2 M sodium chloride pH 3.5 made using HiPerSolv NaCl and HCl reagents: (BDH, Poole, UK).

2.1.2. Microarray Printing

1. Lyphilized oligosaccharides (10–1,000 μM: ~0.025–2.5 mg/mL).
2. Oligosaccharide printing solvent: 1.5 M Betaine (*N,N,N*-trimethylglycine, stored tightly sealed at 4°C) (Sigma, St Louis, USA) prepared in the fume hood by dissolving 176 mg of the solid in 1 mL formamide (toxic).
3. 386 V-well microtitre plate (Genetix, New Milton, UK).
4. Centrifuge with holder for 386 well microtitre plates.
5. Aminosilane non-barcoded glass slides (GAPS II, Corning, NY, USA).
6. Microspot 2500 pins and MicroGrid II compact contact arrayer (Genomic solutions Ltd, Huntingdon, UK).
7. Commercial 800 W microwave oven.
8. PapJar slide holder (ReceptorTech, UK).
9. Air compressor.
10. Silica gel Chameleon® c1-3 mm (VWR, Lutterworth, UK).

2.1.3. Surface Blocking

1. Slides prepared as described in **Subheading 3.1.2**.
2. 22 × 25 mm LifterSlip (Erie Scientific, Portsmouth, USA).
3. Forceps.
4. PapJar slide holders (ReceptorTech, UK).

2.1.4. Protein Binding Profiling

Examples of protein solutions for oligosaccharide interaction interrogation:

1. his/vsv-tagged phage display anti-heparin scFv antibody AO4BO8 (the supernatant from bacteria culture) and mouse anti-vsv antibody supernatant (P5D4) (both provided by Prof. van Kuppevelt, University of Nijmegen, Netherlands) were frozen in aliquots using liquid nitrogen for storage at –20°C. For array interrogation the stock solutions were thawed on ice and diluted fivefold or twofold using pPBST/1% BSA or pPBS2T/2% BSA solutions, respectively.
2. Lyophilised recombinant human fibroblast growth factor 1 (FGF1), expressed in *E. coli* (232-FA: R & D systems, UK) and goat anti-human FGF1 (AF232: R & D Systems, UK) was reconstituted as 18 μg/mL (1 μM) or 100 μg/mL in sterile sPBS, respectively, frozen as aliquots with liquid nitrogen and stored at –80°C. For array interrogation the stock solutions were thawed on ice and diluted to 100 nM and 1 μg/mL, respectively using pPBST/1% BSA.
3. Lyophilised human glial cell line-derived neurotrophic factor (GDNF) precursor, expressed in mouse myeloma cell line NSO (212-GD: R & D systems, UK) and goat anti-human GDNF (AF-212-NA: R & D systems, UK) was reconstituted at 60 μg/

mL (~3 μM) or 100 μg/mL in sterile sPBS, respectively, frozen as aliquots with liquid nitrogen and stored at –80°C. For array interrogation the stock solutions were thawed on ice and diluted to 100 nM and 1 μg/mL, respectively, using pPBST/1% BSA.

4. Fibroblast growth factor receptor 2-IgG1-Fc (FGFR2-Fc) recombinant protein was expressed and purified as described previously (*4*), frozen as aliquots in sPBS with liquid nitrogen and stored at –80°C. Lyophilised goat anti-human IgG Fc-horse radish peroxidase conjugate (ImmunoPure®: cat no 31416, Pierce, Rockford, USA) was resuspended at 0.8 mg/mL in 50% glycerol and stored at –20 C. For array interrogation the stock FGFR-Fc solution was thawed on ice and diluted to 100 nM, whilst goat anti-human Fc was diluted to 0.8 μg/mL both using pPBST/1% BSA.
5. Lyophilised avidin (ImmunoPure®: cat no 21121, Pierce, Rockford, USA) was resusupended at 1 mg/mL (~15 μM) in sterile pPBS and Alexa Fluor® 546 biocytin at 250 μg/mL in sterile deionised water. Both were frozen as aliquots with liquid nitrogen and stored at –20°C. For array interrogation the stock solutions were thawed on ice and diluted to 500 nM and 1 μg/mL, respectively using pPBST/ 1% BSA.
6. Detecting fluorescent anti-mouse antibody solution: Alexa Fluor® 546 goat anti-mouse IgG (H + L) conjugate (Molecular Probes, Eugene, USA) was diluted to 1 mg/mL and 50% glycerol and stored at –20°C in the dark. For array detection of mouse antibodies the 1 mg/mL stock solution was diluted to 1 μg/mL using pPBST/1% BSA solution.
7. Detecting fluorescent anti-goat antibody solution: Alexa Fluor® 546 donkey anti-goat IgG (H + L) (Molecular Probes, Eugene, USA) was stored at 4°C in the dark. For array detection of goat antibodies the 1 mg/mL stock solution was diluted to 1 μg/mL using pPBST/1% BSA solution.

2.1.5. Scanning and Evaluation

1. Slides as prepared in **Subheading 3.1.4**.
2. GenePix 4000A scanner (Molecular Devices, Wokingham, UK) operated with GenePix Pro 3.0 software.
3. Data evaluation software (E.g. Microsoft Excel).

2.2. Gold Surface Glass Slide Microarrays

General solvents used for gold surface glass slide microarray generation and utilization are tabulated.

1. Phosphate buffered saline (pH 7.3) (PBS) (Oxoid, Basingstoke, UK) (1 tablet/100 mL) was dissolved in 0.01% Tween (Sigma-Aldrich, St Louis, USA).
2. Bovine serum albumin (BSA) (Fraction V: Sigma-Aldrich Inc, St Louis, USA) (3%) was dissolved in tenfold diluted PBS buffer with 0.01% Tween.

2.2.1. Synthesis of MHDA and Adipic Dihydrazide Construct

1. Adipic dihydrazide (Aldrich, Milwaukee, USA).
2. 16-mercaptohexadecanoic acid (MHDA) (Aldrich, St Louis, USA).
3. 1-ethyl-3-(3-dimethylaminopropyl) carbodiimide (EDC) (Pierce, Rockford, USA).
4. Dimethyl sulfoxide (DMSO) for coupling reactions.

2.2.2. Synthesis of MHDA and Ethanolamine Construct

1. MHDA (Sigma-Aldrich, St Louis, USA).
2. Ethanolamine (Aldrich, Sigma-Alrich, St Louis, MO).
3. EDC (Pierce, Rockford, USA).
4. DMSO for coupling reactions.

2.2.3. Formation of Mixed Hydrazide Displaying SAM on Gold Surface

1. Gold-coated microarray slides (100 nm Au, 10 nm Cr or Ti) were obtained from Nunc (Denmark), and EMF (NY, USA).
2. Ethanol.
3. Nitrogen for drying.

2.2.4. Microarray Printing

1. Oligosaccharides: Heparin decasaccharides (**Subheading 3.1.1**, isolated decasaccharide after strong anion exchange (*5, 6*)), α1-3, α1-6-D-mannopentaose (Man5), and α1-6-Mannobiose (Man2) (Dextra Laboratories, Reading, UK).
2. Oligoaccharide printing solvent: 1 M Betaine (Sigma-Aldrich, St Louis, USA) was prepared by dissolving 117 mg of the solid with 1 mL deionised water.
3. Slides prepared as **Subheading 3.2.3**.
4. MicroGrid II compact pin-type contact arrayer (Genomic solutions Ltd, Huntingdon, UK).

2.2.5. Surface Blocking

1. 10 mg/mL α-methoxy-ω-formyl poly(ethylene glycol) (MW 750) (PEG-aldehyde) (Rapp Polymere GmbH, Germany) was prepared by dissolving 10 mg of the solid in 1 mL deioinised water.
2. 3% BSA blocking solution in PBS buffer with 0.01% Tween-20.
3. 0.01% Tween-20.
4. Tenfold diluted PBS in 0.01% Tween-20.
5. Nitrogen for drying.

2.2.6. Protein Binding Profiling

1. Alexa Fluor® 546 goat anti-mouse IgG (H + L) conjugate 2 mg/mL (Molecular Probes, Eugene, USA) was diluted 100-fold using 3% BSA solution.
2. Alexa Fluor® 546 donkey anti-goat IgG (H + L) 2 mg/mL (Molecular Probes, Eugene, USA) was diluted 1,000-fold using 3% BSA solution.

3. Lyophilised recombinant human fibroblast growth factor 1 (FGF1) and goat anti-human FGF1 were obtained from R & D Systems, UK, and prepared as described in **Subheading 2.1.4**.
4. Glial cell line-derived neurotrophic factor (GDNF) (60 μg/mL) and its corresponding antibody (0.1 mg/mL) (R & D Systems, UK) were diluted 30- and 100-fold using 3% BSA solution, respectively.
5. Fibroblast growth factor receptor 2-IgG1-Fc (FGFR2-Fc) recombinant protein was expressed and purified as described previously (*4*) and anti-IgG Fc (0.1 mg/mL) (Pierce, USA) were diluted 10- and 100-fold, respectively, using 3% BSA solution.
6. Sonic hedgehog (5 μg/mL) and anti-sonic hedgehog (2 μg/mL) were obtained from Abcam (UK), the protein solutions were prepared from lyophilized solids using 3% BSA solution.
7. Concanavalin A (Con A; 0.1 mg/mL) and anti-Con A antibody (0.1 mg/mL), snowdrop *Galanthus Nivalis* lectin (1 mg/mL), and anti-*Galanthus Nivalis* (0.1 mg/mL) (Vector, Burlingame, USA) were diluted 50-fold using 3% BSA solution.

2.2.7. Protein Binding Profiling

1. Genepix 4000A laser microarray scanner (Molecular Devices, Wokingham, UK).

3. Methods

3.1. Aminosilane Glass Slide Microarrays

Arrays can be prepared on commercial aminosilane glass slide surfaces, offering a facile saccharide array method for interrogating binding of proteins on a large scale. A schematic outlining the fluorescence-based approach described here is given in **Fig. 1**.

3.1.1. Oligosaccharide Probe Preparation

1. Heparin saccharides were generated by partial nitrous acid or partial heparitinase I digestion of porcine mucosal heparin (PMH) or bovine lung heparin (BLH) according to established procedures (*1*). PMH and BLH were partially digested using heparinase I (1 mU/mg heparin) in 100 mM sodium acetate, 0.1 mM calcium acetate (pH 7), at 37°C for different time intervals before heating at 100°C for 2 min to terminate the reactions. For partial nitrous acid digestion, 100 mg/mL BLH or PMH solutions were diluted with an equal volume of 250 mM nitrous acid solution. The reaction was kept on ice and aliquots removed at specific time points and neutralized by the addition of 1 M sodium hydrogen carbonate.

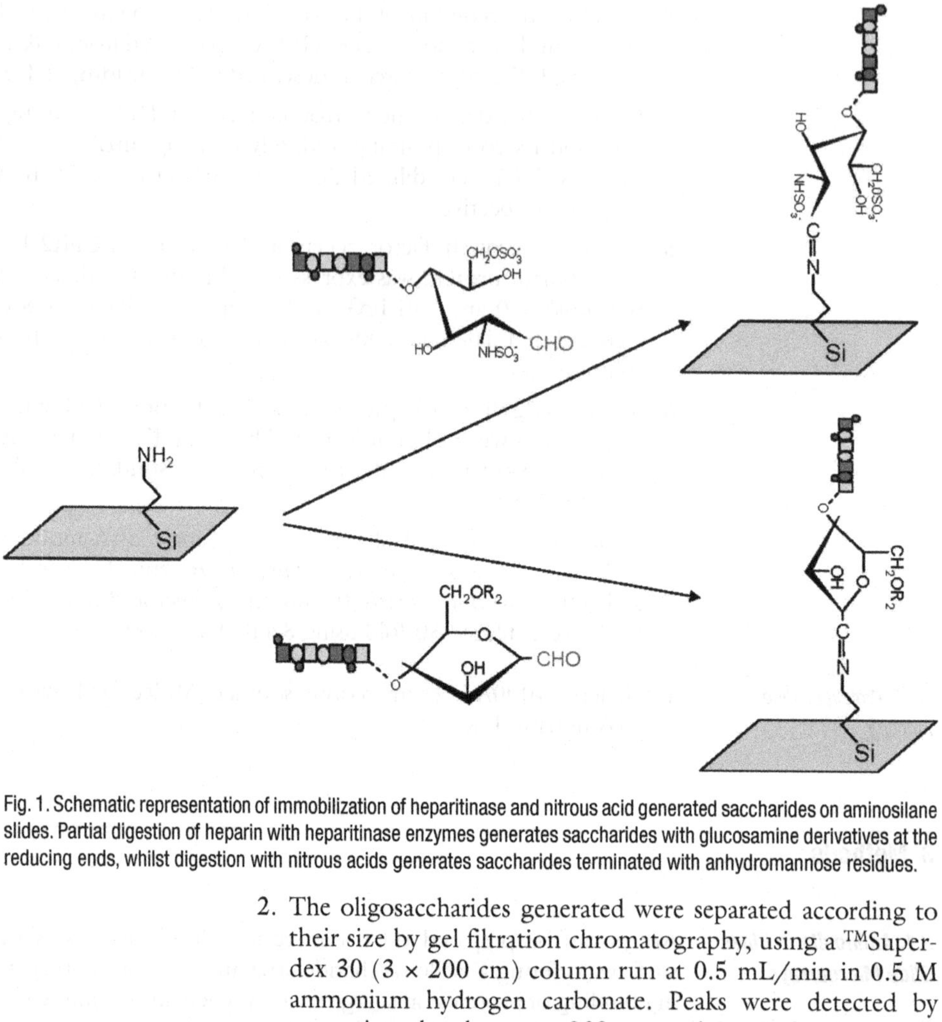

Fig. 1. Schematic representation of immobilization of heparitinase and nitrous acid generated saccharides on aminosilane slides. Partial digestion of heparin with heparitinase enzymes generates saccharides with glucosamine derivatives at the reducing ends, whilst digestion with nitrous acids generates saccharides terminated with anhydromannose residues.

2. The oligosaccharides generated were separated according to their size by gel filtration chromatography, using a ™Superdex 30 (3 × 200 cm) column run at 0.5 mL/min in 0.5 M ammonium hydrogen carbonate. Peaks were detected by measuring absorbance at 232 nm and 1 mL fractions were collected.
3. Fractions containing size-defined oligosaccharides were pooled and desalted using ™HiPrep 26/10 desalting columns, monitoring elution using absorbance at 232 nm, before lyophilization. Note that saccharides prepared by different methods probably have differential reducing-end reactivities (*see* **Note 1**).
4. Size-defined fractions were further separated by strong anion exchange (SAX) chromatography using a Propac PA1 column (0.4 × 25 cm) and HPLC. Saccharides were eluted with a 0-1 M linear gradient of sodium chloride over 90 min. After each chromatography step, elution fractions were pooled and desalted using ™HiPrep 26/10 desalting columns, eluting with water before lyophilization. For each chromatography step, elution was monitored at 232 nm.

3.1.2. Microarray Printing

1. Wearing gloves at all times, oligosaccharide printing solutions were prepared by lyophilizing 15 μL of saccharide at the desired concentration and resuspending in freshly prepared 15 μL oligosaccharide printing solvent (formamide supplemented with 1.5 M betaine) immediately before each sample was transferred to the correct well of a 386 V-well microtitre plate.
2. The plate was centrifuged at 120g for 30 s before immediately spotting saccharides onto freshly opened aminosilane glass slides. The following printing variables were used: 0.4 mm pitch (distance between adjacent spot centers), humidifier controlled humidity of 57–70%, 18°C temperature, 60 spots per source visit, seven wash cycles with a 1 s fill time, 5.5 s drain time, and 8 s delay between cycles. It is important to consider potential spotting variation in the experimental design (*see* **Note 2**).
3. A template showing the area of the slide where the arrays were printed was drawn. Printed slides were placed in a commercially available 800 W microwave oven and subjected to microwave heating for 5 min before removing from the microwave into a plastic slide holder to cool for 10 min in the dark (*5*). This procedure was repeated a total of three times.
4. Oligosaccharides that were not bound after spotting and heating were removed by washing the slides three times for 1 min with distilled water in a sealed PapJar slide holder and dried for storage using an air compressor. Immobilized oligosaccharides were stable to storage for at least several months when stored at room temperature in a sealed plastic bag containing silica gel.

3.1.3. Surface Blocking

1. Immediately before protein interrogation, slides were rinsed three times for 1 min with de-ionised water (pH ~7).
2. 10–15 μL of pPBS/ 1% BSA blocking solution was pipetted under a 22 × 25 mm LifterSlip placed carefully over the array area on the slides using forceps.
3. The blocking solution was incubated with the slide in a slide incubation box, humidified using wet paper towels, for 30 min at room temperature.
4. Slips were gently lifted from the slide were soaked by immersion in a beaker of deionised water (pH ~7) before being washed with deionised water (pH ~7) with a gentle lateral rotation in sealed PapJar slide holders for 1 min five times before rapidly drying slides using an air compressor.
5. A second blocking step was performed, as described above, using the pPBST/ 1% BSA blocking solution (*see* **Note 3**).

3.1.4. Protein Binding Profiling (See Note 4)

1. Heparin binding proteins and their corresponding detection antibodies or reagents (*see* **Note 4**) were diluted to a final concentration in pPBST/1% BSA or pPBS2 T/ 2% BSA and

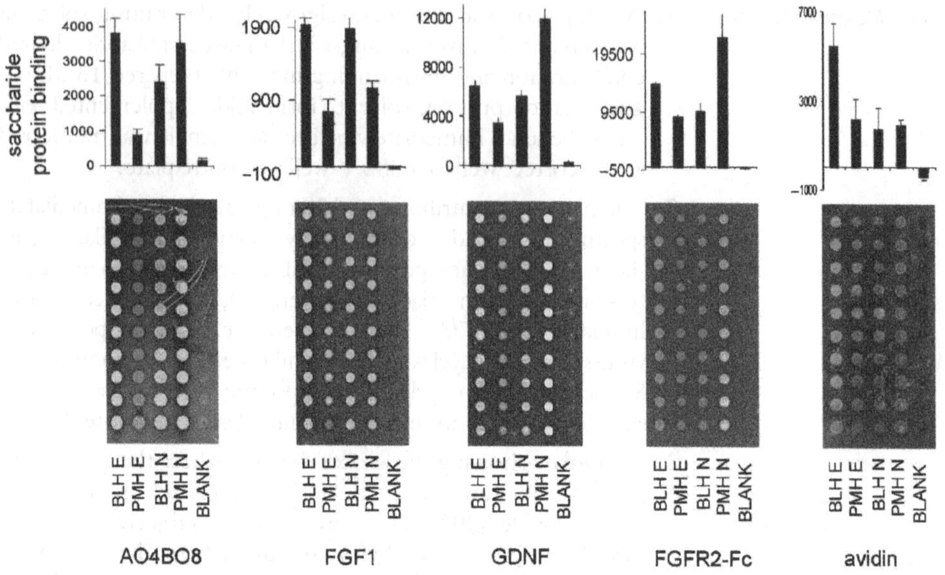

Fig. 2. Interrogation of enzyme and nitrous acid generated heparin saccharide aminosilane microarrays with diverse proteins. Binding of diluted A04B08 scFv phage display antibody supernatant, 100 nM FGF1, GDNF and FGFR2-Fc or 500 nM avidin to bovine lung heparin (BLH) and porcine mucosal heparin (PMH) 8 mer saccharides generated by *Heparitinase I* cleavage (E) and nitrous acid cleavage (N). Protein binding was detected using antibodies to the heparan sulfate binding proteins and Alexa Fluor® 546 secondary antibody or ligand conjugates and slide scanning at 532 nm. For quantification of protein binding to the saccharides the mean of the difference between the median spot and background fluorescence intensity at 532 nm for ten spots was calculated using Genepix Pro 3.0 software and Excel. Error bars represent the standard deviation of the mean of protein binding to the replicate spots.

incubated under LifterSlips placed over the array area of the slide. It is important to consider potential protein spreading variation in the experimental design (*see* **Note 5**).

2. Arrays were incubated with reagents at room temperature for 30 min for each step, except for the heparin binding proteins, which were left 1 h.
3. After each protein incubation, slides were washed and dried for the blocking step. Final incubation was with detecting florescent antibody solutions containing Alexa Fluor® 546 goat anti-mouse IgG (H + L) or Alexa Fluor® 546 donkey anti-goat IgG (H + L), depending on the secondary antibody. For examples of protein binding interrogation using this method *see* **Fig. 2**.

3.1.5. Scanning and Evaluation

1. The fluorescence signals of the microarrays were read at 10 μm resolution and 532 nm using a GenePix 4000A scanner operated with GenePix Pro 3.0 software. The photomultiplier tube intensity was adjusted to the maximum permitted without saturation of any signals.
2. Fluorescent intensity analysis with GenePix Pro 3.0 software used features fitted by an eye to spots so that the diameter

matched that of the spot as closely as possible. The difference between the median feature and background intensities was calculated as the saccharide binding intensity to allow for any variability in protein spreading across the array (hence, background binding) affecting the comparative intensity in binding to individual saccharides in the array. The median feature and background intensities were used instead of the mean to circumvent the effect of 'rogue' pixels on the mean intensity calculation.

3. The mean and standard deviation of the spot diameter and saccharide binding intensity, as well as coefficient of variances for repeat sample spots, were calculated from the GenePix Pro 3.0 software fluorescence intensity data using Excel.
4. Spot diameters were commonly in the range 170–230 μm.

3.2. Gold Surface Glass Slide Microarrays

Arrays can be prepared on gold surfaces using SAMs and hydrazide chemistry. This approach offers a controlled surface which is engineered for low non-specific protein binding and can be exploited using a variety of read-out techniques. A schematic outlining the fluorescence-based approach described here is given in **Fig. 3**.

3.2.1. Synthesis of MHDA and Adipic Dihydrazide Construct (See *Note 6*)

1. MHDA (2.0 mg) and adipic dihydrazide (35 mg) were mixed in 10 mL DMSO, and sonicated to completely dissolve the mixture, sonicated .
2. EDC (30 mg) was then added and the reaction mixture was stirred at ambient temperature for 6–8 h. The reaction product was used directly without further purification.

3.2.2. Synthesis of MHDA and Ethanolamine Construct (See *Note 7*)

1. MHDA (2.0 mg) and ethanolamine (10 μL) were mixed in 10 mL DMSO and sonicated to completely dissolve the mixture.

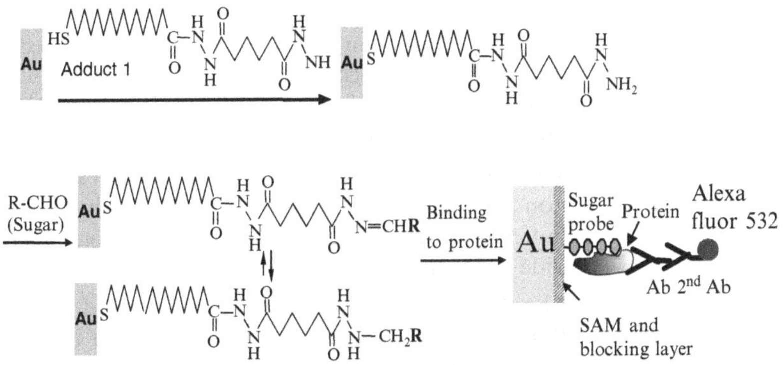

Fig. 3. Schematic representation of the surface modification process and fabrication of an oligosaccharide microarray. The approach is based on direct attachment of unmodified oligosaccharides onto hydrazide-derivatized, SAM covered gold surfaces. The binding of the proteins to the microarray was detected using specific antibodies and visualized by fluorescence-labelled secondary antibodies.

2. EDC (30 mg) was then added and the reaction mixture was stirred at ambient temperature for 6–8 h. The reaction product was used directly without further purification.

3.2.3. Formation of Mixed Hydrazide Displaying SAM on Gold Surface (See Note 8)

1. The gold-coated glass slides were coated with a mixed SAM of construct of MHDA and adipic dihydrazide and construct of MHDA and ethanolamine, by soaking the slides for 16–24 h in a mixture of construct according to the pre-set ratio (e.g. 90:10; *see* **Note 7**).
2. The SAM covered slides were washed with ethanol and sonicated for 5 min in ethanol, and dried with a nitrogen stream.

3.2.4. Microarray Printing

1. Oligosaccharide printing solutions were prepared in 1 M betaine, and spotted on hydrazide-derivatized gold-coated glass slides using a MicroGrid II compact pin-type contact arrayer in 65% relative humidity.
2. 1 M betaine was added to the samples to prevent water evaporation from the droplets.
3. The oligosaccharides were typically arrayed as a tenfold dilution serious with starting pickup solution concentration of 1 mg/mL for ten replicate spots (with approximately 1 nL per spot delivered by contact of the pins with the surface). The distance between the centers of adjacent spots was 400 μm.
4. The printed slides were incubated overnight at 18°C in a closed environment (a plastic dish sealed with a parafilm).
5. Oligosaccharides that were not bound after spotting and incubation were removed by washing the slides twice with distilled water. Immobilized oligosaccharides were stable to storage for at least several weeks when stored at 4°C in a sealed container.

3.2.5. Surface Blocking (See Note 9)

1. The printed slides were treated in 0.5 mL of PEG-aldehyde (10 mg/mL) in water and incubated for 1 h at room temperature.
2. The slides were then rinsed with water and treated with 0.5 mL of 3% BSA blocking solution in PBS buffer with 0.01% Tween-20 for 30 min, and washed again with 0.01% Tween-20 solution for 2 min.
3. Thereafter, the binding proteins and the corresponding detection antibodies (each of 100 μL, diluted with 3% BSA in PBS buffer in 0.01% Tween-20) were added sequentially on the slide and incubated for 40 min for each step.
4. The slide was rinsed three times with tenfold diluted PBS in 0.01% Tween-20 solution for a total period of 10 min between each binding step and after the final step.
5. After the last step of the fluorescence-labelled antibody binding, the slide was rinsed with water and dried at room temperature under a stream of nitrogen.

3.2.6. Protein Binding Profiling

Sulphated oligosaccharide microarrays can be probed with heparin-binding proteins (e.g. FGF1, GDNF, FGFR2-Fc, Sonic hedgehog) whereas binding to microarrays produced with mannoses can be tested using lectins (e.g. ConA and snowdrop *Galanthus Nivalis*); for examples *see* **Fig. 4**.

1. The microarray-bound specific proteins were probed with appropriate cognate antibodies, followed by a fluorescence (Alexa Fluor® 546)-labelled secondary antibody. For each binding step, an incubating step of 40 min was used (*see* **Note 10**).
2. HybriWells (Grace Biolabs, USA) were used to cover the slides during the incubation in order to prevent evaporation.
3. The slides were rinsed with water and dried at room temperature under a stream of gaseous nitrogen It is worth noting that chip regeneration is not recommended (*see* **Note 11**).

3.2.7. Scanning and Evaluation

1. The fluorescence signals of the microarrays were read using a Genepix 4000A laser microarray scanner with PMT voltage set at 800 V and laser power at 100%.
2. Signals at 532 nm were quantified using the GenePix Pro 3.0 image analysis software package.

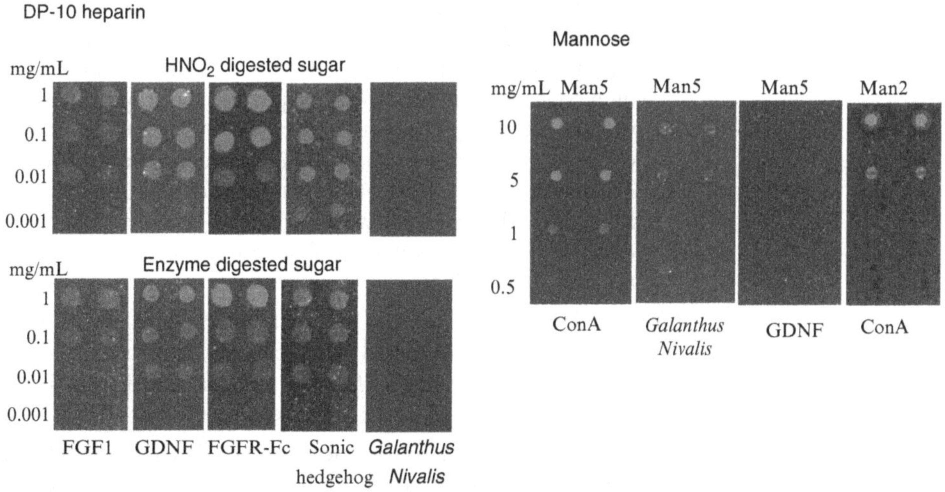

Fig. 4. Binding of carbohydrate-binding proteins to immobilized sugars in microarray format. Interaction of heparin-binding proteins (FGF1, GDNF, FGFR-Fc, and Sonic hedgehog) and lectins (ConA and *Galanthus Nivalis*) to heparin decasaccharides and mannose microarrays fabricated on gold. Detection was achieved using primary antibodies specific for each protein, followed by an Alexa Fluor® 546-labelled secondary antibody. Rows of duplicate spots for each dilution are shown. The concentrations of sugar probes in the pickup solutions are shown at the left side of the panels. The left upper panel shows the microarray of heparin oligosaccharide (10 mer) produced by partial HNO_2 digestion. The left lower panel shows the microarray of heparin oligosaccharide produced by partial heparitinase / enzymatic digestion. *Galanthus Nivalis* lectin was used as a negative control. The right panel shows the binding of lectins (5 µg/mL ConA and 100 µg/mL *Galanthus Nivalis*) to Man5 and Man2 microarrays. A non-lectin (GDNF) was used as the negative control. Note the smaller size of the spots of Man5 and Man2 compared to that of negatively charged heparin saccharides; this is likely due to the lower surface tensions of the spotted solutions containing the non-charged Man5 or Man2.

3. Mean signal intensities of pixels selected from the centre of triplicate spots were used for data analysis.

4. Notes

1. Differential reducing end reactivities: Heparinases have differential substrate specificities (*7*), hence are likely to generate saccharides with different reducing end-terminating glucosamine derivatives if used to produce saccharide libraries from tissue-derived HS. The identity of the reducing end residue probably effects the reactivity of the reducing end aldehyde to amines and hence possibly the level of immobilization of saccharides. As a result the signal obtained for a protein binding to particular saccharide may reflect differential amounts of saccharide immobilization on the aminosilane surface as well as a differential ability of proteins to bind saccharide fractions, hence this should be considered when interpreting results.
2. Spotting variation: To minimize the effect of systematic signal variation stemming from pin-pin spot volume deposition, variation and potential sample carry-over on results interpretation, dilution series of samples were printed from low to high concentration with the same pin and spotting the solvent without the sample was printed before and after the sample concentration series to confirm negligible sample carry over. To minimize the effect of decreasing spot size immediately following a filling of pins with the sample (known as the 'rush effect'), slides possessing the first ten spots following a pin refill were discarded.
3. Addition of 0.1% Tween to the second blocking and subsequent protein solutions is required for facile spreading of solution across the BSA blocked array surface but not for spreading of initial blocking solution across the unblocked aminosilane surface.
4. Both enzyme and nitrous acid derived heparin saccharides were shown to bind as arrays to heparin binding proteins such as growth factors, anti-HS antibodies, receptors and miscellaneous HSBPs detected with sequential antibodies and directly using fluorescently labelled ligands. Where applicable, proteins expressed in eukaryotic cell lines should be used as the *N*-glycosylation of HSBPs that is likely to affect binding to HS saccharides in terms of affinity (*8*) and possibly specificity.
5. Protein spreading variation: Early experiments indicated that the uniform application of sequential protein solutions to the array area of the aminosilane glass slide is important. Firstly,

with poor application, gaps in the BSA blocking solution result in intense fluorescent signal which hinders signal from saccharide-protein interactions from being accurately quantified. We observed that the facile spreading of low volumes of protein solution using various chambers or coverslips was more difficult following the incubation of the array area with the blocking solution. Empirical studies of different methods of sample incubation with the arrays suggested that the use of LifterSlips was the most robust method for applying aqueous protein solutions to aminosilane slides. Quantitative analysis of spot signal with position in the array indicated signal variation probably due to a gradient of protein spreading from the corner of the LifterSlip where the sample is applied. Comparison of spot intensity should therefore only be performed for spots within the same subarray (i.e. printed in close vicinity using the same pin).

6. The microarray chip surface modification was prepared by synthesizing an MHDA-adipic dihydrazide construct (mass 444) in the solution phase (in DMSO) using carbodiimide coupling chemistry, and then present it onto the gold surface forming a SAM, thus creating a single-step procedure to make the hydrazide surface (*see* **Fig. 3**). A higher yield for the synthesis of the construct was achieved by allowing the reaction to occur in dimethyl sulfoxide (DMSO) which overcomes the hydrolysis problems of carbodiimide-formed esters.
7. It was found initially that this method produced lower fluorescent signals for the sugar-antibody binding-based assay. However, experiments on the optimization of the hydrazide density on the chip surfaces by adding an MHDA-ethanolamine construct (10–20%) as a 'dilutor' showed that improved detection sensitivity could be obtained.
8. The hydrazide functional group displayed on the gold surface could efficiently and chemoselectively react with unmodified carbohydrate in aqueous solutions, forming covalent bonds through the reducing end aldehyde moiety. The sugar-hydrazide linkage chemistry included a ring opening process at the reducing end residue forming firstly a hydrazone, followed by conversion of the structure into predominantly the desired native beta-pyranose form, leaving the oligosaccharide structure unchanged (*9*). Thus, the sugar linkage can be made without risk of decomposition of the saccharide structure, and the sugars can be displayed on the surface in a directed orientation analogous to cell surface presentation.
9. A combination blocking scheme involving the use of PEG aldehyde and supplementing BSA in binding protein solutions was found to be the best for minimizing the non-specific background and noise signals on gold surfaces. Note that other commercially available blocking reagents, such as casein and

SuperBlock (Pierce, USA), were found to strongly inhibit the binding of some heparin-binding proteins to the immobilized sugars, and are thus not suitable as blocking agents. Another blocking agent, TopBlock (Sigma, St Louis, USA) was found to be not efficient enough to reduce the background.

10. It should be noted that the assay shown in **Fig. 3** had three protein binding layers (i.e. the target protein and two subsequent antibodies) added consecutively, with the last layer antibody being fluorescence-labelled. Additional assays were performed to compare whether the number of layers was also important for high-sensitivity detection. As expected, experiments using only one layer of a fluorescence-dye tagged ConA showed very weak fluorescence signal, even at very high concentrations. In contrast a three layer protein detection approach demonstrated the most effective detection (data not shown). It is thus very probable that both an efficient insulating SAM layer and a multilayer protein structure (preferably three layers) are essential to provide efficient fluorescence detection on the SAM modified gold surfaces. The use of an antibody-based assay can avoid the problems associated with direct labelling of target proteins that may compromise binding or create artefactual protein characteristics.
11. Proteins bound to the gold chip surfaces, once dried, could no longer be removed by the use of typical chip regeneration protocols, e.g., 0.1 M glycerin-HCl, pH 2.0. This fact, combined with the ease of generation of multiple slide-based microarrays indicates that re-use of chips is neither recommended nor required.

Acknowledgements

This research was supported by the UK 'Glycoarrays' Consortium funded by the Research Councils UK Basic Technology Programme (to JET); a Biotechnology and Biological Sciences Research Council project grant (to AKP and JET) and a grant to the Liverpool Centre for Bioarray Innovation (Prof. Andy Cossins), and a Medical Research Council Senior Research Fellowship (to JET). The Authors would like to thank Yassir Ahmed, and Lorraine Wallace for technical assistance in the production of saccharides and PAGE analysis, Abdelmadjid Atrih for mass spectrometry analysis of saccharides and Margaret Hughes for instruction and advice in operation of the microarray robot. We also thank Rob Field, Jonathan Blackburn, Marianne Brille, Tim Rudd, Ed Yates and Dave Fernig for useful discussions.

References

1. Shin I, Park S, Lee MR. (2005). Carbohydrate microarrays: an advanced technology for functional studies of glycans. *Chemistry* 11, 2894–2901
2. Ratner DM, Adams EW, Disney MD, Seeberger PH (2004). Tools for glycomics: mapping interactions of carbohydrates in biological systems. *ChemBiochem* 5, 1375–1383
3. Zhi ZL, Powell AK, Turnbull JE (2006). Fabrication of carbohydrate microarray on gold surface: direct attachment of non-derivatized oligosaccharides to hydrazide-derivatized self-assembled monolayer. *Anal. Chem.* 78, 4786–4793
4. Powell AK, Fernig DG, Turnbull JE (2002). Fibroblast growth factor receptors 1 and 2 interact differently with heparin/heparan sulfate – implications for dynamic assembly of a ternary signaling complex. *J. Biol. Chem.* 277, 28554–28563
5. Yates EA, Jones MO, Clarke C, Powell AK, Johnson SR, Porch A, Edwards P, Turnbull JE (2004). Microwave enhanced reaction of carbohydrates with amino-derivatised labels and glass surfaces. *J. Mater. Chem.* 13, 2061–2063
6. Skidmore MA, Turnbull JE (2005). Separation & sequencing of heparin and HS saccharides. In: *Chemistry and Biology of Heparin and Heparan Sulphate*, Garg H (ed.). Elsevier Press, New York, pp. 181–203
7. Linhardt RJ, Turnbull JE, Wang H, Loganathan D, Gallagher JT (1990). Examination of the substrate specificity of heparin and heparan sulphate lyases. *Biochemistry* 29, 2611–2617
8. Duchesne L, Tissot B, Rudd TR, Dell A, Fernig DG (2006). N-glycosylation of fibroblast growth factor receptor 1 regulates ligand and heparan sulfate co-receptor binding. *J Biol Chem* 281, 27178–27189
9. Flinn NS, Quibell M, Monk TP, Ramjee MK, Urch CJ (2006). A single-step method for the production of sugar hydrazides: Intermediates for the chemoselective preparation of glycoconjugates. *Bioconjug. Chem.* 16, 722–728, 2005

Chapter 23

Glycosaminoglycan Characterization Methodologies: Probing Biomolecular Interactions

Vikas Prabhakar, Ishan Capila, and Ram Sasisekharan

Summary

Interactions between glycans and proteins are central to many of the regulatory processes within biology. The development of analytical methodologies that enable structural characterization of glycosaminoglycan oligosaccharides has fostered improved understanding of the specificity of these biomolecular interactions. This facilitates an appreciation in understanding how changes in GAG structure can regulate physiology as well as pathology. While there are various techniques for studying the interaction of GAGs with proteins, in this chapter we focus on two approaches. First, an integrated analytical methodology, surface non-covalent affinity mass spectrometry (SNA–MS), is described to isolate, enrich, and sequence tissue-derived GAGs that bind to specific proteins. The broad applicability of this powerful platform offers an insight into how changes in cell-surface and extracellular GAG composition and sequence influences the ability of cells and tissues to dynamically alter responses to signaling molecules. Thus, this approach provides a window into understanding how changes at a molecular level manifest with respect to cellular phenotype. Second, surface plasmon resonance, or SPR, represents an additional platform for the study of protein–polysaccharide interaction, specifically for measuring the binding between GAG chains and proteins.

Key words: Glycosaminoglycans, Capillary electrophoresis, Mass spectrometry, Surface plasmon resonance, Heparinase, Chondroitinase, Heparin, Heparan sulfate, Chondroitin, Dermatan.

1. Introduction

Glycosaminoglycans (GAGs) on the cell surface or in the extracellular matrix are important mediators of cross-talk between different cells and tissues (*1–5*). They play key roles in multiple activities including cell proliferation, differentiation, and migration. GAGs exert their effects primarily by binding to extracellular "signals" (e.g., growth factors, chemokines) and highly specific cell surface receptors (*6–8*). By facilitating or inhibiting the interaction between

Nicolle H. Packer and Niclas G. Karlsson (eds.), *Methods in Molecular Biology, Glycomics: Methods and Protocols, vol. 534*

DOI: 10.1007/978-1-59745-022-5_23

these signals and their receptors, GAGs can modulate different physiological as well as pathophysiological processes. The interaction between GAG and protein likely involves specific sequences on the GAG chain that bind the protein with high affinity. The most widely studied example of such an interaction is the binding of antithrombin III and heparin, which leads to accelerated inhibition of the coagulation cascade. Studying the interaction of GAGs and proteins, in terms of both specificity and affinity of binding, is an important pursuit that enables a better understanding of the in vivo mechanisms and the underlying biology of these interactions (*3,8–13*).

Various techniques are employed to study the interaction between GAGs and proteins. These include affinity chromatography, isothermal titration calorimetry (ITC), nuclear magnetic resonance (NMR), fluorescence spectroscopy, gel electrophoresis, and more recently, surface plasmon resonance (SPR). These analyses enable an evaluation of GAG-protein interactions in either a qualitative (high/low affinity) or quantitative (kinetics and thermodynamics of binding) manner. Beyond this primary level of physical characterization of the binding interaction are additional aspects of the structural specificity of the interaction. The complexity and heterogeneity inherent in GAG structure suggests the potential for structural signatures along a given chain that specifically bind a particular protein. Therefore, correlating structural analysis along with the interaction of glycosaminoglycans and proteins enables a better understanding of the in vivo roles of these biomolecules (*14–19*). Here, we highlight two techniques that have been used in our laboratory to study specificity and affinity of GAG–protein interactions: (1) Surface Non-covalent affinity–Mass Spectrometry (SNA–MS) (*20*) involves the integration of different analytical methodologies into one procedure that allows direct isolation and identification of GAGs that interact with a protein, and (2) Surface Plasmon Resonance (SPR) (*21*) allows the evaluation of kinetics of GAG–protein interactions.

The overall strategy for SNA–MS involves protein immobilization on a thin hydrophobic film or on a metal surface, and subsequent addition of an aqueous mixture of oligosaccharides that is allowed to bind to the protein (**Fig. 1**). Following a low salt wash to remove the non-specific binders, a synthetic basic peptide, $(\text{Arg-Gly})_{15}\text{Arg}$ ($[\text{RG}]_{15}\text{R}$), is added in the matrix solution to chelate specific binders. The oligosaccharides are detected as $(\text{RG})_{15}\text{R}$ peptide–oligosaccharide complexes using MALDI-MS. Using enzymatic or chemical methods, these oligosaccharides can also be subjected to depolymerization before chelation with $(\text{RG})_{15}\text{R}$, thereby enabling their direct sequencing. SPR is a powerful technique for studying molecular interactions (*22, 23*). With this platform, one of the interacting species is immobilized on the surface of a chip and its binding partner is made to flow over this chip. The resulting interaction changes the refractive index of the chip, which is measured as a change in the intensity and angle of light reflected from the chip's surface.

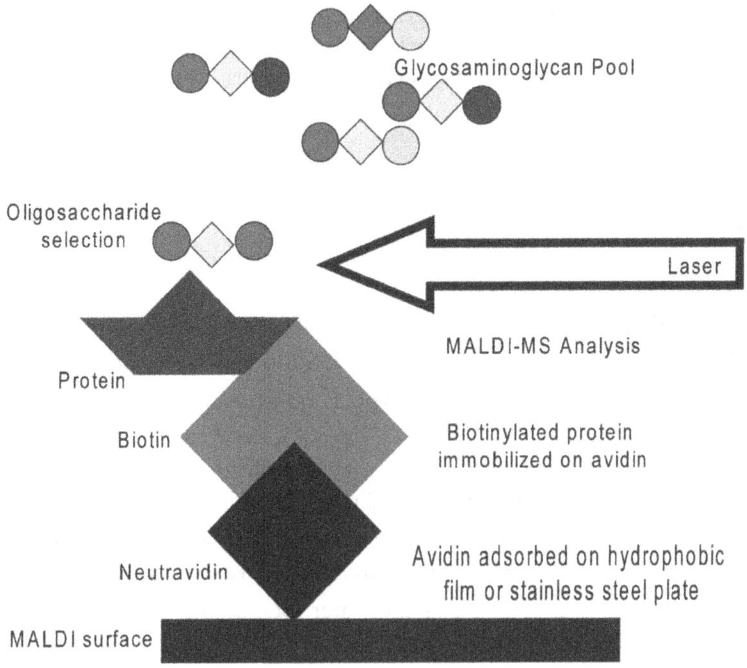

Fig. 1. Overview of SNA–MALDI–MS. SNA–MS schematic for protein immobilization, oligosaccharide binding and selection, and MS analysis.

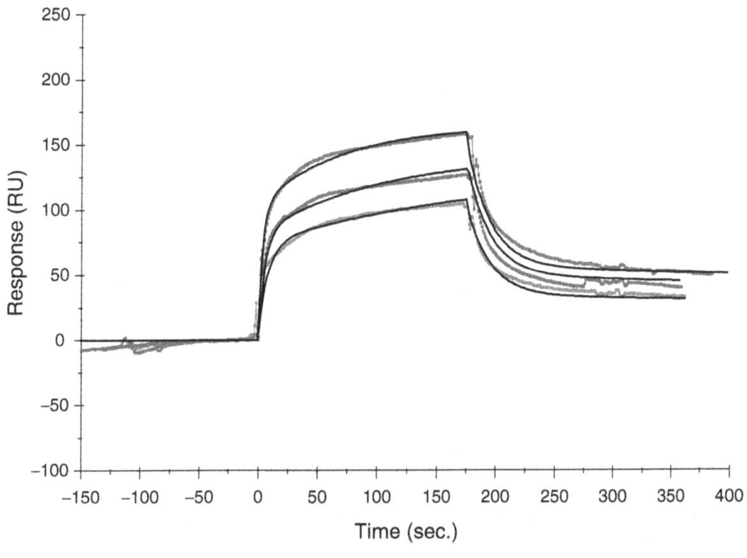

Fig. 2. Surface plasmon resonance. Sensorgrams for the interaction of antithrombin III with biotinylated heparin, immobilized on a streptavidin chip. The three different curves correspond to different concentrations of AT-III flowing over the chip surface.

The magnitude of this change is directly proportional to the amount of the binding partner being bound, affording a real-time measurement of association and dissociation rates from which the dissociation constant can be calculated (**Fig. 2**). Signals are easily obtained from sub-microgram quantities of material. While this technique can provide important information on binding kinetics, it suffers from potential experimental artifacts that arise from ligand immobilization (*24*).

2. Materials

2.1. Protein Preparation and Immobilization for SNA–MS

1. Protein of interest (e.g., FGF-2 or AT-III)
2. EZ-Link® Sulfo-NHS Biotin (Pierce, Rockford, IL).
3. Canon NP Type E transparency film, taped to a MALDI sample plate.
4. Spin columns with a molecular weight cutoff (MWCO)= 10,000 Da (Millipore, Billerica, MA).
5. Porcine intestinal mucosal heparin (Celsus Labs, Cincinnati, OH), Neutravidin (Pierce, Rockford, IL).
6. 1 M NaCl and water for washing of heparin from MALDI-spots.

2.2. Saccharide Binding, Selection and Analysis for SNA–MS

1. Bovine aortic smooth muscle cells.
2. T75 Flasks (VWR, West Chester, PA).
3. Phosphate buffer saline (PBS – 10 mM sodium phosphate, 150 mM NaCl pH = 7.4).
4. Heparinase I or III (recombinantly produced in our laboratory).
5. DNase and RNase (Roche, Mannheim, Germany).
6. DEAE filter (Millipore).
7. Elution buffer: 10 mM sodium phosphate 1 M NaCl pH 6.0.
8. Spin columns with a molecular weight cutoff (MWCO) = 3,000 MWCO membrane (Millipore).
9. Selection buffers: NaCl solution ranging from 0.2 to 1 M.
10. Matrix solution: Caffeic acid matrix in 30% acetonitrile/water (saturated solution) with 2 pmol/μL $(RG)_{15}R$ prepared as described in the chapter "The Structural Elucidation of Glycosaminoglycans."

2.3. SNA–MS Saccharide Characterization Using Digestion by Heparinase I or III After Binding Selection

1. Samples spotted as described in **Subheading 3.2**.
2. Heparinases I or III in water.
3. Water.
4. Matrix solution: Caffeic acid matrix in 30% acetonitrile/water (saturated solution) with 2 pmol/μL $(RG)_{15}R$ prepared as in the chapter "The Structural Elucidation of Glycosaminoglycans."

2.4. Immobilization of Glycosaminoglycans on a Sensor Chip for Surface Plasmon Resonance

1. Biotinylated heparin (Celsus Labs, Cincinnati, OH).
2. SA sensor chip (Streptavidin sensor chip) from Biacore (Piscataway, NJ).
3. Pre-treatment solution: 50 mM NaOH in 1 M NaCl.
4. HEPES running buffer (HBS): 10 mM HEPES, 150 mM NaCl, 3.4 mM EDTA, pH 7.4 containing 0.005% (v/v) P-20 surfactant.
5. 2 M NaCl.
6. Biacore® 3000 system from Biacore.

2.5. Kinetic Measurement of Glycosaminoglycan–Protein Interactions Using SPR

1. SA sensor chip (Streptavidin sensor chip) with immobilized biotinylated heparin as described in **Subheading 3.4**.
2. Protein of interest (0.25–1.5 μM of protein in buffer compatible with the protein).
3. Regenerating buffer: 50 mM HEPES, pH 7.4 containing 0.1 mM DTT and 0.5 M NaCl.

2.6. Immobilization of Protein on a Sensor Chip and Kinetic Measurement of Glycosaminoglycan–Protein Interactions

1. CM5 sensor chip (carboxy-methylated dextran sensor chip) from Biacore.
2. *N*-hydroxysuccinimide(NHS)and *N*-ethyl- *N*-[(dimethylamino) propyl] carbodiimide (EDC) or "Amine-coupling kit" from Biacore.
3. 10 mM sodium acetate buffers at pH ranging from 4 to 6.
4. GAG binding protein of interest: 100 μg/mL in acetate buffer.
5. Blocking solution: 1 M ethanolamine.
6. Regeneration buffer: 2 M NaCl.

3. Methods

3.1. Protein Preparation and Immobilization for SNA–MS

1. The protein of interest (e.g., FGF-2 or AT-III) is incubated overnight with porcine intestinal mucosal heparin to occupy/block the heparin binding sites.
2. Protein is then biotinylated with EZ-Link® sulfo-NHS biotin according to the instructions provided by the supplier (Pierce).
3. Excess biotin is removed by spin column.
4. Canon NP Type E transparency film is taped to the MALDI sample plate and used as a protein immobilization surface.
5. Protein is immobilized by first drying 4 μg neutravidin in water (1 μL) on the film surface, then adding 30 pmols biotinylated protein in water (1 μL) to the neutravidin spot.

6. Heparin is removed by washing ten times with 1 M NaCl and ten times with water.

3.2. Saccharide Binding, Selection and Analysis for SNA–MS

1. Bovine aortic smooth muscle cells are grown to confluence in T75 flasks.
2. Cells are washed twice with PBS and then 200 nM heparinase I or III is added for 1 h. The supernatant is heated to 50°C for 10 min to inactivate heparinase, and the sample is filtered.
3. To remove polynucleotide contamination the samples are treated with excess DNase and RNase (Roche) at room temperature overnight.
4. Partly degraded heparan sulfate oligosaccharides are isolated from the sample by binding to a DEAE filter (Millipore), washing away unbound material, and eluting with 10 mM sodium phosphate 1 M NaCl pH 6.0.
5. The material is then concentrated using a 3,000 MWCO membrane (Millipore) and buffer exchanged into water. The retentate containing the oligosaccharides is lyophilized and reconstituted in water.
6. Saccharides are bound to immobilized proteins by spotting 1 μL aqueous solution of 5–20 pmol/μL on the protein spot for at least 5 min.
7. Unbound saccharides are removed by washing with water fifteen times. For selection experiments, the spot is washed ten times with NaCl concentrations ranging from 0.2 to 1 M, followed by ten water washes.
8. Caffeic acid matrix in 30% acetonitrile (saturated solution) with 2 pmol/μL $(Arg\text{-}Gly)_{15}Arg$ is added to the spot before MALDI analysis.
9. All saccharides are detected as non-covalent complexes with $(Arg\text{-}Gly)_{15}Arg$ using MALDI parameters as described in the chapter "The Structural Elucidation of Glycosaminoglycans."

3.3. SNA–MS Saccharide Characterization Using Digestion by Heparinase I or III after Binding Selection

1. Saccharides selected for protein binding are digested with heparinases I or III by spotting 8 ng of enzyme in water (1 μL) after selection is completed (*see* **Note 1**).
2. Matrix with 2 pmol/μL $(RG)_{15}R$ is added to the spot for MALDI analysis. This enables direct sequence analysis of the selected binding saccharides.
3. Mass spectrometric (MALDI) analysis is performed as described in the chapter "The Structural Elucidation of Glycosaminoglycans."

3.4. Immobilization of Glycosaminoglycans on a Sensor Chip for Surface Plasmon Resonance

1. The procedure described below assumes the use of a Biacore® 3000 system from Biacore (Piscataway, NJ).
2. A streptavidin sensor surface (SA sensor chip) is pre-treated with three 5 μL injections of 50 mM NaOH in 1 M NaCl, to remove any non-specifically bound contaminants.
3. A 5 μL injection of reducing-end biotinylated heparin (10 μg/μL) in HBS, followed by a 10 μL injection of 2 M NaCl.
4. One other flow cell of the sensor chip is similarly treated, however, without the heparin injection, only to serve as a control.

3.5. Kinetic Measurement of Glycosaminoglycan–Protein Interactions Using SPR

1. Typically, a 15 μL injection of protein (at a concentration range of 0.25–1.5 μM in buffer compatible with the protein) is made at a flow rate of 5 μL/min.
2. At the end of the sample plug, the same buffer (without added protein) is flowed past sensor surface to allow dissociation (*see* **Note 2**).
3. After a suitable dissociation phase, the sensor surface is regenerated for the next protein sample using a 10 μL pulse of regenerating buffer (50 mM HEPES, pH 7.4 containing 0.1 mM DTT and 0.5 M NaCl) (*see* **Note 3**).
4. The same solution is flowed past all 4 flow cells on the chip in sequence and the response is monitored as a function of time (sensorgram) at 25°C.
5. The control cell is used to subtract the contribution of non-specific interaction with the dextran matrix itself (*see* **Note 4** for exception).
6. Kinetic parameters are evaluated using the BIA Evaluation software according to the manufacturer's methods (*see* **Note 5**).

In **Fig. 2**, a sensorgram of antithrombin III with biotinylated heparin immobilized on a streptavidin chip is shown.

3.6. Immobilization of Protein on a Sensor Chip and Kinetic Measurement of Glycosaminoglycan–Protein Interactions

In cases where it is not possible to measure the GAG–protein interaction with the glycosaminoglycan immobilized on the chip (due to non-specific interaction of the protein with the streptavidin chip), different chip immobilization strategies can be explored that allow immobilization of the protein on the chip surface.

1. A protein biochip can be prepared by covalent attachment of the protein of interest to the sensor surface through its primary amino groups. The chip used for this case is a carboxymethylated dextran surface chip (CM5 sensor chip).
2. The chip is first activated with an injection pulse (7 min, 35 μL) of an equimolar mix of *N*-hydroxysuccinimide (NHS) and *N*-ethyl-*N*-[(dimethylamino) propyl]carbodiimide (final concentration of 0.05 M; *see* **Note 6**).

3. The protein solution, 80 μL, is then injected manually (100 μg/mL in citrate buffer at pH ranging from 4 to 6). Typically four different pH values (4.5, 5, 5.5, and 6) are examined and the pH that provides best immobilization is chosen for subsequent experiments with this protein.
4. Excess unreacted sites on the sensor surface are blocked with a 35 μL injection of 1 M ethanolamine, and the surface is cleaned with the desired buffer (pH 7.4).
5. A 15 μL injection of the GAG (different concentrations from 50 to 750 nM are evaluated initially) in desired buffer (pH 7.4) is made at a flow rate of 5 μL/min.
6. At the end of the sample plug, the same buffer is passed over the sensor surface to facilitate dissociation. After a suitable dissociation time, the sensor surface is regenerated for the next sample with a 10 μL pulse of 2 M NaCl.
7. The response is monitored as a function of time (sensorgram) at 25°C. Kinetic parameters are evaluated with BIA Evaluation software (*see* **Note 5**).

4. Notes

1. The spot is kept wet for the duration of the desired digestion time by adding water as necessary.
2. All buffers should be degassed prior to their use in SPR analysis to minimize any background interference while collecting data.
3. In some cases, due to re-binding during the dissociation phase, a clear dissociation curve is not observed. To observe dissociation kinetics and to minimize rebinding effects the addition of minor amounts of heparin in the dissociation buffer has been utilized (*25*).
4. In many cases it is likely that the protein of interest may exhibit non-specific binding to the streptavidin surface. If this is observed then the chip selection needs to be modified. A CM5 chip can be used and a procedure for preparing and immobilizing a heparin–albumin conjugate on the chip surface has previously been described (*21*). Alternately, the protein of interest can be immobilized on the chip surface and heparin can be flowed over the surface (**Subheading 3.2.3**).
5. Various models can be used to fit the GAG–protein binding interaction. Since curve fitting is a mathematical process it is important to choose a model that accurately reflects a possible biological interaction scenario for GAGs. Since GAGs present multiple binding sites, "heterogeneous ligand"

binding models often work well. In cases where the GAG is flowed over an immobilized protein surface, "bivalent analyte" models may work best. The simple 1:1 Langmuir binding models should also be evaluated.

6. The NHS–EDC equimolar solution should be prepared fresh and mixed just before injection to obtain effective activation of the surface and thereby good immobilization.

Acknowledgements

The authors thank Dr. Karthik Viswanathan for helpful discussions. Funding was provided by the National Institutes of Health Grant GM57073. Vikas Prabhakar was the recipient of the National Institutes of Health Biotechnology Training Grant (5-T32-GM08334).

References

1 Esko, J. D. (1999) Glycosaminoglycan binding proteins. In: Essentials of glycobiology (Varki, A.,). Cold Spring Harbor Laboratory Press, Cold Spring Harbor, New York, pp. 145–159

2 Sanderson, R. D., Yang, Y., Suva, L. J. and Kelly, T.. (2004) Heparan sulfate proteoglycans and heparanase – partners in osteolytic tumor growth and metastasis. *Matrix Biol* 23, 341–352

3 Liu, D., Shriver, Z., Venkataraman, G., El Shabrawi, Y. and Sasisekharan, R.. (2002) Tumor cell surface heparan sulfate as cryptic promoters or inhibitors of tumor growth and metastasis. *Proc Natl Acad Sci U S A* 99, 568–573

4 Liu, J., Shriver, Z., Pope, R. M., Thorp, S. C., Duncan, M. B., Copeland, R. J., Raska, C. S., Yoshida, K., Eisenberg, R. J., Cohen, G., Linhardt, R. J. and Sasisekharan, R.. (2002) Characterization of a heparan sulfate octasaccharide that binds to herpes simplex virus type 1 glycoprotein D. *J Biol Chem* 277, 33456–33467

5 Bernfield, M., Gotte, M., Park, P. W., Reizes, O., Fitzgerald, M. L., Lincecum, J. and Zako, M.. (1999) Functions of cell surface heparan sulfate proteoglycans. *Annu Rev Biochem* 68, 729–777

6 Trowbridge, J. M. and Gallo, R. L. (2002) Dermatan sulfate: new functions from an old glycosaminoglycan. *Glycobiology* 12, 117R–125R

7 Sugahara, K., Mikami, T., Uyama, T., Mizuguchi, S., Nomura, K. and Kitagawa, H.. (2003) Recent advances in the structural biology of chondroitin sulfate and dermatan sulfate. *Curr Opin Struct Biol* 13, 612–620

8 Shriver, Z., Sundaram, M., Venkataraman, G., Fareed, J., Linhardt, R., Biemann, K. and Sasisekharan, R. (2000) Cleavage of the antithrombin III binding site in heparin by heparinases and its implication in the generation of low molecular weight heparin. *Proc Natl Acad Sci U S A* 97, 10365–10370

9 Shriver, Z., Raman, R., Venkataraman, G., Drummond, K., Turnbull, J., Toida, T., Linhardt, R., Biemann, K. and Sasisekharan, R.. (2000) Sequencing of 3-O sulfate containing heparin decasaccharides with a partial antithrombin III binding site. *Proc Natl Acad Sci U S A* 97, 10359–10364

10 Trowbridge, J. M., Rudisill, J. A., Ron, D. and Gallo, R. L.. (2002) Dermatan sulfate binds and potentiates activity of keratinocyte growth factor (FGF-7). *J Biol Chem* 277, 42815–42820

11 Tovar, A. M., de Mattos, D. A., Stelling, M. P., Sarcinelli-Luz, B. S., Nazareth, R. A. and Mourao, P. A. (2005) Dermatan sulfate is the predominant antithrombotic glycosaminoglycan in vessel walls: implications for a possible physiological function of heparin cofactor II. *Biochim Biophys Acta* 1740, 45–53

12 Powell, E. M., Fawcett, J. W. and Geller, H. M. (1997) Proteoglycans provide neurite guidance at an astrocyte boundary. *Mol Cell Neurosci* 10, 27–42

13 Popp, S., Andersen, J. S., Maurel, P. and Margolis, R. U. (2003) Localization of aggrecan and versican in the developing rat central nervous system. *Dev Dyn* 227, 143–149

14 Desaire, H. and Leary, J.A. (2000) Detection and quantification of the sulfated disaccharides in chondroitin sulfate by electrospray tandem mass spectrometry. *J Am Soc Mass Spectrom* 11, 916–920

15 Kuberan, B., Lech, M., Zhang, L., Wu, Z. L., Beeler, D. L., and Rosenberg, R. D. (2002) Analysis of heparan sulfate oligosaccharides with ion pair-reverse phase capillary high performance liquid chromatography-microelectrospray ionization time-of-flight mass spectrometry. *J Am Chem Soc* 124, 8707–8718

16 Perez, S. and Mulloy, B. (2005) Prospects for glycoinformatics. *Curr Opin Struct Biol* 15, 517–524

17 Sturiale, L., Naggi, A. and Torri, G. (2001) MALDI mass spectrometry as a tool for characterizing glycosaminoglycan oligosaccharides and their interaction with proteins. *Semin Thromb Hemost* 27, 465–472

18 Zaia, J., McClellan, J. E. and Costello, C. E.. (2001) Tandem mass spectrometric determination of the 4S/6S sulfation sequence in chondroitin sulfate oligosaccharides. *Anal Chem* 73, 6030–6039

19 Zaia, J. and Costello, C. E.. (2001) Compositional analysis of glycosaminoglycans by electrospray mass spectrometry. *Anal Chem* 73, 233–239

20 Keiser, N., Venkataraman, G., Shriver, Z. and Sasisekharan, R. (2001) Direct isolation and sequencing of specific protein-binding glycosaminoglycans. *Nat Med* 7, 123–128

21 Zhang, F., Fath, M., Marks, R. and Linhardt, R. J.. (2002) A highly stable covalent conjugated heparin biochip for heparin-protein interaction studies. *Anal Biochem* 304, 271–273

22 Pattnaik, P. (2005) Surface plasmon resonance: applications in understanding receptor–ligand interaction. *Appl Biochem Biotechnol* 126, 79–92

23 Rich, R.L. and Myszka, D.G. (2000) Advances in surface plasmon resonance biosensor analysis. *Curr Opin Biotechnol* 11, 54–61

24 Osmond, R. I., Kett, W. C., Skett, S. E. and Coombe, D. R. (2002) Protein–heparin interactions measured by BIAcore 2000 are affected by the method of heparin immobilization. *Anal Biochem* 310, 199–207

25 Delehedde, M., Lyon, M., Gallagher, J. T., Rudland, P. S. and Fernig, D. G. (2002) Fibroblast growth factor-2 binds to small heparin-derived oligosaccharides and stimulates a sustained phosphorylation of p42/44 mitogen-activated protein kinase and proliferation of rat mammary fibroblasts. *Biochem J* 366, 235–244

Chapter 24

Development and Characterization of Antibodies to Carbohydrate Antigens

Jamie Heimburg-Molinaro and Kate Rittenhouse-Olson

Summary

Antibodies to carbohydrate antigens are critical for the study of bacteria, tumors, blood groups, and cell-cell adhesion interactions; for the analysis of viral, hormone, and toxin receptors; and, finally, for analysis of the glycosylation of recombinant proteins. However, antibodies to carbohydrate structures are more difficult to develop because of the T-cell-independent response to carbohydrates. This can result in the production of low affinity and difficult to work with IgM antibodies to these molecules. Screening technologies that include IgM antibodies can cause selections of antibodies with low-affinity binding sites because of the net avidity enhancement. Unfortunately, the low-affinity binding site can also have a similar affinity for unwanted structures. Production of antibodies using cellular extracts can result in antibodies that react with multiple related structures, and therefore the resultant bioassays have sensitivity or specificity problems. Protein conjugates of saccharides for the production of polyclonal and monoclonal antibodies to carbohydrate structures can be used to solve these problems. For monoclonal antibody development to oligosaccharides, mapping with closely related saccharides allows the determination of the areas of the saccharide to which the antibody binds so that conclusions can be made concerning which saccharide structures will cross-react. Determination of the reactivity of the produced antibodies with related saccharide structures is essential prior to utilization.

Key words: Polyclonal antibody, Monoclonal antibody, Carbohydrate antigens, Protein carrier, Enzyme immunoassay, Epitope mapping, Specificity analysis.

1. Introduction

Antibody development to carbohydrate antigens is important because carbohydrate antigens play a role in many biological processes. They are bacterial antigens, but in addition they are blood group antigens and tumor-associated antigens. Toxin,

Nicolle H. Packer and Niclas G. Karlsson (eds.), *Methods in Molecular Biology, Glycomics: Methods and Protocols, vol. 534*

DOI: 10.1007/978-1-59745-022-5_24

hormone, and viral receptors are often carbohydrates, and carbohydrates act as adhesion molecules involved in cellular interactions and homing (*1, 2*). If appropriate immunizations are performed, a powerful and unique antibody response to the carbohydrate antigen can be created. Antibodies of a desired specificity can then be chosen for many uses, including the detection of the antigen on cell surfaces, analysis of the glycosylation of recombinant proteins, and the stability of vaccines. Important antibodies toward tumor antigens, bacterial antigens, viral antigens, and others can be produced that are useful for the recognition of these antigens as well as for therapy. Monoclonal antibodies are also critically important for many techniques, including immunohistochemistry, western blots, affinity columns, and immunotherapy. Antibodies to carbohydrates can have exquisite specificity and sensitivity. These antibodies can require up to six sugars for their optimal reactivity but can be developed with as few as one sugar (*1, 2*). Important in the binding of antibodies to the saccharide structure can be the presence of each sugar in the oligosaccharide (i.e., both monosaccharides in the disaccharide Galβ1-3GalNAc), the identity of each sugar (i.e., Gal rather than GalNAc or Glc) and the lack of substitution of the sugar (i.e., Galβ1-3GalNAc rather than NANA-Galβ1-3GalNAc) (*3, 4*). An immunized animal will produce antibodies of different specificities, and these are represented to varying extents as polyclonal antibodies in the serum of the immunized animal. It is interesting and important to note that the immune response to saccharide structures is often oligoclonal rather than polyclonal, and therefore the simpler production of an antiserum rather than a monoclonal antibody may suffice (*5, 6*). When long-term use of antibody is anticipated and the defined specificity is important, production of monoclonal antibodies is preferred. Specificity of the produced antibody depends on the quality of the immunizing antigen and the presence of related structures in the vaccinated animal. The presence, for instance, of the A blood group in a human would prevent the production of antibody to this carbohydrate structure in that individual, but allows the production of a very specific oligoclonal response to the B blood group, and these antigens differ only by the presence of a terminal galactose rather than an *N*-acetyl galactosamine (*1*).

The immune response to saccharides is T-cell independent and predominantly results in the production of the IgM antibody, with little memory and little affinity maturation of the antibody (*1, 2*). In order to form a better T-cell-dependent response, saccharides are chemically coupled to carrier proteins and this conjugate is used to create the response (*5–8*). When the conjugate is made, if the reducing sugar of the antigen is important, care must be taken so that the reducing sugar remains in its cyclic hemiacetyl form, without opening it to its free carbonyl form.

Antibody affinity relies on fit, and so changing the shape of any part of the antigen will ultimately affect the response. Methods that open the reducing sugar include reductive amination with cyanoborohydride, which should be avoided in cases where the reducing sugar is critical. The reducing sugar is less important if the structure is longer than seven sugars, or if the structure contains repeats of the required epitope. Methods that do not affect the reducing sugar include use of a diazotization reaction of an aminophenyl derivative of the saccharide, the use of the hetero bifunctional linker *m*-maleimidobenzoyl *N*-hydroxysuccinamine ester, or preparation of a hetero bifunctional linker, 1,2 *O*-cyanoethylidene derivative (*7–10*). When a linker region is used, controls for binding a different, unrelated saccharide attached to the protein via the same linker should also be performed. Attempts using glycosylated amino acids as building blocks in peptide synthesis may in the future yield the most natural saccharide conjugates for vaccination (*8*). The carrier proteins predominately used for human vaccination are derivatives of diphtheria or tetanus toxoid. These are chosen because most humans would have received previous vaccinations with these antigens, and so a larger booster response is seen (*8*). Animals used for antibody production will generally not have been previously vaccinated with these toxoids, so this higher booster effect is not seen. These toxoid carrier proteins are also relatively expensive. In our laboratory, we commonly use bovine serum albumin (BSA) as the carrier protein. We chose BSA because it is well established as a successful carrier and it is inexpensive to use in all subsequent assays where controls will be needed.

Specificity of the produced antibody is of critical importance, but the importance will vary depending on your requirements. For example, if studying exposure of Galβ1-3GalNAc (called TF-Ag) in tumor cells, an antibody that does not react with the sialylated TF-Ag seen in some normal tissues would be desired. The ultimate use of the antibody must be planned for when preparing to vaccinate and when setting up the screening and inhibition analysis assays.

The protocol for the development of monoclonal-antibody-producing cells relies on the fusion of the cell making the antibody of choice (the splenic B cell from the mice chosen) with a cell that has the property of immortality (a myeloma cell). The progeny cell, the hybridoma, has the desired antibody-producing characteristics and can live indefinitely. The cells are fused using polyethylene glycol (PEG), which allows the loosening of the lipids in the cell membrane lipid bilayer, and when two cells touch under these conditions, they can fuse and become one. Successful fusion is a matter of the use of the appropriate two cell populations, their ratio, the amount of time in the PEG, and how gently the cells are treated during this time when there is increased fluidity of their membranes. After the fusion is performed, it is important to select the progeny hybridoma cells, which secrete the antibody of the desired specificity. The myeloma cells that are used in this protocol are nonsecretory, indicating that they do not secrete immunoglobulin of their own. In addition, these cells do not have a functional hypoxanthine-guanine phosphoribosyl transferase gene, nor do they have a functional thymine kinase gene. Therefore they cannot synthesize DNA and replicate if the medium contains aminopterin, a compound that blocks both salvage pathways for DNA production. The spleens cells have no such defect, so hybridoma cells that contain a spleen cell and a myeloma cell grow in the medium containing aminopterin. The spleen cells, by themselves, can only replicate a limited number of times, so these cells will not be selected for tissue culture growth. Aliquots of the medium from the growing hybridoma cells are taken when the medium has just become yellow, and then these aliquots are tested for the desired antibody through the enzyme immunoassays. Selected cells are expanded, subcloned, and expanded further (*11–13*).

Enzyme-linked immunosorbent assays (EIAs) are a very valuable and widely utilized technique for measuring specific antibodies to antigens. Many variations of this technique are used, but the most common and the easiest to implement are indirect EIAs. The indirect method allows easy screening of antisera and hybridomas without the need for labeled primary antibodies or very pure antigen (*11–14*). The assay is sensitive, reproducible, and relatively quick, and does not require radioactive materials. In screening for antibodies to carbohydrate antigens, our laboratory employs use of a secondary antibody that detects only IgG not IgM. IgM in either a polyclonal or monoclonal antibody has a lower specificity. Although IgM may be useful if IgG cannot be developed, IgG is certainly preferred.

Sometimes it is important to determine the isotype of the antibody produced, in order to determine whether the antibody would be cytotoxic or complement binding, or whether it would pass through the placenta. The possible subclasses of monoclonal antibodies produced are IgG_1, IgG_{2a}, IgG_{2b}, or IgG_3. Commercial isotyping kits are now available, which allow rapid and easy determination. One example is the Roche Applied Science IsoStrip kit (Roche Diagnostics, Indianapolis IN), which allows fast, easy, reliable, and specific isotype determination.

Monoclonal antibody production to bacterial saccharides can allow the mapping of protective epitopes, enable vaccine stability studies, allow epidemiologic analysis as to which strains are causing the infections, and may be useful for passive transfer of antibody to reduce morbidity and mortality of the infection. Antibody development to carbohydrate tumor-associated antigens can create tools for diagnosis, prognosis, and possibly immunotherapy. Antibodies to toxin, hormone, and viral receptors may help to develop therapeutic regimes for associated illnesses. Antibodies to adhesion molecules may help in the development of anti-metastatic therapies, as well as in immunomodulatory modalities. With increased interest in the use of recombinant proteins, and the renewed understanding of the importance of the glycosylation of these proteins, monoclonal antibodies may help in the determination of the extent of glycosylation of the produced glycoproteins.

2. Materials

2.1. Antigen Preparation/ Conjugation

1. Bovine serum albumin (BSA, 95% or higher purity).
2. *O*-Nitrophenyl derivative of the desired saccharide.
3. Platinum oxide.
4. Hydrogen gas.
5. HCl (0.1 M).
6. NaOH (0.5 M).
7. Sodium nitrate (0.05 M).
8. Sodium chloride (0.15 M).

2.2. Immunizations

1. Antigen of interest.
2. Adjuvant of choice (Complete Freund's Adjuvant (CFA)*, and Incomplete Freund's Adjuvant (IFA)). (*use protective equipment – potent inflammatory agent).
3. Animals (2–10 Balb/c mice for monoclonal, two rabbits for polyclonal).
4. Glass syringes (1–5 mL) with Luer-lock tips, sterile (minimum 2), Luer-lock connector, 20G or 22G needles.
5. Tuberculin plastic syringes (1 mL), 20G needle.
6. *Blood collection by retro-orbital bleeds*: Capillary tubes, Eppendorf tubes (serum separator tubes optional), anesthesia machine, gauze pads.

2.3. Enzyme Immunoassay (EIA), Indirect Using Alkaline Phosphatase Detection

1. The antigen conjugate of interest.
2. BSA.
3. Unrelated saccharide antigen similarly coupled to BSA.
4. Coating buffer (0.1 M Na_2CO_3 pH 9.6).

5. Phosphate buffered saline (PBS) dilution buffer (1.9 mM NaH_2PO_4, 8.1 mM Na_2HPO_4, 0.15 M NaCl, pH 7.2).
6. Wash buffer (PBS/0.1%Tween/0.01%NaN_3, pH 7.2).
7. Antisera diluent (5% BSA in wash buffer).
8. Conjugate buffer (1%BSA in wash buffer).
9. Substrate buffer (0.05 M Na_2CO_3 + 0.001 M $MgCl_2$, pH 9.8).
10. Flat-bottom 96-well microtiter plates (Immulon or Nunc (Rochester, NY)).
11. Antibody test samples (serum, supernatant).
12. Anti-mouse or rabbit IgG alkaline phosphatase conjugate, heavy chain specific.
13. *p*-Nitrophenyl phosphate substrate tablets.
14. Microtiter plate washer or plastic squirt bottles, microtiter plate reader capable of reading at 405 nm, multichannel pipettes and tips if available.

2.4. Fusion

1. All media and plasticware for the fusion and subsequent cell culture must be sterile; all procedures after the removal of the spleen should be performed in a tissue culture hood.
2. Myeloma cell line (X63Ag8.653, ATCC, Manassas, VA).
3. Complete medium (DMEM supplemented with 10% fetal bovine serum, L-glutamine, sodium pyruvate, nonessential amino acids, and vitamins).
4. Spleen cell counting solution (1% acetic acid, 150 mM NaCl).
5. DMEM (not supplemented).
6. PEG-DMSO (dimethylsulfoxide) Hybri-Max (50% w/v PEG, 10% w/v DMSO) (Sigma).
7. Hybridoma cloning factor (BioVeris, Gaithersberg, MD).
8. HAT medium (complete medium supplemented with 1× HAT and 1% hybridoma cloning factor).
9. HT media (complete medium supplemented with 1× HT and 0.5% hybridoma cloning factor).
10. Subcloning medium (complete medium + 1% hybridoma cloning factor); primed mice (see above).
11. Sterile flasks (175 cm^2).
12. Sterile microscope slides with frosted ends.
13. Conical polypropylene tubes (50 ml); 96-well microtiter plates, sterile.
14. Anesthesia for euthanizing mice; surgical tools for removal of spleens.

15. Hemacytometer.
16. Trypan blue.
17. Hybridoma screening – see EIA above.

2.5. Antibody Characterization

1. IsoStrip isotyping kit (Roche Diagnostics).
2. BSA.
3. Unrelated saccharide–BSA conjugate.
4. Related sugars that have substitutions at various hydroxyls.
5. Related sugars that have modifications from immunizing sugar: that is, a different linkage or a different sugar (*see* **Table 1** and **Fig. 1** for an example).
6. Biological substances known to contain the saccharide.

Table 1
Rabbit Ab to Galβ1-3GalNAc-BSA (T antigen) (data for Table 1 and Fig. 1 contained in ref. 3)

Inhibition	
Galβ1-3(GlcNAcβ1-6)GalNAc	The C-6 hydroxyl of the GalNAc is not involved in binding
GlcNAcβ1-6Galβ1-3(GlcNAcβ1-6)GalNAc	The C-6 hydroxyl of the Gal is not involved in binding
Galβ1-3GalNAcα is better than Gal β 1-3GalNAcβ	α Anomer is better binding than β
No inhibition	
Galactose *N*-Acetyl galactosamine	Both saccharides are needed for reaction
GlcNAc β1-3Gal β1-3GalNAc NANAα1-3Gal β1-3GalNAcβ	C-3 Hydroxyl of the Gal is involved in binding
Gal β1-3GlcNAc	C4 Axial hydroxyl of the GalNac is involved in binding

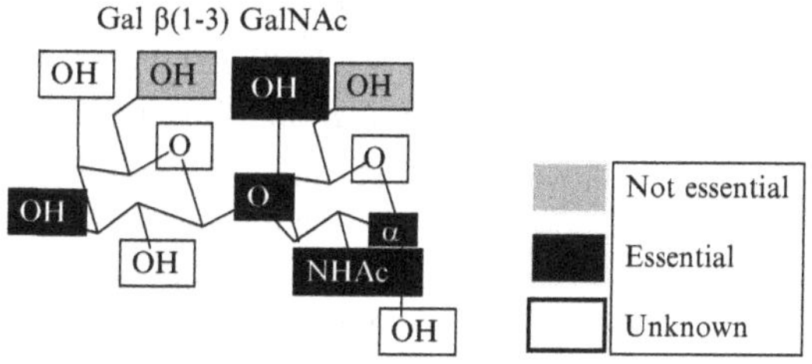

Fig. 1. Shading indicates importance of the hydroxyls for antibody binding, as indicated in the key.

3. Methods

The antigen conjugation method chosen preserves the cyclic hemiacetyl form of the reducing sugar and is derived from the method of McBroom et al. (*7*). This method utilizes a nitrophenyl derivative of the saccharide, which is often commercially available. BSA is chosen as the carrier protein because of its immunogenicity and the ease of separating immune reactions to the saccharide from immune reactions to this carrier. The sugar content of the conjugate is measured by the phenol sulfuric method (*15*). The protein content is measured by the Lowry method (*16*).

The immunization material should be prepared fresh on the day of injection. Generally, the antigen/adjuvant emulsion is prepared in glass syringes and can be transferred to 1-mL plastic syringes for easier and more accurate injection with sterile needles. Proper animal handling is important with this technique. Immunization of either rabbits or mice follows very similar protocols, varying only in the antigen concentration, number of injection sites, and volume (*11*).

The methods describe a standard indirect EIA using enzymatic detection with alkaline phosphatase-conjugated secondary antibody and a colorimetric substrate for detection. Assay optimization may be needed for coating the antigen (concentration and buffer), diluting the primary antibody, and blocking. Optimal assay conditions are necessary for obtaining accurate antibody detection (*11–14*).

3.1. Antigen Preparation and Conjugation (See Notes 1–4)

1. *Aminophenyl derivative*
   (a) One hundred milligrams of the *o*-nitrophenyl saccharide derivative (e.g., *o*-nitrophenyl 2 acetamido-2-deoxy-3-*O*-D-galactopyranosyl-b-D-galactopyranoside) and 50 mg of platinum oxide in methanol are shaken under hydrogen at 345 kPa pressure for 4 h at room temperature.
   (b) The reaction mixture is filtered to remove the catalyst, and the filtrate evaporated under pressure to yield the aminophenyl derivative as a solid.
2. *Preparation of the p-diazophenyl derivative*
   (a) The aminophenyl derivative is immediately linked to BSA by the method of McBroom et al. (*7*). Seventy milligrams of the aminophenyl derivative is dissolved in 4 ml of ice-cold 0.1 M HCl, followed by dropwise addition of 5 mL of sodium nitrite (0.05 M) with stirring (*see* **Note 2**).
   (b) The reaction proceeds for 30 min on ice.

3. *Preparation of the saccharide–BSA conjugate*
   (a) The *p*-diazophenyl derivative is mixed with BSA (20 mL at 25 mg/mL) in 0.15 M NaCl, pH 9 (pH adjusted with 0.5 M NaOH). Keep at pH 9 with 0.1 HCl.
   (b) The product is separated from the reactants by dialysis.
4. *Molar ratio of haptenation*
   The haptenation is determined using the Dubois method for molar saccharide concentration and the Lowry method for protein concentration. **Note 3** and **4**. For example, Galβ1-3GalNAc was attached to BSA at 55:1 ratio, and we typically find ratios ranging from 4 to 55.

3.2. Immunizations (See Notes 5–9)

1. For each rabbit, 50–200 μg protein conjugate in 0.3 mL PBS is added and mixed with an equal volume of Freund's complete adjuvant for injection. For each mouse, 5–50 μl protein in 500 μL of PBS is added and mixed with an equal volume of Freund's adjuvant for injection. Extra volume should be made to compensate for loss in the mixing vessel **Note 5**.
2. To mix, transfer the antigen and Freund's adjuvant back and forth between two syringes attached to each other by a Luer-lock device. Emulsify the mixture into a thick, stable white emulsion (*see* **Notes 6, 7, 8** and **9**).
3. Transfer to 1 mL glass or plastic syringe and attach a 20G needle.
4. Inject 0.6 mL into each rabbit, splitting the injections into six sites, so that each drains into different regional lymph nodes. Do not inject over the backbone. Inject 0.1 mL of the emulsion subcutaneously into each mouse, depressing the syringe slowly and carefully to prevent the needle from dislodging from the syringe and to prevent leakage out of the injection site. Split the 0.1 mL into at least four injection sites along the back of the mouse, so that each site drains into different regional lymph nodes. Do not inject over the backbone.
5. Ten to fourteen days after the primary injection, boost the mice or rabbits with the same dose of antigen, mixed with IFA rather than with CFA (*see* **Notes 5–8**). Repeat the booster injections every 2 weeks for a total of four injections.
6. Prior to the fusion, inject the antigen intraperitoneally into the mice selected. This antigen is not mixed with the adjuvant.
7. Determine antibody titers before fusion, so that the mice with maximum response are used. Blood should be taken from the mice just prior to each booster by retro-orbital bleed under anesthesia (or the institution's preferred method), the serum separated from the red cells, and the serum tested by EIA (*see* **Subheading 2.3**). Blood should be taken from the ear vein of the rabbit.

3.3. EIA (See *Notes 10–18*)

1. Coat each well of the microtiter plate with the antigen. As a guide, 1–10 μg/mL of carbohydrate–BSA conjugate can be coated in 0.1 M sodium bicarbonate buffer, pH 9.6. One hundred microliters of the antigen in buffer should be added using a multichannel pipette into each well. Coat a second plate with the control of either the carrier protein alone (BSA) or a different, unrelated saccharide linked to the protein in a similar fashion (*see* **Notes 10–13**).
2. Seal the plates with parafilm or plate sealers and place at 37°C for 3 h or at 4°C overnight before use.
3. Wash the antigen from the wells using an automated plate washer or squirt bottle containing wash buffer; perform three washes. Tap the plate upside down on several paper towels to remove residual wash buffer (block step optional, *see* **Note 12**).
4. Add 100 μL of the proper primary antibody dilutions to each well, in triplicate. If using sera, a 1:100 dilution would be an appropriate starting dilution if enough serum was obtained. For sera, start with twofold serial dilutions to 1:6,400. For monoclonal antibody supernatant testing, use a 1:20 final dilution from the hybridoma tissue culture fluid in which the cells have been allowed to grow until the medium just turns yellow.
5. For each serum or supernatant dilution, use both the carbohydrate–protein conjugate coated plates and a negative control of the carrier protein (BSA) alone.
6. Dilute the antiserum preparations in a buffer containing 5% BSA to block the reactivity to the BSA carrier protein.
7. Cover the plate and incubate at 37°C for 1 h for antibody binding.
8. Wash plate three times as in **step 3** to remove the unbound antibody.
9. Add 100 μL of a dilution of labeled secondary antibody (use EIA dilution guidelines as suggested by the manufacturer, for example 1:1,000, anti-mouse IgG, alkaline phosphatase labeled, Sigma) in conjugate buffer to each well using a multichannel pipette.
10. Cover the plate and incubate at room temperature for 1 h for secondary antibody binding.
11. Wash the plate three times as in **step 3** to remove the unbound antibody.
12. To measure bound antibody, add 100 μL of *para*-nitrophenyl phosphate substrate to each well (1 mg/mL solution in substrate buffer, pH 9.8 *see* **Note 14, 15, 16, 17, 18**).

13. After 30–60 min of color development, read the optical density (OD) on a microplate reader at 405 nm.
14. Positive sera and supernatants will be positive on the carbohydrate-BSA plate but not (or much less) on the BSA control plate.

3.4. Fusion (See *Notes 19 and 20*)

1. Prepare the myeloma cells 1 week before the fusion by thawing and expanding in the complete medium.
2. Boost the primed animal 3 days before fusion as described above.
3. Prepare materials 1 day before fusion, including PEG, and split the myeloma cells into fresh medium.
4. On the day of the fusion, you should have roughly 5×10^7 myeloma cells per spleen, or 1×10^8 for a typical two-spleen fusion. This will give an excess of myeloma cells. Splitting the myeloma cells on the day prior to the fusion is important in order to have the cells in a growth phase (*see* **Notes 19** and **20**).
5. On the day of the fusion, check the myeloma cells for growth characteristics and lack of contamination, and harvest and count. The optimal cell density is $6.0–8.0 \times 10^5$/mL.
6. To harvest the spleen cells, sacrifice the mice as per the approved protocol.
7. Remove blood via heart puncture and separate the serum to serve as a control for the EIAs.
8. Spray the mouse with 70% ethanol, place with left side up, and make an incision to visualize spleen.
9. Change to clean scissors and forceps and remove the spleen to a Petri dish containing a small amount of plain DMEM medium.
10. Separate the spleen into a single cell suspension, first by cutting into small pieces with scissors and then by further disaggregating by rubbing the tissue gently between the frosted ends of two sterile microscope slides.
11. Place harvested cells in a 15-mL tube. Allow the aggregates to settle, and move the harvested cells to a fresh tube.
12. Wash the myeloma cells and spleen cells in DMEM for a total of three washes, centrifuging for 10 min at 1,000 rpm (235*g*) between each wash.
13. Count and combine in a five spleen cells/myeloma cell ratio after the second wash. Ratios from 4:1 to 12:1 have been successful. Count the spleen cells using the 1% acetic acid/150 mM NaCl solution.

14. After the third wash, discard medium, tap up the cell pellet, and slowly (timing so that the addition takes 60 s) add 1 mL of the prepared PEG. Swirl the tube for 60 s.
15. Slowly (taking 2 min) add 2.5 mL DMEM.
16. Add 17.5 mL DMEM.
17. Centrifuge and resuspend in HAT medium to 2×10^6 spleen cells/mL.
18. Add 100 μL of the cells to each well of flat-bottom, sterile microtiter plates
19. In addition, add 100 μL of HAT medium to the wells
20. Approximately 5–8 plates per spleen will be required. Place in a 37°C CO_2 incubator.
21. Check the next morning for signs of contamination.
22. Check the wells for clones after 7 days. Begin to screen after 10 days to 2 weeks. Collect the medium when it is yellow orange into another microtiter plate, replacing the medium in the tissue culture plate as it becomes exhausted.
23. After a number of supernatants have been collected, use the supernatants that have been stored in the microtiter plate to test in EIA as directed.
24. Expand the wells that contain the selected clones by transferring them first into 24-well plates, then transferring into HT media, then as they expand, transferring the selected clones into flasks.
25. After 1 week in HT medium (at least two changes), clones can go into regular culture media.

3.5. Subcloning

Selected clones should be subcloned for long-term stability and use.

1. Harvest cells from a flask, count, and dilute to 3 cells/mL in the subcloning medium.
2. Add 100 μL of the subcloning medium to each well of a 96-well plate and incubate to per warm. Plate 100 μL of cells per well. Less than 30% of the wells should have colonies, as statistically this should confirm the count of 0.3 cells/well and little chance of >1 cell/well.
3. After 4–5 days, the wells should be observed with an inverted microscope, and wells containing 1 cluster/well are marked for retesting in EIA.
4. The subclone with the best growth characteristics and highest OD in EIA should be expanded for use and frozen for long-term storage and use.

3.6. Specificity Analysis

1. Serially dilute the hybridoma supernatant to determine the concentration needed to develop an OD of 1.0 in the EIA.
2. Dilute the supernatant to twice this concentration (for example if a 1:1,000 dilution gave an OD of 1.0, prepare your monoclonal supernatant at 1:500).
3. Prepare serial dilutions of the saccharide and related saccharides (*see* **Note 16**)
4. Mix with equal volumes of each dilution of your saccharide with the 2× hybridoma supernatant. Incubate for 1 h and then add to the saccharide-BSA conjugate coated EIA plate.
5. Incubate the mixtures for 1 h on the plates, and then wash three times.
6. Wash the antigen from wells using an automated plate washer or squirt bottle containing wash buffer; perform three washes. Tap the plate upside down on several paper towels to remove the residual wash buffer.
7. Add the secondary antibody, wash, and add the substrate as in the general indirect EIA protocol.
8. Measure the OD.
9. Calculate the percentage inhibition. Determine the amount of each saccharide required for 50% inhibition.

4. Notes

1. A wonderful review that outlines many conjugation techniques is *Neoglycoproteins: preparation and applications*, edited by Y.C. Lee and R.T. Lee (*8*).
2. During the diazotization reaction, check for $NaNO_2$ excess by placing a drop on a starch iodide paper. The indicator paper should be blue.
3. When performing the Dubois method, be sure to use glass tubes, as plastic will melt. Add the sulfuric acid quickly. Use tubes larger than needed for the volume, as the heat will cause boiling.
4. While the Lowry method for protein determination is given here, any protein assay will suffice that would include the Bio-Rad assay based on the Bradford method or the Bicinchoninic acid (BCA) assay.
5. Whereas CFA and IFA are the most common adjuvants in use, others exist that are useful for different scenarios. A common systemic adjuvant is alum and derivatives of

aluminum. This compound is used to adsorb antigens, allowing slow release and therefore prolonged exposure.

6. When preparing the CFA or IFA mixture by transferring the antigen solution into the CFA syringe and injecting back into the antigen syringe repeatedly to emulsify the mixture, a thick, stable emulsion should be formed. This is characterized by the lack of dispersion of a drop of the mixture placed in cold water. If the droplet disperses, continue to emulsify the mixture by repeatedly passing through the syringe. Other mixing methods can also be used, including the use of small blender or sonic mixing.
7. While CFA is used commonly, it does have side effects in the animals. Multiple injections with CFA can cause necrotic areas and granulomas at the injection sites. Therefore, IFA is used for all subsequent injections to decrease the risk of tissue damage in the injection areas. Proper care should also be taken by those preparing the CFA mixture to prevent skin or eye exposure, by wearing gloves and masks or goggles.
8. Often, the number of animals used for immunizations can be varied along with the immunization schedule. Immunizations every 2 weeks are very common; however, 1 and 3 weeks are also feasible. Multiple injections are recommended. Balb/c variants are most commonly used and usually give the highest fusion efficiency; however, other strains of mice such as C3H or C57/Bl6 can be used if good immune response is not seen with Balb/c mice.
9. When preparing the antigen for immunization, be sure that the glass syringes are large enough to hold the volume of both the antigen and the adjuvant being used. If whole cells are used as immunogens rather than protein/peptides, wash the cells three times to remove media components before injecting.
10. In the indirect enzyme immunoassay, flat-bottomed Nunc or Immunolon microtiter plates are suggested. Other manufacturers' plates can also be used. Plates are available in medium or high binding. To determine the best plate to use for further testing, after a positive serum is available, make a serial dilution and test it on coated plates of each type.
11. To determine the specificity of the reaction, find the concentration of the serum that gives a reaction with an OD of 1.0 in the plate. Use twice this concentration, mix with excess antigen, and incubate prior to incubation on the plate. Measure the OD and compare it with that obtained with the same final concentration of serum that was not inhibited. The percentage inhibition equals (1 – OD of the inhibited sample divided by the OD of the uninhibited sample). The percentage inhibition should be over 90%. If it is not, a blocking step can be added.

12. In the indirect enzyme immunoassay method described above, a blocking step is not included, as BSA is present in the dilution buffer for the antibody, which serves to block some nonspecific interactions. However, if high backgrounds are seen, adding a blocking step may be warranted. Traditionally, BSA and/or Tween 20 are used to block the residual binding sites on the plate after coating with the antigen. Additional blocking agents are gelatin and non-fat dry milk. Use caution when using non-fat dry milk as it contains saccharide structures and may inhibit or mask your reaction. The typical blocking conditions are 30–60 min at room temperature or at 37°C. When the blocking time is completed, the buffer can be removed by usual washing (three times with wash buffer), or the block can be dumped out of the wells and the plate tapped onto paper towels to remove the residual fluid. Select the plate that allows the highest sensitivity and the lowest background.
13. The amount of coating antigen required can also be optimized. Coat the chosen plate with 0.1–10 μg/mL of the antigen. Use the selected sera at the concentration that yielded an OD of 1.0 under previous testing, and plate on the differently coated wells. Plot the values. A peak should be seen in which higher concentrations of coating the antigen do not increase the OD, and this concentration should be used. The optimization of coating the antigen is important when the antigen is costly. Incubation times for coating, blocking, primary antibody, and secondary antibody may also be varied.
14. Several types of enzyme substrates can be used for the reaction visualization. Enzymes include alkaline phosphatase, horseradish peroxidase, and β-galactosidase, in the order of their usage. Alkaline phosphatase is a rapid and stable enzyme with several available substrates, both chromogenic and fluorogenic. Both substrates are easy to assay; however, fluorescence readers are more expensive than traditional spectrophotometers used to read the chromogenic output. In order to decide which enzyme to use in this step, one must consider whether one will be assaying fluids or materials which contain endogenous sources of these enzymes or endogenous inhibitors of these enzymes, and choose accordingly.
15. Several controls should be included in the EIA protocol. An antibody known to bind the antigen serves as a positive control, while preimmunization sera or isotype control antibodies serves as a negative control. Further, wells tested with secondary antibody alone (no primary antibody) also provide a control to zero the spectrophotometer.

16. For specificity analysis, choose saccharides that will help you determine whether your antibody will have the reactivity you desire. If you are making antibody to *Streptococcus pneumoniae* serogroup 1, you may want to test against serogroups 3, 4, 5, 6B, 7F, 9V, 14, 18C, 19F, and 23F. If you want to make an antibody that reacts with the type 1 chain related trisaccharide, galactose β1-3*N*-acetylglucosamine β1-3 galactose, test against the type 2 chain related trisaccharide Galβ1-4GlcNAcβ1-3Gal. Run the inhibition analysis with the associated monosaccharides. If you are using a branched saccharide, use linear saccharides derived from this. Some related saccharides will not be available, and, of course, we all must work within these limits. See **Table 1** and **Fig. 1** for an example of how related saccharides are used to define the reactivity of an antibody.
17. If the IsoStrip kit (Roche) is not used, the standard procedure is a sandwich EIA for isotype determination. This process is similar to the indirect EIA above; however, isotype-specific capture antibodies are coated directly on the microtiter plate followed by the addition of the test sample. If the specific isotype is present, a signal will be detected. EIA kits are also available for isotype determination from BD, Bio-Rad (Hercules, CA), Invitrogen (Carlsbad, CA), and Pierce (Rockford, IL). A detailed protocol can be found in *Current Protocols in Immunology* (*11*).
18. Ten to 50 μg of antibody is produced by hybridoma cells under these conditions. For increased concentrations of the antibody, bioreactors can be used (producing 0.5–1.0 mg Ab/mL), or if the ascites production is possible, the cells can be grown as ascites, and up to 1 mg/mL of the antibody will be produced (*11*).
19. Grow the myeloma cells in T-75 flasks. You will need at least four T75 flasks. We grow in eight T75 flasks and pick the flasks in which the cells look the healthiest.
20. An average spleen from an immunized animal will yield 1.5 × 10^8 cells. This number varies with how much response was made to the antigen.

Acknowledgments

The authors would like to thank Sue Morey for her expertise in monoclonal antibody production, help in the experiments concerning carbohydrate antigens for the last 11 years, and help in editing the chapter.

References

1. Cunningham, R. (1994) Immunochemistry of Polysaccharide and Blood Group Antigens, in *Immunochemistry* (van Oss, C.J. and van Regenmortel M.H.eds.), Marcel Dekker Inc., New York, NY, pp. 319–337
2. Berzofsky, J.A. and Berkower, I.J. (2003) Immunogenicity and Antigen Structure, in *Fundamental Immunology*, 5th edition (Paul, W.E., ed.), Lippincott Williams and Wilkins, Philadelphia, PA, pg 631–683
3. Diakun, K.R., Yazawa, S., Abbas, S.A., and Matta, K.L. (1987) Synthetic Antigens as Immunogens. Part II. Antibodies to Synthetic T Antigen. *Immunol Investig* 16(2):151–163
4. Rittenhouse-Diakun,K.,Xia,Z.,Pickhardt,D., Morey, S., Baek, M.G., and Roy, R. (1998) Development and Characterization of Monoclonal Antibody to T-Antigen: GalB1–3GalNAc-alpha-O. *Hybridoma* 17(2):165–173
5. Insel, R.A. and Anderson, P. (1986) Oligosaccharide-Protein Conjugate Vaccines Induce and Prime for Oligoclonal IgG Antibody Response to the *Haemophilus influenzae* b Capsular Polysaccharide in Human Infants. *J. Exp. Med.* 163:262–269
6. Insel, R.A., Kittleberger, A., and Anderson, P. (1986) Isoelectric Focusing of Human Antibody to the *Haemophilus influenzae* b Capsular Polysaccharide: Restricted and Identical Spectrotypes in Adults. *J. Immunol.* 135:2810–2816
7. McBroom, C.R., Samanen, C.H., and Goldstein, I.J. (1972) Carbohydrate Antigens: Coupling of Carbohydrates to Proteins by Diazonium and Phenylisothiocyanate Reactions. *Methods Enzymol.* 28:212–219
8. (1994) Lee, Y.C. and Lee, R.T. (eds.) *Neoglycoconjugates: Preparation and Applications.* Academic Press, San Diego, CA
9. Eby, R. (1979) The Synthesis of Isomalto-Oligosaccharide Derivatives and Their Protein Conjugates. *Carbohydr. Res* 70:75–82
10. Davis, B.G. (2002) Synthesis of Glycoproteins. *Chem. Rev.* 102:579–601
11. (2001) Coligan, J.E., Kruisbeek, A.M., Marguiles, D.H., Shevach, E.M., and Strober, W. (eds.) *Current Protocols in Immunology*, Wiley and Sons, New York, NY
12. 1980 edn. Hudson, L., and Hay, F.C. (eds.) *Practical Immunology..* Blackwell Scientific Publications, , Oxford, England
13. Campbell, D.H., Garvey, J.S., Cremer, N.E., and Sussdorf, D.H. (eds.) (1963) *Methods in Immunology: A Laboratory Text for Instruction and Research*, W.A. Benjamin inc., New York, NY
14. Engvall, E. and Perlman, P. (1972) Enzyme-linked immunosorbent assay, Elisa 3. Quantitation of Specific Antibodies by Enzyme-Labeled Anti-Immunoglobulin in Antigen-Coated Tubes. *J. Immunol.* 109:129
15. Dubois, M., Gilles, A., Hamilton, K., Rebers, P.A., and Smith, F. (1956) Colorimetric Method for Determination of Sugars and Related Substances. *Anal. Chem.* 28:350–356
16. Lowry, O.H., Rosebrough, N.J., Farr, A.L., and Randall, R.J. (1951) Protein Measurement with the Folin-Phenol Reagents. *J. Biol. Chem.* 193:265–275

Chapter 25

Inhibition of Glycosyltransferase Activities as the Basis for Drug Development

John Schutzbach and Inka Brockhausen

Summary

Glycosyltransferases are involved in the biosynthesis of protein-bound glycan chains that have multiple and important biological functions in all species. In this protocol, we describe methods to assess the inhibition of glycosyltransferase activities. The kinetic mechanisms of the enzymes, information from structural studies and preliminary inhibition studies can aid in designing appropriate inhibitors. The inhibition of β4-Gal-transferase can be studied with GlcNAc derivatives that act as alternative acceptor substrate analogs and are expected to dock in the acceptor binding site of the enzyme. The inhibition of core 2 β6-GlcNAc-transferase can be studied with compounds that may compete with binding of the acceptor or glycosyl-donor substrate. Another example is the use of a class of amino acid specific reagents as inhibitors that help to obtain information about amino acid residues at or near the active site of dolichol-phosphate-mannose synthase or those involved in the enzyme mechanism. These inhibitors can be useful for studies of glycan functions, and have potential as therapeutic drugs for a number of diseases involving glycosylation.

Key words: Glycosyltransferases, Glycoproteins, Enzyme kinetics, Enzyme inhibition, Enzyme mechanism.

1. Introduction

All species synthesize complex glycans which play important biological roles. In mammals, for example, the glycan chains of glycoproteins participate in cellular communication and in the immune system (*1, 2*). Glycans have been shown to be altered in many diseases, and they contribute to the antigenicity and invasive behavior of cancer cells and in the metastatic process (*3*). In cystic fibrosis and other lung diseases, the extensive glycosylation

Nicolle H. Packer and Niclas G. Karlsson (eds.), *Methods in Molecular Biology, Glycomics: Methods and Protocols, vol. 534*

DOI: 10.1007/978-1-59745-022-5_25

of mucus glycoproteins contributes to the pathological obstructions of airways and bacterial colonization (*4, 5*). Glycosyltransferases involved in the synthesis of glycan chains are thus potential targets for therapeutic drug development.

Inhibitors of glycosyltransferases that can block the biosynthesis of specific carbohydrate structures could be of value in investigations of biological functions of glycans, and in the reduction of inflammatory diseases and cancer. They could potentially block cell adhesion events, alter activities of the immune system, prevent cancer metastasis, reduce the production of highly viscous mucus and prevent bacterial adhesion during infections.

In this protocol, we describe studies of enzyme inhibition with a rational design approach that examines the kinetics of inhibition by individually designed compounds. Knowledge of the detailed substrate specificity of the enzyme is the basis for the rational design of inhibitors. In addition, information on the protein structure and amino acids involved in substrate binding and catalysis aid in designing compounds that fit well into the binding site and may form non-productive complexes.

Three glycosyltransferases will be used as examples. The Golgi enzymes β4-Gal-transferase 1 (β4GalT1), involved in glycan chain extension, and core 2 β6-GlcNAc-transferase (C2GnT1) that synthesizes a branch of *O*-glycans, have crucial roles in the immune system and in cancer(*3, 4*). Purified β4GalT1 is readily available and its inhibition can be studied with synthetic acceptor substrate analogs, based on GlcNAc (*6, 7*). C2GnT1 binds UDP-GlcNAc as the donor substrate. Its acceptor substrate is Galβ1-3GalNAc-R, and the enzyme is known to be absolutely specific for most of the substituents of the sugar rings (*8–10*). GlcNAc derivatives may thus compete with UDP-GlcNAc binding to the enzyme. Dolichol-phospho-mannose synthase (Dol-P-Man synthase) synthesizes essential lipid-bound glycosylation intermediates in ER membranes, and utilizes GDP-Man as the donor substrate and dolichol-phosphate (Dol-P) as the acceptor substrate. Inhibition of Dol-P-Man synthase has the potential of eliminating functionally important *N*-glycans, as well as glycosylphosphatidyl-inositol (GPI) anchors and protein *O*-mannosyl linkages (*11, 12*).

The kinetic mechanism of the enzymes, and information from structural and preliminary inhibition studies aid in designing appropriate inhibitors. Detailed methodology for the kinetic analysis of enzyme reactions and mechanisms is beyond the scope of this protocol however, and expert texts(*13, 14*) should be consulted for such information.

In order to develop drugs for a number of biological and pathological situations, inhibitors should have defined potency, stability, solubility, and hydrophobicity. They should have the ability to enter cells to reach the subcellular site of glycosylation and be specific for defined glycosyltransferases. Compounds may be designed as biosynthetic precursors that are converted in vivo to achieve accurate enzyme targeting. High concentrations of alternative substrates inhibit by competition with the natural substrates. An example is the presence of synthetic GalNAcα-benzyl in cultured cells(*15*) that causes *O*-glycan chains to be built up on GalNAcα-benzyl rather than the natural glycoproteins. Thus, natural substrates and substrate analogs have the ability to reduce glycoprotein glycosylation. Another possibility for inhibition is the use of blocked derivatives (for example per-*O*-acetylated GlcNAcβ1,3Galβ-*O*-naphthalenemethanol (*16*)) that are converted inside cells to substrate analogs. Compounds that are substrates and contain reactive groups may bind to the enzyme upon activation (*9*), thereby preventing substrate binding and catalytic activity of the enzyme. If a substrate derivative binds much tighter to the enzyme than the natural substrate, it will block substrate binding and thus activity. Finally, many non-substrate inhibitors bind non-specifically to enzymes or target specific amino acids that are critical for the enzyme protein folding, substrate binding or catalysis.

1.1. Glycosyltransferases and Enzyme Assays

Before discussing specifics of enzyme inhibition, it is first necessary to provide some guidelines for basic enzyme assays and for studies of the kinetics of glycosyltransferases. For the characterization of enzyme kinetics it is usually preferable to study purified or homogeneous protein preparations. However, in contrast to the famous commandment of Efraim Racker 'to not waste clean thoughts on dirty enzymes', a great deal of information about enzyme specificity and kinetics can be obtained with crude enzyme preparations provided that the assays are specific to the enzyme under investigation and side reactions do not remove substrates or essential co-factors. Soluble portions of membrane-bound enzymes or recombinant enzymes containing a tag can give useful information and often facilitate the assay procedures. Depending upon the reaction involved, glycosyl-donors and acceptors are often only available in limited quantities or are quite expensive, or the assays may be quite cumbersome, thereby limiting the number and designs of kinetic experiments. Assays usually involve the use of radioactive glycosyl-donors or fluorescent acceptors in product determinations at a specific time point although for some reactions, it may be possible to utilize coupled enzyme assays to assess the activity of the glycosyltransferase. With the latter type of assay, however, the enzyme products and reaction intermediates for the coupled reactions, such as UDP, can interfere with determining the effects of inhibitors on the glycosyltransferase. This is usually less of a problem for the glycosyl acceptor, but proper controls must be carried out to assure that the enzyme products have minor or no effects on the test enzymes. Time point determinations of product formation can limit the interpretation of glycosyltransferase assays since these assays do not represent measurements of initial rates but rather approximations of the initial rates. The utilization of substrates below 10% can be acceptable, but the changing conditions within the incubation time must be kept in mind when interpreting the results.

The reaction needs to be approximately linear with time since the rate is dependant both on initial substrate concentration and stability of the enzyme. In addition, the reaction must be proportional to enzyme concentration, and metal ion activators and other co-factors should be present at optimal concentrations. One should work at or near the temperature and pH optimum of the enzyme and ensure that any detergents used to support enzyme activity do not interfere with substrate and/or inhibitor availability. It is essential to have a convenient and accurate assay procedure that is specific for the enzyme under study.

In order to understand the type of inhibition observed, kinetic data of the enzyme are useful. The kinetic mechanism for some glycosyltransferases has been shown to involve a sequential ordered reaction in which the glycosyl-donor or nucleotide sugar binds first followed by binding of the acceptor, then the newly formed saccharide leaves followed by nucleoside diphosphate (*see* **Fig. 1A**). Examples for enzyme mechanisms that have been published include those for β4-galactosyltransferase (*17*), β2-*N*-acetylglucosaminyltransferase II (*18*), β2-*N*-acetylglucosaminyltransferase I (*19*), and mucin β6-*N*-acetylglucosaminyltransferase C2GnT2 (*20*). The sequential ordered mechanism is compatible with the x-ray crystallography of the structures of several glycosyltransferases in which the nucleotide sugars were shown to bind in deep clefts in the enzymes(*7, 21, 22*). However, the kinetic mechanism for Dol-P-Man synthase appears to be random bi bi (*see* **Fig. 1B**) in which either substrate, GDP-Man or Dol-P, can bind first, and then either Dol-P-Man or GDP can leave (Forsee, W.T. and Schutzbach, J.S., unpublished). Dol-P-Man synthase, in contrast to the examples given above, catalyzes a reversible reaction.

Establishing the kinetic mechanism for a new glycosyltransferase requires carrying out numerous assays, varying the concentrations of both substrates, usually in a 5 × 5 matrix. This involves measuring initial rates at five different fixed concentrations of glycosyl-donor while varying the acceptor at five different concentrations. Parallel lines in reciprocal (Lineweaver-Burk) plots ($1/v$ vs. $1/S$) indicate a ping pong mechanism in which the first product leaves the enzyme before the second substrate binds. A series of intersecting lines, however, suggests a sequential reaction as described above. The lines can intersect either above, on or below the x-axis. Inhibitor studies using enzyme products are required, however, to differentiate between the sequential and random mechanisms (*13*). This involves assessing the types of

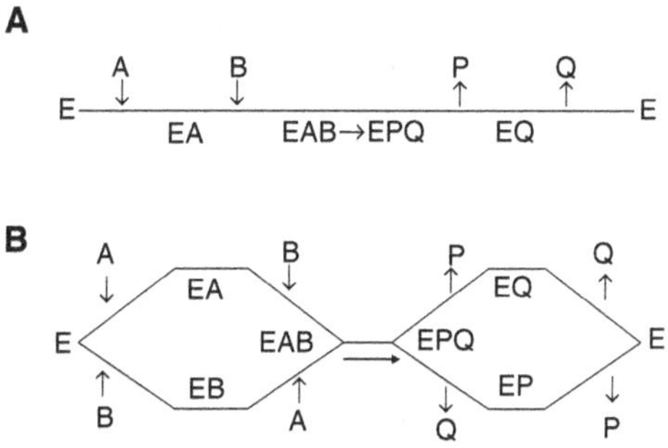

Fig. 1. Common enzyme mechanisms for glycosyltransferases. (**A**) is a representative diagram of an ordered sequential mechanism in which the glycosyl-donor (A) binds to free enzyme before the acceptor (B) binds. Following glycosyltransfer, the newly formed glycoproduct (P) is released prior to release of the nucleoside diphosphate (Q). (**B**) is a representative diagram of a random bi bi reaction in which either glycosyl-donor (A) or glycosyl-acceptor (B) can bind to free enzyme. Following glycosyltransfer, either the newly formed glycoproduct (P) or the nucleoside diphosphate (Q) can be released first.

inhibition- whether competitive, noncompetitive, uncompetitive or mixed competitive- that is obtained with each product versus each substrate. The mechanism should be confirmed with the use of dead-end inhibitors or inhibitors that are not substrates, as well as analogs of the substrates and products.

Since some substrates and inhibitors for glycosyltransferases are quite expensive or must be prepared by chemical/enzymatic synthesis, they are often not available for large numbers of initial assays. Thus, for a first approximation, one should obtain an apparent K_M for both glycosyl-donor and acceptor in a minimum number of assays using one variable substrate and the other substrate at some reasonably high concentration. Thus, the apparent K_M for glycosyl-donor is determined by varying donor concentrations (e.g. from 0.002 to 4 mM) while keeping the acceptor concentration at the highest reasonable level which should be at least twice the expected K_M value (e.g. 2–4 mM). Then the apparent K_M for the acceptor is determined while keeping the donor concentrations above (at least two times) the K_M for this compound. The intervals between the substrate concentrations should be spaced to give approximate equal intervals in Lineweaver-Burk plots. A number of non-linear regression programs are available from commercial sources for determining enzyme kinetic constants but all data should first be manually plotted to detect unusual data, outliers, or curved lines. The substrate that binds second in an ordered sequential reaction often demonstrates substrate inhibition because the acceptor can bind in a non-productive complex at high concentrations, blocking binding of the first substrate or glycosyl-donor. Substrate inhibition can be recognized by many non-linear regression programs.

Finally, considering an ordered sequential mechanism in which nucleotide sugar binds first, compounds that bind at the glycosyl-donor site will be competitive inhibitors, whereas, compounds that bind at the acceptor site will likely give non-competitive or mixed type kinetic patterns.

For first approximations initial studies can be limited to establishing that the test compounds provide inhibition when tested at reasonable concentrations of substrate, i.e. at or near K_M concentrations. Acceptor analog inhibitors usually bind reversibly to the enzyme or the EA complex (*see* **Fig. 1**). By contrast, a number of inhibitors form covalent linkages with the enzyme protein and can bind either near or at the active site. These inhibitors are often specific for amino acids and will usually be kinetically non-competitive since they remove enzyme from the rate equations, thus lowering V_{max} in a non-reversible reaction (under assay conditions). These types of inhibitors have been useful in identifying the active site of Dol-P-Man synthase.

2. Materials

2.1. β4-Gal-Transferase

2.1.1. Enzyme Source

1. Soluble bovine milk β1,4-Gal-transferase T1 (Sigma, Toronto, ON, Canada).

2.1.2. Enzyme Assays

1. 0.5 M MES buffer, pH 7 (Sigma).
2. 0.4% bovine serum albumin (BSA) (Sigma).
3. 0.1 M $MnCl_2$.
4. AG1x8 (Bio-Rad, Hercules, CA).
5. Sep-Pak C18 columns (Waters, Milford CA), methanol.
6. HPLC system (Waters), acetonitrile, reverse-phase C18 column (Phenomenex, Torrance, CA).
7. Table top vacuum centrifuge, Readisolv scintillation fluid (Beckman, Fullerton CA), 7-mL scintillation vials, and scintillation counter.

2.1.3. Substrates

1. 5 mM GlcNAc (Sigma).
2. 5 mM GlcNAcβ-benzyl (Sigma).
3. UDP-[^{3}H]Gal (Perkin Elmer) diluted with non-radioactive UDP-Gal (Sigma) to 10 mM, 2,000 cpm/nmol.

2.1.4. Inhibitors (Table 1)

1. 10 mM 4-deoxy GlcNAcβ-methyl (*23*).

Table 1
Substrates and inhibitors for β4Gal-transferase

Compound	Substrate	Potency as inhibitor	References
GlcNAc	++	+	(*6*)
GlcNAcβ-benzyl	+++	+	(*6*)
6-*S*-GlcNAcβ-benzyl	+	+	(*6*)
4-deoxy-GlcNAcβ-methyl	–	(+)	(*6,23*)
GlcNAcβ-*O*-naphthyl	–	+++	(*6*)
GlcNAcβ-*S*-naphthyl	–	+++	(*6*)
GlcNBuβ-*O*-naphthyl	–	+++	(*6*)
GlcNBuβ-*S*-naphthyl	–	+++	(*6*)
6-*S*-GlcNBuβ-*S*-naphthyl	–	+	(*6*)

2. 5 mM GlcNAcβ-2-naphthyl in methanol (*6, 24*).
3. 5 mM *N*-butyryl-glucosamineβ-thio-2-naphthyl dissolved in methanol (*6*).
4. 10 mM UDP (Sigma).

2.1.5. Enzyme Kinetics

1. Non-linear regression programs (*6*); EZ-Fit (FPerrell@JLC.net), Enzfitter (Sigma-Aldrich).
2. Lineweaver-Burk linear regression analysis.

2.2. Core 2 β6-GlcNAc-Transferase

2.2.1. Enzyme Source

1. Recombinant soluble human C2GnT1, containing a His_6 tag and an enterokinase site, from insect cell supernatants (*10*).

2.2.2. Enzyme Assays

1. MES buffer, 0.5 M, pH 7.
2. 0.1 M AMP (Sigma).
3. AG1x8 (Bio-Rad).
4. Sep-Pak C18 columns (Waters), methanol.
5. HPLC, acetonitrile, reverse-phase C18 column.
6. Table top vacuum centrifuge, Readisolv scintillation fluid, scintillation counter.
7. UV reactor Rayonet RPR-100 with 15 RPR-3500 lamps (Rayonet, Branford, Connecticut).

1. 5mM Galβ1-3GalNAcα-benzyl (Sigma).

2.2.3. Substrates

2. UDP-[^{3}H]GlcNAc (Perkin Elmer) diluted with non- radioactive UDP-GlcNAc (Sigma) to 10 mM, 4,000 cpm/nmol (*10*).

2.2.4. Inhibitors (Table 2)

1. 5 mM Galβ1-3(6-deoxy)GalNAcα-$(CH_2)_8$COOmethyl (*23*).
2. 5 mM Galβ1-3GalNAcα-*p*-nitrophenyl (*9*).
3. 5 mM 4-deoxy-3,6-di-*O*-acetylGlcNAcβ-*O*-methyl (*25*).
4. 10 mM UDP.

2.2.5. Enzyme Kinetics

1. As described in **Subheading 2.1.5**

2.3. Dolichol-Man-Synthase

2.3.1. Enzyme Source

1. Recombinant *Saccharomyces cerevisiae* Dol-P-Man-synthase prepared from *E. coli* homogenates(*26*).

2.3.2. Enzyme Assays

1. 25 mM Tris/acetate, pH 7.5, containing 0.25 mM EDTA, 5 mM $MnCl_2$ and 2.5 mM $MgCl_2$.
2. 2M Dithiothreitol (DTT) (Sigma).
3. Thesit (Sigma).
4. 5,5′-Dithiobis(2-nitrobenzoic acid) (Nbs_2) (Sigma).
5. Ultrasonic cleaning bath (Branson model 1510 or equivalent).
6. Water-saturated 1-Butanol prepared by mixing (shaking) deionized water and 1-butanol in a glass stoppered bottle and then letting the mixture stand until 2 phases are obtained.
7. Xylene based scintillation fluid (Scintilene, Fisher Scientific, Pittsburgh, PA), 5-mL scintillation vials, scintillation counter.

Table 2
Substrates and inhibitors for core 2 β6GlcNAc-transferase

Compound	Substrate	Potency as inhibitor	References
Galβ1-3GalNAcα-benzyl	+++	+	(*9*)
Galβ1-3GalNAcα-pnp	+++	+++(UV)	(*9*)
Galβ1-3(6-deoxy-) GalNAcα-benzyl	–	(+)	(*23*)
4-Deoxy-3,6-di-*O*-acetyl GlcNAcβ-*O*-methyl	–	+	(*25*)

pnp p-nitrophenyl; *UV* needs to be activated by UV-irradiation

Table 3
Reagents useful for titrating specific amino acid residues in enzymes

Reagent	Abbreviation	Amino acid specificity
Lucifer yellow iodoacetamide	LYI	Cysteine
5,5′-Dithiobis(2-nitrobenzoic acid)	Nbs_2	Cysteine
Diethylpyrocarbonate	DPC	Histidine
2,4,6-Trinitrobenzenesulfonic acid	TNBS	Lysine
p-Hydroxylphenylglyoxal	HPG	Arginine

It should be kept in mind that none of the reagents are absolutely specific to one amino acid

2.3.3. Substrates

1. GDP-[^{3}H]mannose (American Radiolabeled Chemicals Inc. St. Louis, Missouri, 50 μCi; 5–15 Ci/mmol), unlabeled 15 mM GDP-Man (Sigma).
2. 0.6mg Dol-P (American Radiolabeled Chemicals Inc.) in $CHCl_3/CH_3OH$ (2:1).

2.3.4. Inhibitors

1. Amino acid specific reagents **(Table 3)** (Sigma), Lucifer yellow iodoacetamide (LYI) (Invitrogen) (*see* **Note 1**).

2.3.5. Enzyme Kinetics

1. As described in **Subheading 2.1.5**.

3. Methods

3.1. Inhibition of β4-Gal-Transferase

3.1.1. Enzyme

1. Soluble β4-Gal-transferase should be diluted with water to approximately 0.2 mg/mL. This solution can be stored at −20°C for several months, or at −80°C indefinitely.

3.1.2. Inhibitors

1. The purity and stability of all inhibitors in solutions have to be established by HPLC and, by mass spectrometry.
2. Inhibitors that are soluble in methanol or other solvents can be used in the assay up to a concentration of 10% methanol, above which enzyme will be inactivated. Solutions in assay tubes can be dried in the vacuum centrifuge, residue is then taken up in solvent, and the assay mixture is added. For higher concentrations, the solubility of the inhibitor has to be con-

firmed in the assay mixture at the temperature of incubation (37°C). If necessary, detergent (e.g. 0.1% Triton-X 100) can be used in the assay.

3.1.3. Enzyme Assays

1. Preliminary studies to ensure linearity should be done first. The optimal conditions should be used from the literature (*4, 6, 27*). Prepare a reagent cocktail. All reagents and conditions should be the same for a set of experiments, except for the variable substrate and inhibitor concentrations. Always include assays without substrate to determine the background radioactivity, and with inhibitor alone to establish whether it is an acceptor substrate or not. As a preliminary approximation, donor substrate is used at 0.5 and 1mM, with 1 and 2mM acceptor and 1 and 2mM inhibitor concentrations. All assays are performed in duplicate determinations. This will suggest the potency of the inhibitor and guide the choice of concentrations used in subsequent more detailed studies.
2. For the standard assay, add substrates, inhibitors and cocktail containing 1μL enzyme to a total volume of 40 μL. If the volume is exceeded, dry solutions in a table top vacuum centrifuge (but not detergents, proteins or volatile reagents). If inhibitors are in methanol solution, add the same amount of methanol in each assay. Leave all assay tubes on ice until ready for incubation.
3. Incubate assay mixtures for a fixed time period (e.g. 30 min) in a water bath at 37°C.
4. Stop the reactions by adding 600 μL of cold water and freezing. At this stage, tubes can be stored at −20°C.
5. Enzyme product can be isolated by ion exchange chromatography (0.4mL AG1x8) that yields product plus free [^{3}H] Gal, or by hydrophobic chromatography using Sep-Pak C18 columns that yields product and substrate after elution with methanol. For further isolation and identification of enzyme product, HPLC analysis can be performed (*27*).

3.1.4. Inhibition Study

1. Choose inhibitor concentrations equal to 0.5, 1.0, 2.0 and 5.0 times the K_M value of the substrate. Inhibitor and substrates are added first, and dried if necessary and taken up in up to 8μL methanol. This is followed by the addition of reagent cocktail and enzyme. If both substrate and inhibitor are acceptor substrates, the individual products should be isolated by HPLC to evaluate a reduction in product formation.

3.1.5. Enzyme Kinetics and Analysis of Inhibition

1. Enzyme kinetics will be evaluated by plotting enzyme product formation versus substrate concentration, by linear regression analysis and using non-linear regression programs.
2. Based on the potency of inhibitor and the type of inhibition observed, design compounds with similar structural features or those containing reactive groups as improved inhibitors.

The enzyme requirement for specific hydroxyls and other structural features of the substrate should be considered, as well as the three-dimensional arrangement and the properties of amino acids in the substrate binding sites (*7*).

3.2. Inhibition of Core 2β6-GlcNAc-Transferase C2GnT1

3.2.1. Enzyme

1. The insect cell supernatant containing soluble C2GnT1 can be used as the enzyme source (*see* **Note 2**) (*25*) Enzyme is stable at −80°C.

3.2.2. Inhibitors

1. The purity of all compounds used as inhibitors should be verified. The use of solvents in assays is discussed in **Subheading 3.1.2.**
2. Inhibitors that resemble the Galβ1-3GalNAc- structure are expected to bind in the acceptor binding site.
3. Analogs of partial structures of UDP-GlcNAc are expected to bind competitively in the nucleotide sugar binding site.
4. Compounds having a UV-activatable group should be preincubated and irradiated with the enzyme in the UV reactor for 10min at 350nm at 30°C (*9*).

3.2.3. Enzyme Assays

1. Standard assays are as described (*4,8–10, 27*), and preliminary studies to ensure linearity and purity, and inhibition experiments, as well as the kinetic analysis of inhibition are as described above in **Subheadings 3.1.3–3.1.5**.

3.3. Inhibition of Dol-P-Man Synthase: Identification of a Cysteine Residue Near the Active Site

3.3.1. Preparation of Enzyme

1. Preparation of Dol-P-Man synthase is as reported *(26)*, except that Nonidet P-40 should be replaced by Thesit (*see* **Note 3**). The concentration of purified Dol-P-Man synthase is determined from the calculated molar extinction coefficient at 280nm of 28,020 $M^{-1}cm^{-1}$. Purified Dol-P-Man synthase preparations are routinely stored in buffers containing 1mM DTT. The enzyme is stable for several days at 0–4°C and stable for months at −20°C (*see* **Note 4**).
2. The reducing reagent (DTT) is removed from the enzyme by chromatography on Sephadex G-50 (superfine) using a procedure similar to that described by Penefsky (*28*). A disposable syringe (3mL) is packed with Sephadex G-50 equilibrated with Buffer B, and excess buffer is removed by centrifugation in a swinging bucket rotor at a setting of 3 for 2min using an International Clinical Centrifuge (Model CL). The gel is washed three times by adding 1.0mL of buffer B, followed by centrifugation using the same procedure. Enzyme (0.5mL) is then layered on the gel and separated from DTT by centrifu-

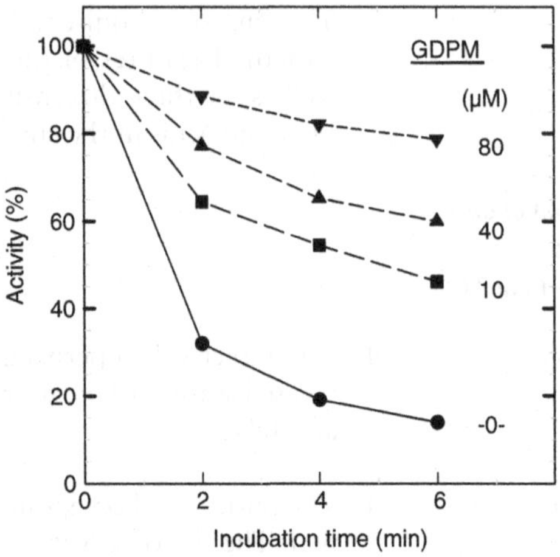

Fig. 2. Protection of yeast Dol-P-Man synthase by GDP-Man against inactivation by Nbs2. Enzyme was incubated on ice in 0.2 mL 25 mM Tris/acetate pH 7.5, 0.25% Nonidet-P-40, 5 mM $MnCl_2$, 2.5 mM $MgCl_2$, and 50 μM Nbs_2. GDP-Mannose concentrations were 0 (*filled circle*), 10 μM (*filled square*), 40 μM (*filled up triangle*), 80 μM (*filled down triangle*). At the indicated times, 5-μL aliquots of enzyme were assayed for activity in the absence of DTT.

gation as described above. All the DTT is retained by the gel, and 70–80% of the applied protein is recovered in the eluate.

3.3.2. Inhibition Study

1. Sulfhydryl reagent-reactive groups in the enzyme are derivatized by the following procedures. Nbs_2 is incubated with enzyme at ice bath temperature. Aliquots of enzyme are withdrawn at the indicated times (**Fig. 2**) and diluted to determine enzyme activity. Since inactivation of the enzyme by Nbs_2 is reversed by reducing agents, DTT is not added to assay mixtures when the enzyme is inhibited with this thiol-reactive reagent.

2. Titration of sulfhydryl residues in Dol-P-Man synthase (*see* **Note 5**):
Inhibition by LYI is carried out in reaction mixtures incubated on ice with enzyme (10 nmol) in 0.5 mL of 0.05M Tris/acetate, pH 7.5, with 0.2% Thesit. Aliquots are taken at 30 min intervals to determine residual activity. The reaction is terminated at 100 min with 4 mM DTT and excess LYI is removed by gel chromatography on a column (1.0×18 cm) of Sephadex G-25 (coarse), equilibrated with 10 mM Tris/acetate, pH 7.5, 200 mM NaCl and 0.1% Thesit. Lucifer-yellow-enzyme conjugate (Lucifer-yellow-carboxamido-methyl-protein derivative) is determined using an absorption coefficient value of

13,000 $M^{-1}cm^{-1}$ at 426 nm (Molecular Probes). Protein values are determined by absorbance at 280 nm and corrected for Lucifer-yellow absorbance at this wavelength. Titration with LYI to the modification of 0.8 mol sulfhydryl/mol enzyme results in 90% inhibition of enzyme activity.

3. Protection of reactive cysteine by substrate. Pure enzyme (0.24 μg) is incubated on ice in 0.2 mL of 25 mM Tris/acetate, pH 7.5, and 0.35% Thesit. The control is incubated in the absence of Nbs_2. Other reaction mixtures contain 25 μM Nbs_2 and GDP-Man at 0, 10, 40 and 80 μM. Aliquots (5 μL) are taken at 0 time and at 2, 4 and 6min and assayed for enzyme activity in the absence of DTT (*see* **Fig. 2**, **Note 6**).
4. Similar procedures are used for the other amino acid specific reagents (*29*).
5. The reduction of enzyme activity reflects the effectiveness of inhibitors.
6. Substrates and substrate analogs will be preincubated with the enzyme, before addition of amino acid reagent. This will suggest whether the amino acids are in or near the substrate binding site, or involved in the catalytic mechanism.

4. Notes

1. Although the listed reagents are specific for individual amino acid residues, all these reagents do exhibit some non-specificity. Therefore, one must always interpret the results with appropriate caution and confirm the conclusions by other methods including site directed mutagenesis if possible. For a more detailed description of potential problems with specificity consult Lundblad (*29*).
2. Inhibitors should be tested for their ability to inhibit cell-bound enzyme in cell homogenates, their specificity by testing other glycosyltransferases, and their ability to enter cells.
3. Thesit is a UV transparent detergent with many of the properties of Nonidet P-40 for the solubilization of membrane enzymes. During synthase purification on hydroxylapatite, Nonidet P-40 is replaced with Thesit with no discernible differences in properties of the enzyme.
4. Enzyme activity in old preparations of enzyme can usually be completely restored by the addition of fresh reducing agent to 5mM and SDS to a final concentration of 0.01%.
5. Yeast Dol-P-Man synthase contains three cysteine residues at positions 93,172, and 259, based on the deduced amino acid

sequence for this protein. Site directed mutagenesis of each of these residues has demonstrated that cysteine is not required for enzyme activity but inhibition studies place Cys93 in or near the active site of the enzyme (*30*).

6. The apparent K_M for GDP-Man is 1.5 μM. Therefore, the concentration of GDP-Man that protects against modification of eysteine by Nbs_2 corresponds to the affinity of the enzyme for this substrate.

Acknowledgments

This work was supported by the NIH (grant to John Schutzbach), the Canadian Cystic Fibrosis Foundation, and NSERC (grants to Inka Brockhausen).

References

1. Lowe, J.B. (2003) Glycan-dependent leukocyte adhesion and recruitment in inflammation. *Curr. Opin. Cell Biol.* 15, 531–538
2. Brockhausen, I. (2006) The role of galactosyltransferases in cell surface functions and in the immune system. *Drug News Perspect.* 19, 401–409
3. Brockhausen, I. (2006) Mucin-type *O*-glycans in human colon and breast cancer: glycodynamics and functions. *EMBO Rep.* 7, 599–604
4. Brockhausen, I., Kuhns, W. and Schutzbach, J.S. (1998) Glycoproteins and their relationship to human disease. *Acta Anat.* 161, 36–78
5. Hollingsworth, M.A., and Swanson, B.J. (2004) Mucins in cancer: protection and control of the cell surface. *Nat. Rev.* 4, 45–60
6. Brockhausen, I., Benn, M., Bhat, S., Marone, S., Riley, G.J., Montoya-Peleaz, P., Vlahakis, J.Z., Paulsen, H., Schutzbach, J.S. and Szarek, W. (2006) UDP-Gal: GlcNAc-R β1,4-Galactosyltransferase – a target enzyme for drug design. Acceptor specificity and inhibition of the enzyme. *Glycoconj. J.* 23, 523–539
7. Qasba, P.K., Ramakrishnan, B., and Boeggeman, E. (2005) Substrate-induced conformational changes in glycosyltransferases. *Trends Biochem. Sci.* 30, 53–62
8. Kuhns, W., Rutz, V., Paulsen, H., Matta, K.L., Baker, M.A., Barner, M., Granovsky, M., and Brockhausen, I. (1993) Processing *O*-glycan core 1, Galβ1-3GalNAcα-R. Specificities of core 2, UDP-GlcNAc:Galβ1-3GalNAc-R (GlcNAc to GalNAc) β6-*N*-acetylglucosaminyltransferase and CMP-sialic acid: Galβ1-3GalNAc-R α3-sialyltransferase. *Glycoconj. J.*10, 381–394
9. Toki, D., Granovsky, M., Reck, F., Kuhns, W., Baker, M.A., Matta, K.L., and Brockhausen, I. (1994) Inhibition of UDP-GlcNAc: Galβ1-3-GalNAc-R (GlcNAc to GalNAc) β6-*N*-acetylglucosaminyltransferase from acute myeloid leukemia cells by photoreactive nitrophenyl substrate derivatives. *Biochem. Biophys. Res. Commun.* 198, 417–423
10. Toki, D., Sarkar, M., Yip, B., Reck, F., Joziasse, D., Fukuda, M., Schachter, H., and Brockhausen, I. (1997) Expression of stable human *O*-glycan core 2 β1,6-*N*-acetylglucosaminyltransferase in Sf9 insect cells. *Biochem. J.* 325, 63–69
11. Burda, P., and Aebi, M. (1999) The dolichol pathway of *N*-linked glycosylation. *Biochim. Biophys. Acta* 1426, 239–257
12. Schutzbach, J. (1997) The role of the lipid matrix in the biosynthesis of dolichyl- linked oligosaccharides. *Glycoconj. J.* 14, 175–182
13. Segel, I.H. (1993) *Enzyme Kinetics; Behavior and Analysis of Rapid Equilibrium and Steady-State Enzyme Systems.* Wiley Classics Library Edition, John Wiley and Sons,| New York
14. Mahler, H.R., and Cordes, E.H. (1971) *Biological Chemistry.* Harper & Row Publisher, New York
15. Delannoy, P., Kim, I., Emery, N. et al. (1996) Benzyl-*N*-acetyl-alpha-D-galactosaminide inhibits the sialylation and the secretion of

mucins by a mucin secreting HT-29 cell subpopulation. *Glycoconj. J.* 13, 717–726

16. Brown, J.R., Fuster, M.M., Li, R., Varki, N., Glass, C.A., and Esko, J.D. (2006) A disaccharide-based inhibitor of glycosylation attenuates metastatic tumor cell dissemination. *Clin. Cancer Res.* 12, 2894–2901
17. Morrison, J.F., and Ebner, K.E. (1971) Studies on galactosyltransferase: kinetic investigations with N-acetylglucosamine as the galactosyl group acceptor. *J. Biol. Chem.* 246, 3977–3984
18. Bendiak, B., and Schachter, H. (1987) Control of glycoprotein synthesis: kinetic mechanism, substrate specificity, and inhibition characteristics of UDP-*N*-acetylglucosamine:α-D-mannoside β1–2 *N*-acetylglucosaminyltransferase II from rat liver. *J. Biol. Chem.* 262, 5784–5790
19. Nishikawa, Y., Pegg, W., Paulsen, H., and Schachter, H. (1988) Control of glycoprotein synthesis: purification and characterization of rabbit liver UDP-*N*-acetylglucosamine:α-3-D-mannoside β1–2 *N*-acetylglucosaminyltransferase I. *J. Biol. Chem.* 263, 8270–8281
20. Ropp, P.A., Little, M.R., and Cheng, P.W. (1991) Mucin biosynthesis: purification and characterization of a mucin β6*N*-acetylglucosaminyltransferase. *J. Biol. Chem.* 266, 23863–23871
21. Ünligil, U.M., and Rini, J.M. (2000) Glycosyltransferase structure and mechanism. *Curr. Opin. Struct. Biol.* 5, 510–517
22. Fritz, T.A., Hurley, J.H., Trinh, L.B., Shiloach, J., and Tabak, L.A. (2004) The beginnings of mucin biosynthesis: the crystal structure of UDP-GalNAc:polypeptide alpha-*N*-acetylgalactosaminyltransferase-T1. *Proc. Natl. Acad. Sci. U.S.A.* 101, 15307– 15312
23. Hindsgaul, O., Kaur, K.J., Srivastava, G., Blaszczyk-Thurin, M., Crawley, S.C., Heerze, L.D., and Palcic, M.M. (1991) Evaluation of deoxygenated oligosaccharide acceptor analogs as specific inhibitors of glycosyltransferases, *J. Biol. Chem.* 266, 17858–17862
24. Chung, S.J., Takayama, S., Wong, C.H. (1998) Acceptor substrate-based selective inhibition of galactosyltransferases. *Bioorg. Med. Chem. Lett.* 8, 3359–3364
25. Benn, M. (2005) MSc thesis, Queen's University, Ontario, Canada
26. Schutzbach, J. (2006) Assay of dolichyl-phospho-mannose synbthase reconstituted in a lipid matrix. In: 'Glycobiology Protocols', *Methods in Molecular Biology*, pp. 31–42
27. Brockhausen, I. (2006) Analysis of the glycodynamics of primary osteoblasts and bone cancer cells. In: 'Glycobiology Protocols', *Methods in Molecular Biology*, pp. 211–236
28. Penefsky, H.S. (1977) Reversible binding of Pi by beef heart mitochondrial adenosine triphosphatase. *J. Biol. Chem.* 252, 2891–2899
29. Lundblad, R.L. (1994) *Techniques in Protein Modification*. CRC Press, Boco Raton
30. Forsee, W.T., McPherson, D. and Schutzbach, J.S. (1997) Characterization of recombinant yeast dolichyl mannosyl phosphate synthase and site-directed mutagenesis of its cysteine residues. *Eur. J. Biochem.* 244, 953–958

Chapter 26

Saturation Transfer Difference NMR Spectroscopy as a Technique to Investigate Protein–Carbohydrate Interactions in Solution

Thomas Haselhorst, Anne-Christin Lamerz, and Mark von Itzstein

Summary

Saturation transfer difference (STD) Nuclear Magnetic Resonance (NMR) spectroscopy is a powerful method for studying protein–ligand interactions in solution. The STD NMR method is capable of identifying the binding epitope of a ligand when bound to its receptor protein. Ligand protons that are in close contact with the receptor protein receive a higher degree of saturation, and as a result stronger STD NMR signals can be observed. Protons that are either less or not involved in the binding process reveal no STD NMR signals. Therefore, the STD NMR method is an excellent tool to investigate how a binding ligand interacts with its receptor molecule. The STD NMR experiment is easy to implement and only small amounts of native protein are required. This chapter comprises a detailed experimental protocol to acquire STD NMR spectra and determine the binding epitope of a ligand bound to its target protein. As representative examples the ligands uridyl-triphosphate (UTP) and uridyl–glucose-diphosphate (UDP-glucose) when bound to the *Leishmania major* UDP-glucose-pyrophosphorylase (UGP) as target protein are examined.

Key words: Nuclear magnetic resonance spectroscopy, Saturation transfer difference NMR, Protein–carbohydrate interactions, Enzyme carbohydrate interaction, Epitope mapping, UTP, UDP-Glc, *Leishmania major* pyrophosphorylase.

1. Introduction

NMR spectroscopy has been utilised to detect protein–ligand interactions for many years, but the recent dramatic increase in reported novel NMR techniques highlight the importance and user-friendly nature of NMR spectroscopy. It is remarkable that

Nicolle H. Packer and Niclas G. Karlsson (eds.), *Methods in Molecular Biology, Glycomics: Methods and Protocols, vol. 534*

DOI: 10.1007/978-1-59745-022-5_26

many of these NMR techniques have been developed while studying protein–carbohydrate interactions, in particular lectin–carbohydrate complexes. Interestingly, one particular method called *Saturation Transfer Difference* (STD) NMR spectroscopy has been shown to be very versatile (*1, 2*). STD NMR was initially designed to screen a library of carbohydrate molecules for binding affinity towards wheat-germ agglutinin (WGA) (*1*) but in this study however it was also realized that STD NMR spectroscopy is an excellent technique for determining the binding epitope of a ligand. The authors observed that ligand protons that are in close contact with the protein showed strong STD NMR signals whereas ligand protons without direct protein contact had STD NMR signals, with only low or even no intensity. Identification of these binding epitopes and pharmacophores of ligands, including substrates and potential inhibitors is of major importance in drug lead optimisation and structure-based drug design.

The STD NMR experiment is based on transfer of saturation from the protein to a bound ligand. Practically, this transfer process can be achieved by selectively saturating the protein resonances with a cascade of Gaussian soft pulses, as shown in **Fig. 1**. The magnetisation is transferred rapidly through the entire protein mediated by spin diffusion. If a ligand resides in the binding site of the protein, saturation may be transferred to this ligand by intermolecular ^{1}H-^{1}H cross relaxation. Protons which are close in space to the protein surface (e.g. proton H_a in **Fig. 1**) receive higher degree of saturation, while protons that are not in intimate contact with the protein (e.g. protons H_b and H_c in **Fig. 1**) receive little or no saturation. Protons from non-binding ligands (e.g. protons H_a', H_b' and H_c' in **Fig. 1**) do not receive any saturation from the protein; hence they are not attenuated in the *on*-resonance spectrum (**Fig.1**). It is therefore logical that only ligand protons that interact with protein are saturated and show decreased intensity in the *on*-resonance spectrum. As it is usually difficult to identify those attenuated signals, Mayer and Meyer suggested the acquisition of an additional *off*-resonance spectrum with a saturation frequency where no protein or ligand signals are detectable, usually around ~30 ppm (*1*). This *off*-resonance reference spectrum represents a normal ^{1}H NMR. Subtraction of the *on*-resonance and *off*-resonance spectrum results in the final saturation transfer difference (STD) NMR spectrum showing only signals from ligands that have binding affinity. Additionally, the STD NMR spectrum may reveal only ligand protons that are in close contact with the protein binding site. STD NMR spectroscopy can be utilised to detect binding ligands with K_D values in the 10^{-2}–10^{-8} M range (*3*). High-affinity ligands that undergo slow chemical exchange typically reside longer within the protein-binding site and are thus not detectable by STD NMR spectroscopy. To quantify STD NMR effects of important ligand protons that interact with the

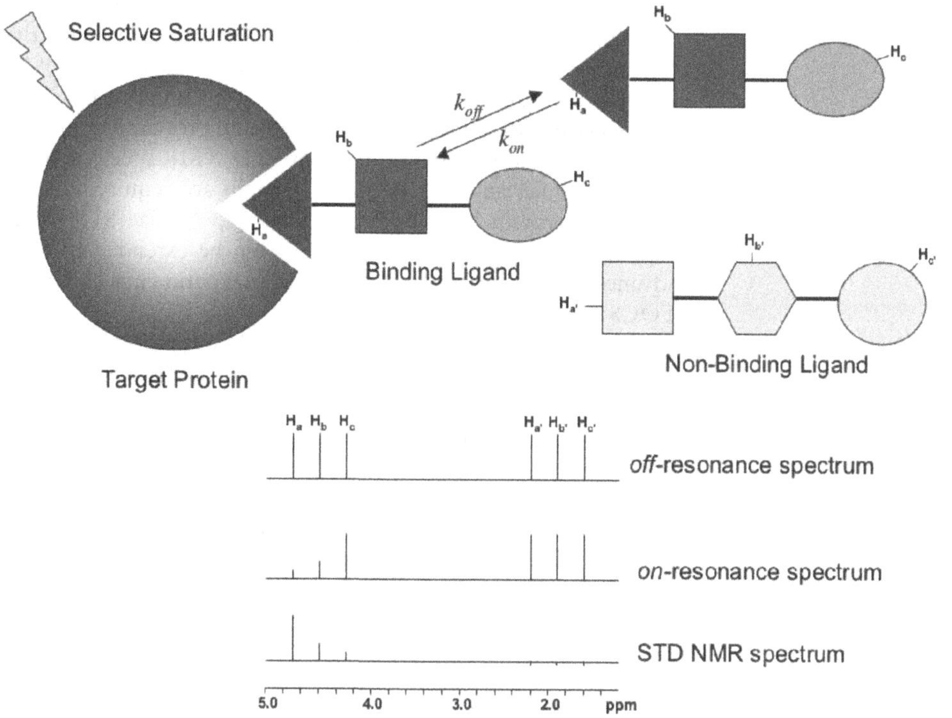

Fig. 1. Scheme showing principle of the STD NMR method. Target protein is saturated with a cascade of selective pulses. Saturation is transferred quickly through the protein mediated by spin diffusion. A ligand binding that is in fast exchange with the target protein receives saturation from the protein. Ligand protons that are in close contact with the protein (H_a) receive more saturation than ligand protons that are solvent exposed (H_c) indicated by the colour scheme of the moiety. Protons from a non-binding ligand (H_a', H_b' and H_c') do not receive any saturation from the protein and are therefore not attenuated in the *on*-resonance spectrum. Subtraction of the *on*-resonance and *off*-resonance spectrum reveals the final difference spectrum (STD) showing only signals from binding ligand protons that are in close proximity with the protein surface. Protons that are exposed to solvent (H_c) show none or weak STD NMR intensity and protons from non-binding ligands (H_a', H_b' and H_c') do not show any STD NMR signals.

protein, Mayer and Meyer introduced the refined group epitope mapping (GEM) that is only possible when the dissociation rate constant (k_{off}) is greater than the T_2 relaxation rate of the bound ligand. As carbohydrates are generally weakly binding ligands (K_D values in the 10^{-3}– 10^{-6}μ range), the corresponding dissociation rate constant k_{off} is rather large and therefore exhibits excellent properties for STD NMR spectroscopy. More recent studies have remarkably shown the rapid development of the STD NMR technique, e.g. the application of STD NMR for the determination of bioactive conformation (*4*) or identification and characterisation of virus-ligand interactions using whole virus particles (*5*).

Claasen et al. introduced the double difference STD NMR experiment (STDD) to remove strong background signals that the authors discovered when investigating the interaction of living cells (human platelets) with ligands (*6*). Overall, STD NMR spectroscopy is a versatile method to study protein-ligand complexes in solution. The advantage of STD NMR spectroscopy is that only small amounts of native protein are required and no expensive protein labelling is necessary (*1, 3*). The STD NMR experiment is easy to implement and can be extended to more advanced two-dimensional STD NMR experiments like STD-TOCSY and STD-HSQC (*7*).

The aim of this chapter is to provide details regarding the acquisition and analysis of STD NMR spectra. The experimental procedure is described using UTP and UDP-glucose as ligands when bound to the UDP-glucose-pyrophosphorylase (UGP) from *Leishmania major*. This enzyme catalyses the formation of UDP-glucose from glucose-1-phosphate and UTP (*8*) (**Fig. 2**) and is essential in the lifecycle of the *Leishmania* parasite ; it is therefore a potential novel target for therapeutic discovery.

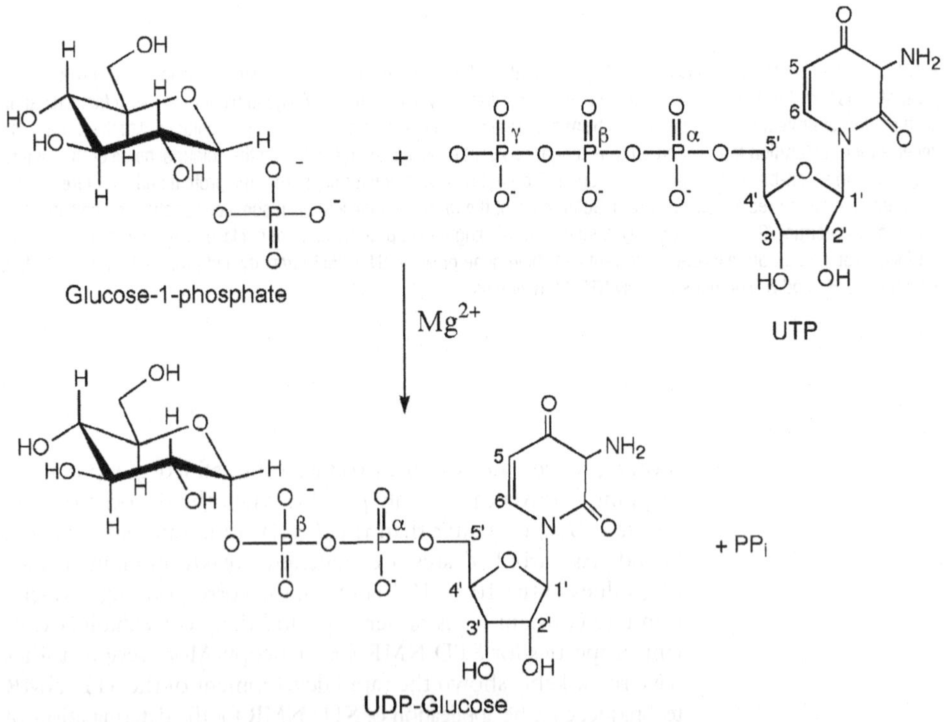

Fig. 2. Scheme showing reaction catalysed by the pyrophosphorylase from *Leishmania major* (UGP) using Glucose-1-phosphate and UTP to form the activated sugar UDP-glucose (*8*).

2. Materials

2.1. Protein Expression

1. The pET-UGP-His plasmid that encodes the full length C-terminally His_6-tagged *L. major* UGP.
2. *E. coli* BL21 (DE3) cells.
3. Luria Bertani medium (Tryptone 10 g, Yeast extract 5 g, NaCl 10 g per L).
4. Carbenicillin.
5. IPTG (Isopropyl β-D-thiogalactopyranoside) (Sigma).
6. Tris-HCl buffer (ICN) supplemented with bestatin (Sigma), pepstatin (Sigma), PMSF (Roche).
7. Branson sonifier.
8. Ni^{2+}-chelating column.
9. Superdex200 HR (Amersham Biosciences).

2.2. NMR Sample Preparation

1. Deuterium oxide (D_2O) (99.9%) (Cambridge Isotope Labs).
2. Deuterated Tris-$_{d5}$ (Sigma).
3. $MgCl_2$ (Sigma).
4. 5 mm 500 MHz 7″ NMR tubes (Kontes Glass).
5. UTP (NMR sample A1).
6. UTP in the presence of UGP (NMR sample A2).
7. UDP-glucose (NMR sample B1).
8. UDP-glucose in the presence of UGP (NMR sample B2).

2.3. NMR Hardware and Software

1. Bruker Avance 600 MHz spectrometer
2. 5-mm TXI probe with xyz-axis gradient
3. TOPSPIN software running on a PC (LINUX) for acquisition and data processing.
4. The temperature unit was set 298 K.
5. The use of a cryoprobe is highly recommended for in correct sensitivity

3. Methods

3.1. Protein Expression

1. For expression of UGP, the pET-UGP-His plasmid was transformed into *E. coli* BL21(DE3) and cells were grown in Luria Bertoni medium containing 200 μg/mL carbenicillin at 15°C to an OD_{600} of 0.6
2. Protein expression was induced with 1 mM IPTG for 20 h.

3. Pellets were lysed using a Branson sonifier.
4. Cell debris was removed by centrifugation and soluble fractions were loaded onto to Ni^{2+}-chelating column.
5. Bound protein was washed with 300 mM imidazole using a step gradient.
6. Fraction containing UGP were pooled and submitted to a size exclusion chromatography on a Superdex200 HR column.

3.2. NMR Sample Preparation

1. NMR buffer preparation: A stock of ~10 mL of 50 mM deuterated Tris containing 10 mM $MgCl_2$ was prepared using D_2O as solvent. The pH was adjusted to pH 7.8.
2. Preparation of NMR sample A1 for spectrum assignment of UTP: A 5 mM UTP solution was prepared using the deuterated NMR buffer. The final volume was adjusted to 600 µL (*see* **Note 1**).
3. NMR sample A2 for acquisition of STD NMR spectrum of UTP in the presence of UGP: A solution containing 6 µM of UGP and 0.6 mM UTP (protein:ligand ratio of 1:100 (*see* **Note 2**)) was prepared using the deuterated NMR buffer. The final volume was adjusted to 600 µL.
4. Preparation of NMR sample B1 for spectrum assignment of UDP-glucose: A 5 mM UDP-glucose solution was prepared using the deuterated NMR buffer. The final volume was adjusted to 600 µL (*see* **Note 1**).
5. NMR sample B2 for acquisition of the STD NMR spectrum of UDP-glucose in the presence of UGP: A solution containing 6 µM of UGP and 0.6 mM UDP-glucose (protein:ligand ratio of 1:100 (*see* **Note 2**)) was prepared using the deuterated NMR buffer. The final volume was adjusted to 600 µL.

3.3. NMR Acquisition

After loading the sample into the 600 MHz NMR spectrometer the sample was locked, tuned and matched according to manufacture's guidelines. The *on*-axis shims were optimised by running the automatic gradient shimming procedure and *off*-axis shims were adjusted manually.

3.3.1. NMR Experiments for Sample A1 and B1

The standard ^{1}H NMR pulse program with pre-saturation was loaded and the following parameters were set:

1. Set number of scans (ns) to 32.
2. Choose a relaxation delay of ~2 s.
3. Set spectral width to a value of 6,000 Hz.

A low power pre-saturation pulse during the relaxation delay for water suppression was applied. Standard ^{1}H NMR acquisition parameters were loaded and 90° high power pulse was determined.

To obtain optimised water suppression, standard ^1H NMR acquisition parameters with water suppression were loaded and the power level for the pre-saturation pulse during the relaxation delay was adjusted to obtain maximal water suppression. The free induction decay (FID) was Fourier transformed and an automatic phase and baseline correction was performed. It is important to note that a complete NMR assignment of the ^1H NMR spectrum can be obtained by acquiring additional 2D NMR experiments (*see* **Note 1**).

3.3.2. STD NMR Experiments Using Sample A2 and B2

The NMR sample was placed into the NMR magnet and temperature set to 298 K (*see* **Note 3**). In the first instance a ^1H NMR reference spectrum was acquired (*see* **Subheading 3.3.1.**). To acquire the STD NMR spectrum, the standard pseudo-2D STD NMR pulse program was loaded as shown in (**Fig. 3**) and the following parameters were set:

1. Set number of scans (ns) to 1 K (1,024).
2. Define frequency list with an *on*-resonance and *off*-resonance frequency (*see* **Note 4**).
3. The STD NMR effect depends on the saturation time t_{sat}. Generally a saturation time of 1.5–2 s is a good choice to perform an initial STD NMR experiment. A train of Gaussian soft pulses defines the overall saturation time t_{sat} and consists of a set of Gaussian pulses with a 50 ms duration each followed by a delay of 100 μs. 40 selective Gaussian pulses were chosen for a total saturation time of 2 s (*see* **Note 5**). For

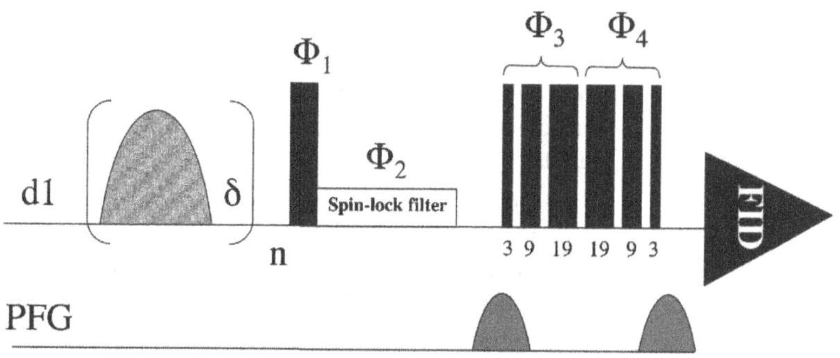

Fig. 3. NMR Pulse sequence showing a STD NMR experiment. The protein is saturated with a Gaussian pulse cascade followed by the 90° high power pulse with phase Φ_1. The additional spin lock pulse with the phase Φ_2 is applied after 90° high power pulse to remove unwanted protein background. STD NMR experiment is set up as a pseudo-2D NMR experiment by switching the *on*- and *off*- resonance frequency of the selective pulse between 7.2 and 33.3 ppm (*see* **Note 5**). The FID of *on*- and *off*-resonance spectra is stored and processed separately. Superimposing and subtracting the *on*- and *off*-resonance spectrum results in the final 1D STD NMR spectrum. Phases are φ_1 = (x,−x), φ_2 = 8(y), 8(−y), φ_3 = 2(x), 2(y), 2(−x), 2(−y). φ_4 = 2(−x), 2(−y), 2(x), 2(y), φ_{rec} = (x, −x) (−x, x). Duration of the selective Gaussian pulse was 50 ms and delay δ was set to 100 μs. For water suppression, the WATERGATE sequence was used by applying a binominal 3–9–19 pulse sandwich. Pulse field gradients (PFG) are of equal intensity and sign (20%).

a complete epitope mapping it is recommended to acquire STD NMR spectra at different saturation times ranging from 0.5 to 4.0 s.

4. Apply a spin lock filter to suppress residual protein background. It is recommended the spin lock pulse is applied after the excitation pulse to suppress unwanted protein background signals. Usually, spin lock pulses of 10–20 ms and strength of 5 kHz are sufficient to achieve protein signal removal.
5. Use the Watergate 3–9–19 pulse train to suppress residual water signals with gradients gpz1 and gpz2 of 20% strength for each.

3.3.3. STD NMR Control Experiments

It is recommended that a number of control experiments are acquired to ensure that the obtained STD NMR signals are genuine effects:

Thus,

1. Prepare an identical STD NMR sample (as for sample A2 or B2) but in the absence of the protein. PeForm a STD NMR experiment using the same parameters as selected for the original STD NMR experiment. As no protein is present in the NMR sample, no saturation transfer occurs and hence the obtained final STD NMR spectrum reveals no signals. In case the saturation frequency was set too close to the ligand signals this control experiment would result in a STD NMR spectrum revealing ligand signals. If this occurs it is recommended the saturation frequency is changed (*see* **Note 4**).
2. It may be possible that subtraction artefacts occur. To identify potential subtraction artefacts in the *on*- and *off*-resonance spectrum, the saturation frequency can be set to the same value (*on/on* or *off/off*). A STD NMR spectrum is then acquired using samples A2 and/or B2. For this control experiment a perfect subtraction is found in a STD NMR spectrum with no signals.
3. If the protein sample itself shows strong signals, it is recommended to acquire a control STD NMR spectrum with a sample containing the protein but in the absence of ligands. The obtained STD NMR spectrum may reveal residual protein background signals and can then be used for a potential double difference STD NMR spectrum (STDD) (*7*).

3.4. STD NMR Processing and Quantification

Firstly, the *on*- and *off*-resonance spectra have to be extracted from the pseudo-2D STD NMR experiment. After data extraction it is important that the *on*- and *off*-resonance spectra are identically processed. Both spectra can then be superimposed and the difference spectrum can be obtained by subtracting the *on*-resonance from the *off*-resonance spectrum. To ensure high digital resolution it is also possible to form the difference of the *on*- and *off*-resonance in the time domain by subtracting individual FIDs. This difference spectrum represents the final STD NMR spectrum and can be separately stored. Relative STD effects are calculated according to the equation $A_{\mathrm{STD}} = (I_0 - I_{\mathrm{sat}})/I_0 = I_{\mathrm{STD}}/I_0$ by comparing the intensity of the signals in the STD NMR spectrum (I_{STD}) with signal intensities of a reference spectrum (I_0) (*2*) (*see* **Note 3**). Practically the quantification of this procedure is as follows: The intensity of the STD NMR spectrum (I_{STD}) is evaluated against the reference ^{1}H NMR spectrum (I_0) (*see* **Note 3**). A convenient way of comparing these spectra is to superimpose the STD NMR spectrum with the reference ^{1}H NMR spectrum in the form of a dual display. The strongest STD NMR signal can then be adjusted to the identical signal intensity in the reference ^{1}H NMR spectrum. The STD NMR signal with the strongest intensity can be set to 100% and relative STD NMR effects for all other observable signals can then be calculated.

Figure 4 shows the ^{1}H NMR spectrum of UTP (a) and UDP-glucose (c) and the corresponding STD NMR spectra in complex with the UGP enzyme for UTP (b) and UDP-glucose (d). It is obvious that UTP and UDP-glucose bind strongly to UGP due to the presence of STD NMR signals in spectrum **Fig. 4B,D**. The H1 proton of the ribose (H1 (Rib)) from UTP and UDP-glucose generates the strongest STD-NMR signal for both ligands (100%). The H5 proton of the nucleotide base (H5

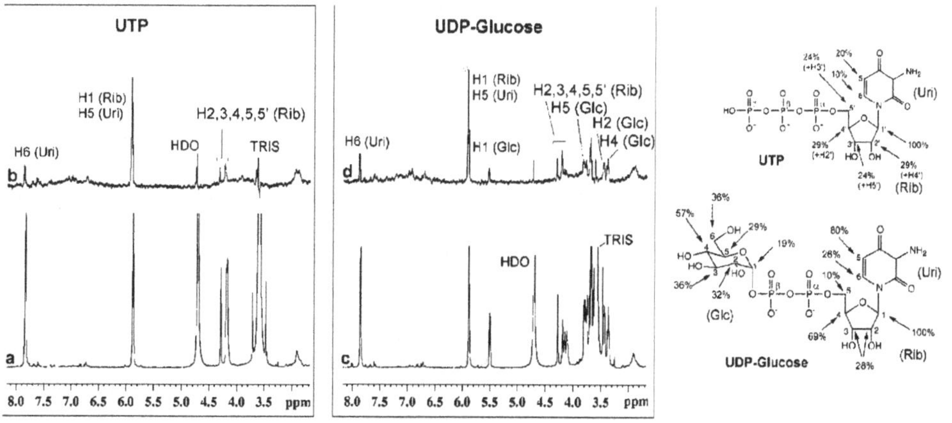

Fig. 4. 600 MHz ^{1}H NMR spectra (a, c) and STD NMR spectra (b, d) of 0.6 mM UTP (a, b) and 0.6 mM UDP-glucose (c, d) in the presence of 6 μM *L. major* UGP (*9*). The protein:ligand ratio is 1:100. Saturation of the protein was achieved with a Gaussian pulse cascade (40 Gaussian pulses of 50 ms duration each with a delay of 100 μs in between each pulse) resulting in a total saturation time of 2 s. The protein resonances (aromatic and amide NH protons) were saturated at 0.7 ppm and the *off*-resonance frequency was set to 40 ppm. Structures on the right show relative STD NMR effects [%] of UTP and UDP-glucose. STD effects were calculated according to the equation $A_{\mathrm{STD}} = (I_0 - I_{\mathrm{sat}})/I_0 = I_{\mathrm{STD}}/I_0$(*2*) by comparing intensity of the signals in the STD NMR spectrum (I_{STD}) with signal intensities of a reference spectrum (I_0).

(Uri)) and the H4 proton of the ribose (H4 (Rib)) in the product UDP-glucose also show strong contacts. All STD NMR effects of the glucose sugar-ring protons show only moderate STD NMR effects. Interestingly, the strongest contact for the carbohydrate moiety was obtained for the H4 proton, which is the substrate defining position of the sugar ring. The pivotal role of this position in binding to the active site was further highlighted by the virtual absence of binding capability of the activated epimer UDP-Galactose (*9*). In summary, STD NMR spectroscopy has become a powerful tool in accessing important information regarding the binding epitope of a ligand and determining important moieties and protons of ligand. Knowledge about this interaction is crucial in design process of novel therapeutics..

4. Notes

1. Standard NMR experiments for chemical shift assignment are routinely run in NMR laboratories, e.g. COSY (*Co*rrelated *S*pectroscop*y*) and TOCSY (*To*tal *c*orrelated *s*pectroscop*y*) that are homonuclear through bond correlation to identify *J*-coupled networks and HSQC (*H*eteronuclear *S*ingle *Q*uantum *C*oherence) that correlates ^{1}H and ^{13}C spins. For users with less NMR experience the book by Braun, Kalinowski and Berger (*9*) can be used as a guideline to perform these standard NMR experiments. Obtained chemical shifts can then be compared with published NMR. The online SUGABASE databank combines the CarbBank Complex Carbohydrate Structure Data (CCSD) with proton and carbon chemical shift values (*11*).
2. STD NMR experiments are usually performed at high molar excess of ligand over protein. Usually a ratio of 1:100 (protein:ligand) is a good choice. This high molar excess of ligand over protein results in stronger STD NMR signals, as it prevents ligands that have already received saturation, from re-entering the protein binding site.
3. Temperature can significantly affect STD NMR signals intensities and it is recommended to acquire a series of STD NMR spectra at different temperatures. At lower temperatures carbohydrates tend to have less conformational flexibility resulting in stronger STD NMR effects for flexible groups compared with STD NMR spectra acquired at higher temperature (*12, 13*).
4. Saturation frequency (*on*-resonance) has to be set to a value where only protein resonances and no ligand signals are located. It is recommended the protein saturation frequency at >700 Hz is placed away from the closest ligand resonance.

If the ligand shows no resonances in the aromatic region (6–8 ppm) usually this is a good choice for protein saturation. If the ligand has aromatic groups the saturation frequency to saturate the protein has to be shifted to 11–12 ppm or around –1 ppm, because no ligand protons resonate in this spectral region, whereas signals from the large protein show significant intensity in this region.

5. Quantification of STD NMR signal intensities exhibit complex dependence on relaxation times, correlation times, exchange rates and the molecular topology of the ligand when binding to the protein. It has been suggested that only when shorter saturation times (t_{sat} < 1 s) are used, STD NMR intensities reflect more closely the exact distances between protein and ligand protons. Yan et al. (*14*) have suggested measurement of the T_1 relaxation times of the complex and the selection of the saturation time for the STD NMR experiment shorter than the T_1 time for improving potential epitope mapping.

References

1. Mayer, M., Meyer, B. (1999) Characterization of ligand binding by saturation transfer difference NMR spectroscopy. *Angewandte Chemie, International Edition,* 38, 1784–1788
2. Mayer, M., Meyer, B. (2001) Group epitope mapping by saturation transfer difference NMR to identify segments of a ligand in direct contact with a protein receptor. *Journal of the American Chemical Society* 123, 6108–6117
3. Meyer, B., Peters, T. (2003) NMR spectroscopy techniques for screening and identifying ligand binding to protein receptors. *Angewandte Chemie, International Edition* 42, 864–890
4. Jayalakshmi, V., Biet, T., Peters, T., Krishna, N.R. (2004) Refinement of the conformation of UDP-galactose bound to galactosyltransferase using the STD NMR intensity-restrained CORCEMA optimization. *Journal of the American Chemical Society* 126, 8610–8611
5. Benie, A.J., Moser, R., Bauml, E., Blaas, D., Peters, T. (2003) Virus–ligand interactions: identification and characterization of ligand binding by NMR spectroscopy. *Journal of the American Chemical Society* 125, 14–15
6. Haselhorst, T., Garcia, F.-r Islam, T., Lai, J.C.C., Rose, F.J., Nicholls, J.M., Peivis, F.S.M. Von Itzstein, M., Avian influence H5-containing Virus-like pertides (VLPs) host-cell receptor sepcificity by STONNR Spectroscopy, Angewendle Chemie Intructional Edition, 47(10), 1910–1917.
7. Claasen,B.,Axmann,M.,Meinecke,M.,Meyer,B. (2005) Direct observation of ligand binding to membrane proteins in living cells by a saturation transfer double difference (STDD) NMR spectroscopy method shows a significantly higher affinity of integrin a(IIb)b3 in native platelets than in liposomes. *Journal of the American Chemical Society* 127, 916–919
8. Vogtherr, M., Peters, T. (2000) Application of NMR based binding assays to identify key hydroxy groups for intermolecular recognition. *Journal of the American Chemical Society* 122, 6093–6099
9. Lamerz, A.-C., Haselhorst, T., Bergfeld, A. K., von Itzstein, M., Gerardy-Schahn, R. (2006) Molecular cloning of the *Leishmania major* UDP-glucose pyrophosphorylase, functional characterisation and ligand binding analyses using NMR spectroscopy. *Journal of Biological Chemistry,* 281(24), 16314–16322
10. Braun, S., Kalinowski, H.-O., Berger, S. (1998) 150 and more basic NMR experiments, 2 ed. Vol. 2nd expanded edition, Weinheim: New York: Chicester: Brisbane: Singapore: Toronto: Wiley-VCH
11. SUGABASE http://www.boc.chem.uu.nl/sugabase/sugabase.html
12. Haselhorst, T., Wilson, J.C., Thomson, R.J., McAtamney,S.,zMenting,J.G.,von Itzstein,M. (2004) Saturation transfer difference (STD) 1H-NMR experiments and in silico docking

experiments to probe the binding of *N*-acetylneuraminic acid and derivatives to *Vibrio cholerae* sialidase. *Proteins: Structure, Function, and Bioinformatics* 56, 346–353

14. Groves, P., Kövér, K.E., André, S., Bandorowicz-Pikula, J., Batta, G., Bruix, M., Buchet, R., Canales, A., Cañada, F.J., Gabius, H.J., Laurents, D.V., Naranjo, J.R., Palczewska, M., Pikula, S., Rial, E., Strzelecka-Kilizzek, A., Jiménez-Barbero, J., (2007) Temperature dependence of ligand-protein complex formation as reflected by saturation transfer difference NMR experiments. Magn Regon Chem 45(9), 745–748

14. Yan, J., Kline, A.D., Mo, H., Shapiro, M.J., Zartler, E.R. (2003) The effect of relaxation on the epitope mapping by saturation transfer difference NMR. *Journal of Magnetic Resonance* 163, 270–276

INDEX